Vitreous Materials at Amarna

The production of glass and faience in 18th Dynasty Egypt

Andrew J. Shortland

BAR International Series 827
2000

Published in 2016 by
BAR Publishing, Oxford

BAR International Series 827

Vitreous Materials at Amarna

ISBN 978 1 84171 038 9

BAR Publishing is the trading name of British Archaeological Reports (Oxford) Ltd. British Archaeological Reports was first incorporated in 1974 to publish the BAR Series, International and British. In 1992 Hadrian Books Ltd became part of the BAR group. This volume was originally published by Archaeopress in conjunction with British Archaeological Reports (Oxford) Ltd / Hadrian Books Ltd, the Series principal publisher, in 2000. This present volume is published by BAR Publishing, 2016.

Printed in England

BAR
PUBLISHING

BAR titles are available from:

BAR Publishing
122 Banbury Rd, Oxford, OX2 7BP, UK
EMAIL info@barpublishing.com
PHONE +44 (0)1865 310431
FAX +44 (0)1865 316916
www.barpublishing.com

for Anja

Acknowledgements

This book was submitted as part of the requirement for the degree of DPhil at teh University of Oxford. I would like to thank Prof Michael Tite for his continuing help and supervision throughout that project and Dr P.R.S. Moorey for his advice on the archaeological elements of it. The members of the Materials Group, especially Chris Doherty and Chris Salter, all gave valuable assistance particularly in the early stages. The project would not have been possible without the generous donation of samples from a number of sources and the help of many museum directors and curators, especially Mrs Rosalind Janssen (Petrie Museum, UCL), Dr Helen Whitehouse, Mr Mark Norman (Ashmolean Museum), Mr Vivian Davies, Dr Jeffrey Spencer (British Museum), Prof Elizabeth Slater (Liverpool University), Prof Dr Dietrich Wildung and Prof Dr K.-H. Priese (Aegyptisches Museum, Berlin). The use of analytical facilities at the British Museum Research Laboratory, kindly arranged by Dr Ian Freestone, was also critical to the completion of the work within the required time frame.

I would also like to thank Dr Paul Nicholson for his advice and for all the assistance that he gave to this project, including giving me access to samples from the recent EES expeditions to Amarna and for allowing use of the ICP-MS at Cardiff, but especially for inviting me to join him on site at Amarna for the September 1998 field season.

Many people at Christ Church helped to keep my limited free time fully occupied and I would make a special mention of the Boat Club Mafia, who contrived to bring me back from a well deserved retirement at least five times during the three years of the project. I would also like to thank Richard Haworth, James Fotheringham, Alexander Wagner, Douglas Chamberlain, Thomas Krebs and many others too numerous to name, for long inspirational cups of coffee and entertaining conversations late into the night.

I owe a special thanks to Dr Derek Jewell, Mr Michael Kettlewell FRCS and Miss Anne-Margrethe Gardner. Without their expertise and care I would never have been able to return to Oxford and this work would never have been written.

My greatest thanks go to my Mother and Father and to Anja. Without their continual support and encouragement throughout the project none of this would have been possible.

Contents

Chapter 1: Introduction

Chapter 2: Analytical Methodology

Chapter 3: Glass, frits and faience

Chapter 4: Ceramics associated with vitreous materials

Chapter 5: Raw materials of ancient production

Chapter 6: Methods of production

Chapter 7: The archaeology of production

Chapter 8: Conclusions

Chapter 1: Introduction

AIM AND SCOPE

The site of Amarna is in Middle Egypt, roughly half way between the ancient capitals of Memphis (Cairo) and Thebes (Luxor) as shown in Figure 1-1. Amarna was the capital city of the 18th Dynasty king, Akhenaten (1352-1336 BC) and provides some of the earliest evidence of workshops for the manufacture of vitreous materials. The manufacture of vitreous materials in Dynastic Egypt reached its zenith in terms of artistic and technical accomplishment in the 18th Dynasty, a level of sophistication not repeated until the Roman periods. There has been much debate over the source of these technological advances, whether some of the vitreous materials were imported or manufactured locally and if they were manufactured locally, the source of their raw materials.

This work investigates the technological processes involved in the making of ancient vitreous materials concentrating on the site of Amarna. The entire process is examined, from the selection of raw materials, preliminary processing and eventual firing right through to the distribution of the finished objects. Analysis of the finished objects and the waste materials of the production sequence by scanning electron microscope and other techniques forms the principal source of evidence, supported by close examination of the archaeological context of both finished products and waste. Information concerning the overall social and political climate of the City of Amarna and other New Kingdom towns is also considered where this might shed light on the conditions of craftsmen in vitreous materials or the overall control of the industry.

This account is divided into eight chapters. This chapter details the aims of the project and provides an introduction to those areas of material science and archaeology that are necessary in order to understand the succeeding chapters. Firstly, what is meant by the term glass is defined and the chemical and physical nature of glass discussed. This is important because all the vitreous materials contain glass phases to a greater or lesser extent, and their physical properties are governed at least partly by the composition and atomic structure of this glass. Each of the vitreous materials referred to is next defined, since in the literature the terminology used for them is confused. The historical development of ancient glass and its place in ancient societies is then covered, concentrating on Amarna, which is placed in its archaeological and historical context.

The second chapter details the analytical procedures followed in all the scientific techniques used. The results of the analysis are then given in the next two chapters, Chapter 3 being the analytical results for the glass, frits and faience and Chapter 4 for the ceramics of the vessels and kiln walls. Interpretations of these results then follow in the next two chapters, Chapter 5 discussing the nature and sources of the raw materials for the vitreous materials and Chapter 6 the methods by which these raw materials are processed and turned into the finished objects. Chapter 7 covers all the work on the archaeological context of the production and distribution of vitreous materials, including identifying the location of the workshops that were manufacturing the vitreous material within the City of Amarna, the control and organisation of these workshops and the distribution of the finished products.

The final chapter draws together the conclusions of the preceding work and goes onto detail what these conclusion imply about the exchange of information between the various crafts at Amarna and the discovery and first use of glass and certain colorants in New Kingdom Egypt.

DEFINITION AND NATURE OF GLASS

By definition, all vitreous materials contain glass to a greater or lesser extent. The term 'glass' refers to a physical structure and atomic arrangement defined by Brill (1962) by reference to kinetic theory. He states that glass lies outside the three "classical states" of matter, since it has a mechanical rigidity that is typical of crystals, but the random and disordered structure that is typical of liquids. However, under close examination the network of interlinking of bonds found within the molecular structure of a glass does not really resemble a liquid. It is closer instead to the "polymerised state", found in materials such as plastics, rubber and dough. The "polymerised state" refers in general to disordered structures consisting of long chains of molecules more or less entangled with each other. The glassy state is a special case of the polymerised state where the individual chains have in themselves great rigidity due to a high degree of cross-linking between the chains. In general, all polymerised materials go through a glassy phase if sufficiently cooled.

Chemical structure

A nearly infinite number of different chemical compositions can be made into glasses and the chemical composition greatly affects the properties of the glass. A common formula for both ancient and modern glass is soda-lime-silica. Pure quartz melts at 1710ºC, which was impossible to achieve in antiquity and is still uneconomically high for most applications at the present day. To reduce this melting point, soda is added to the pure silica and this is often accompanied by lime, which lessens the vulnerability of the glasses to attack by water. The combination of soda-lime-silica gives a minimum liquidus temperature at the ternary eutectic of 725°C, for a composition of 21.9% Na_2O, 5% CaO and 73.1% SiO_2 (Newton and Davison, 1989). A soda-lime-silica composition is common in Roman, Egyptian and other ancient glasses, with the notable difference that these glasses have relatively high levels of impurities (Turner, 1956b) due to the difficulties involved in refining the raw materials. These impurities can, and often do, impart a strong colour to the glass and lead to a complex chemistry compared to modern relatively pure glasses.

The way in which the chemistry of the glass effects its physical structure can be considered in terms of the role played within the glass of the various elements, whether they are 'network formers', 'network modifiers' or colorants and opacifiers. Network formers are those elements within the glass that combine to form the extensive three dimensional network of the glass. The network former in most ancient glasses is silica, but other network formers include the oxides of boron, phosphorous and lead, all of which are either relatively rare or unknown in ancient glasses. Crystalline silica consists of tetrahedra with a silicon atom at its centre and oxygen at each of the corners, which combine in a regular way to create a lattice that repeats itself at regular intervals (Figure 1-2a). However, if molten silica is rapidly cooled, it lacks the time to order itself in regular way and instead the resultant solid has an internal structure that is irregular and non-crystalline (Figure 1-2b). This irregular, amorphous structure occupies a greater volume than the crystalline one, so glasses are less dense than their crystalline counterparts (Newton and Davison, 1989).

Each network forming silicon atom in a glass is surrounded by four oxygen atoms, some of which are shared by two silica tetrahedra ('bridging oxygens') and some are attached to only one ('non-bridging oxygens'). In an ideal crystalline silica lattice, all the oxygens are bridging oxygens, but glasses have silica chains with common spaces between them and these spaces are filled by 'network modifiers' (Figure 1-3), cations that lie in the spaces between the chains (Brill, 1962). The most common network modifiers in most ancient glasses are the monovalent alkali metals sodium (Na^+) and potassium (K^+) and the divalent metal ions calcium (Ca^{2+}) and magnesium (Mg^{2+}) and the presence or absence of these ions leads directly to physical characteristics of the glass. The monovalent ions are relatively loosely held within the network, bonding with only one non-bridging oxygen and as such, they lower the melting point of the structure considerably. If the glass is placed in water, the monovalent ions can move into the water, their places being taken by H^+ ions, so high levels of monovalent ions can therefore lead to a soluble glass. Divalent ions are bonded to two non-bridging oxygens tending to form new links in the network and this means they are bonded more tightly to the structure and do not move around so easily. Under normal circumstances their presence counteracts in part the solubility caused by the monovalent ions.

The colours of ancient glasses were produced in three different ways (Newton and Davison, 1989), either by dissolving oxides of the transition metals, typically Co, Cu, Mn, Ni and Fe in the batch, or encouraging the development of colloidal dispersions of insoluble particles in the glass (as in gold and copper ruby glasses, for instance), or by the use of opacifiers to reduce the transparency of the glass. The complex nature of the composition of ancient glasses leads to complications in the production of colours in the glass. The production of a particular colour not only depends on the presence of a particular transition metal oxide, but also the temperature and oxidation state of the kiln, the nature of the other oxides in the batch and the basic batch composition. For example, under reducing conditions, the presence alone of the Fe^{2+} ion in the batch leads to a blue colour in the glass, but under completely oxidising conditions the Fe^{3+} ion is created imparting a brown or yellow colour to the glass. The usual pattern in ancient glasses is for a mixture of the two forms to be present, leading a green colour. This picture is further complicated by the presence of other ions in that Fe^{3+} can be colourless in the presence of fluorides and phosphates. A further example can be demonstrated in a glass where both Fe and Mn are present, a situation which is quite common in Egyptian glasses. The formula for the equilibrium of the oxidation states of these ions is given below;

$$Fe^{2+} + Mn^{3+} \Leftrightarrow Fe^{3+} + Mn^{2+}$$

The Fe^{3+} and Mn^{2+} are the more stable states, so the equilibrium tends to move to the right. Given a kiln with fully reducing conditions when melting, the equilibrium is forced to the left giving a deep blue ferrous ion plus a colourless manganese leading to an overall blue glass. In fully oxidising conditions, the ferric ion gives a brown or yellow colour and the manganese a purple, leading to a brownish purple glass. In between, a whole range of colours can be produced from greens and yellows to pinks and so forth. Even colourless glass can be produced when the purple from the manganese just compensates for the yellow from the iron. The amount of transition metal ion needed to impart a colour varies depending on the ion involved. Some colorants are very strong, for example only about 0.10% CoO is required to produce a blue glass, and 0.025% CoO will produce a slightly bluish tinge. In contrast, to produce a blue from using FeO would require some 0.5-1.0% FeO and from CuO fully 1-2%.

Colloidal dispersions use a different technique for the production of colour. Here, minute particles are present in the glass (as opposed to ions which are fully dissolved as discussed above). Most commonly gold and copper are used to produce glasses of red and orange colours. The first recorded use of this technique is the twelfth century AD for copper and the seventeenth century AD for gold. They are therefore considerably later than the current study and will not be considered further.

Opaque glasses can be produced by incorporating tiny bubbles or other dispersed materials into the melt. The most common way of producing an opaque glass in antiquity was to add specific opacifying agents (Turner, 1959) which result in opaque glasses by producing tiny particles within the glass that scatter the light. In pre-Roman and Roman glasses, the opacifiers are calcium antimonate ($Ca_2Sb_2O_7$) for opaque white and lead antimonate ($Pb_2Sb_2O_7$) for opaque yellow. These are often combined with copper in Egyptian glasses to produce opaque blue and green respectively. Post-Roman and medieval glasses frequently use tin oxide (SnO_2) and lead stannate as an alternative to calcium and lead antimonates.

Table 1-1: Viscosities in poises of various materials at room temperature (Brill 1962).

Viscosity	Material
10^{-2}	water
10^{0}	SAE motor oil
10^{1}	molasses
10^{8}	cheddar cheese
10^{9}	solid pitch
10^{11}	lead metal
10^{15}	aluminium metal
10^{19}-10^{22}	most glasses

Table 1-2 : Viscosities of various significant points in the working of glass (Brill 1962).

Viscosity	glassworking term
10^{4}	working point
$10^{7.65}$	softening point
$10^{13.5}$	annealing point
$10^{14.5}$	strain point

Physical properties

At room temperature, glass tends to be brittle and breaks with a typical conchoidal fracture, but at higher temperatures glass can flow and be moulded and drawn into different shapes. Its physical properties can be defined in terms of viscosity, thermal expansion, density and hardness.

An important difference between the physical properties of a glass and a crystal is that crystals have a very sharp melting point and glasses do not. This is due to the uniform nature of the chemical bonds within a crystal compared to the distorted and stressed bonds of a glass which often depart from their natural and stable configurations. This means that glasses, instead of having sharp melting points, soften gradually as the temperature is raised and finally become fluid. This gradual softening, often over several hundred degrees centigrade, makes it convenient to discuss glasses in terms of viscosities (Brill, 1962), which, in a qualitative sense, is the resistance that a liquid offers to flow. The viscosities at room temperature of various common materials are shown in Table 1-1.

As a glass is heated, the viscosity drops as the inflexible molecular network breaks down into smaller units. When the viscosity falls below 10^7 poises, the glass becomes deformed under its own weight, whilst at 10^4-10^3 poises, the glass will flow into a mould. If heated strongly enough the viscosity may fall to 10^0 poises. Several different viscosities within this series are given special terms (listed in Table 1-2) to make the comparison of the properties of different glasses possible.

At some point in this range of viscosities, glasses change from the "glassy state" into the true liquid state, but there is no clear-cut constant point of transition. It should theoretically be possible to convert any liquid into a glass if it is cooled fast enough, but in practice it is only liquids with high viscosities close to their freezing points (such as silicates) that are easily converted into glasses. This is because, in general, liquids with high viscosities have rather large and complex molecules, that require more time to arrange themselves into complicated crystal lattices. Even at very rapid rates of cooling, smaller molecules such as water still manage to arrange themselves into perfect crystal lattices.

Glasses, along with most materials, show thermal expansion when heated. This expansion can be measured experimentally or calculated from the chemical composition of the glass and most glasses have a coefficient of linear expansion of between 5×10^{-6} to 10^{-5} per °C. Glasses with high silica have a lower expansion than low silica glasses because the coefficient of linear expansion for silica is low.

Glass densities have a very wide range, anywhere between 2400 and 5900 kgm^{-3}, as is to be expected from their highly variable compositions. Table 1-3 shows the densities for glasses and glass related materials. Hardness is a difficult property to define since it depends on other properties in the material such as brittleness or elasticity. The Mohs scale of hardness is used, with glasses varying between 4.5 and 6.5. For comparison, apatite is 5, orthoclase is 6, and quartz is 7 on the same scale.

DEFINITION AND NATURE OF OTHER VITREOUS MATERIALS

The terminology used for ancient vitreous materials is confused and in several instances the same term is used by different authors to define different materials. It is therefore important to give a clear indication of exactly what material each of the terms used refers to and to be consistent in their use throughout. In addition to 'glass', defined above, the terms 'faience', 'frit' and 'Egyptian blue' are all used frequently in nearly all the chapters of this work.

The most common form of vitreous material at Amarna, and indeed through most of Egyptian dynastic period, is faience, a ceramic with a ground quartz or sand body and an alkaline glaze. An SEM photomicrograph of a cross-section through typical piece of faience is illustrated in Figure 1-4, showing that faience has a layered structure, consisting of a coarse grained core with little glass, a quartz-free glaze and an interaction layer where the quartz grains are cemented together by glass. The thickness of the glaze and interaction layer and the amount of glass in the core varies according to the method of production of the faience and the conditions of temperature and humidity that were prevalent during the preparation of the raw materials (Vandiver, 1998).

The term "faience" stems from the popular medieval tin-glazed products of Faenza in Italy, but has spread to encompass considerably more than this. It was initially applied to objects from Egypt or Egyptian objects found in contemporary foreign contexts, so the term "Egyptian faience" is therefore frequently used. However faience has widespread occurrence, and some authors have argued that the other countries of the Near East were perhaps more dominant producers than Egypt (Tite, 1987). In this work, the term 'faience' is used to describe a glazed ceramic with a ground quartz core and does not imply anything about the possible place of manufacture of the object.

Lucas and Harris (1962) divided Egyptian faience into a number of different types, the most common of which was a blue glazed faience with a white or near white core which they termed 'ordinary faience'. In addition to this ordinary faience, they went on to propose six variants, given the letters A to F. Variant A is the same as ordinary faience, except that a thin layer of fine grained ground quartz or sand, similar in some ways to a slip on a clay based ceramic, was applied between the coarser core and the glaze (Figure 1-5). The next three variants referred mainly to colour variations, Variant B being black faience, Variant C red faience and Variant D deep blue faience with a deep blue, hard vitreous core. Variant E was the name term Lucas and Harris gave to glassy faience, which since it had no glaze, other authors (Kaczmarczyk and Hedges, 1983) have dismissed as not faience at all. The final category, Variant F, refers to faience with a lead glaze as opposed to an alkali glaze. This variant was also dismissed by Kaczmarczyk and Hedges (1983) since they were unable to find "a single specimen having other than an alkaline glaze". The Lucas and Harris classification therefore classifies faience by varying criteria, sometimes in terms of microstructure, sometimes colour, sometimes glaze composition. It misses out some types of faience, like white faience, altogether and some of its variants are disputed. It is therefore of mixed usefulness, but still widely referred to in the literature, so it is retained and used here, but with some reservations.

The term 'frit' is one of the most misused and confusing in the terminology used for vitreous materials. It has been used by some authors to describe any sintered material, sometimes including Egyptian blue (Tite, 1987), and by others to describe an intermediate stage in the production of a glaze. However, its most common use is to describe an intermediate stage in the production of glass. In theory, the primary heating of the raw materials of the glass to a temperature usually around 700°C drives off unwanted carbon dioxide which could otherwise make the resultant glass cloudy and the alkali, which otherwise would be soluble, immobilised. This stage in glass production has been termed 'fritting' and the material derived from the process 'frit'. This frit is then crushed, purified and melted to give the finished glass. Therefore to some authors (Petrie 1894, Lucas and Harris 1962 and others), the term implies more about the chain of production in which the material is perceived to be involved than about the physical nature of the material itself. Since in archaeological material it is often not known what exactly the chain of production was nor the place of a material within it, then the usefulness of the term is limited. However, due to the fact that the term frit is so common in the early literature it is retained in this work, but used only to describe the physical nature of the material, that is to say quartz grains cemented in a glass, and not to imply a possible link to the manufacture of any other material.

'Egyptian blue' was the first synthetic pigment, and one of the first ancient materials to be subjected to a scientific analysis. It is a blue, sintered, polycrystalline material consisting of a calcium-copper tetrasilicate (cuprorivaite, $CaCuSi_4O_{10}$) and quartz embedded in a glass matrix. Here the term 'Egyptian blue' is solely used to describe this bulk material, although in the literature it is often used for both the bulk material and for the cuprorivaite mineral itself. Although Egyptian blue comes in a number of shades of blue (ranging from very pale to dark) and a range of hardnesses, the variation within the material is not relevant for this work and therefore all these variations are simply termed 'Egyptian blue'.

HISTORICAL DEVELOPMENT OF VITREOUS MATERIALS

The earliest vitreous materials in Egypt are glazed quartz and steatite and these are known from the Predynastic Badarian period (c.4000 BC, Vandiver 1983a) onwards. It is possible that the glazing on quartz and steatite led on to the production of faience, but the earliest faience is also from the Predynastic period (although usually dated slightly later) and the difficulties involved in firmly dating Predynastic material means that this sequence cannot conclusively be proved (Vandiver 1983a). However what is clear is that faience very soon becomes the most abundant vitreous material and by the Middle Kingdom, glazed steatite and quartz are relatively rare. Egyptian blue is known from the 4th Dynasty (mid-third millennium BC, Lucas and Harris 1962), so appears to be later than the production of the glazed materials.

18th Dynasty innovations

Following its initial discovery, the development of the techniques for the working of faience was slow, and by the Middle Kingdom, this technology had stabilised into particular methodologies following which for several centuries there was very little change. However, around the start of the New Kingdom (1550 BC), several innovations appear within a very short period of time, including the use of new colorants, the development of a vitreous faience with a hard core termed 'Variant D' and the earliest controlled use of glass in the form of glass vessels. These innovations in vitreous materials fit into a larger pattern of innovations throughout the Egyptian crafts, including the first regular use of bronze and the development of chariotry and new weaponry. Little work has been done on exactly which of these innovations is the earliest and whether there is any links between them. The new colorants are a deep cobalt blue, lead antimonate yellow and calcium antimonate white, and all of these occur on the earliest securely dated glass vessels (discussed in more detail below), and cobalt blue is also common in vitreous faience. These new materials and colours continue to be used throughout the New Kingdom.

The discovery of glass

It is generally thought that true glass developed out of a background of faience working (Bimson and Freestone 1987, Oppenheim et al 1970). Barag (1962, 1970) states that North Mesopotamia is the most likely area in which true glass was first made and places the first glass-making in the sixteenth-fifteenth centuries BC, when the first core formed glass vessels appear. Bimson and Freestone state that the techniques involved in the development of faience were antecedent to the subsequent development of true glass. It is therefore widely accepted that there is a causal relationship between faience working and the development of core formed glass vessels. Unfortunately, beyond the apparent intellectual link between the two materials there is very little hard evidence to link the processes involved.

Peltenburg (1987) argues that if the development of glass is solely through the faience-working tradition, then the most likely time for the discovery of glass is during a period of intensive faience working, particularly a period when larger items are being produced. Since the very earliest glass vessels show an impressive array of colours then it would be logical to conclude that the faience-working should carry polychrome glazes. Unfortunately, there are relatively few faience vessels in the relevant Middle Bonze Age of such sites as Tell Brak, Chagar Bazar and Tell al-Rimah, the centre of the distribution of the first core formed vessels. Alalakh in Northern Syria has some of the earliest core formed glass vessels that may be 16th century BC, along with some of the earliest occurrences of polychrome glass beads in Level VII of this deposit, although the dating is somewhat problematical. However, the pre-Level VII deposits, whilst they are badly preserved, still give little indication of a pre-glass development "surge" in faience production as expected by Peltenburg. This pattern is common elsewhere in Northern Mesopotamia where there are apparently no developments in faience technology which presage the emergence of glass.

While it should be stressed that the quality of the evidence is poor, Bimson and Freestone (1987) believe that there does not appear to be any pronounced move toward the production of glass in any of the relevant faience industries. Studying the earliest glass artefacts, there are often problems with the dating and the provenancing. Peltenburg (1987) presents a table of the earliest occurrence of pre-1500BC small glass finds and a significant number of these finds are probably the result of the re-use of an accident that occurred in the firing of a faience. Peltenburg points out that there is no increase in the general number of finds through time up to 1500BC. He attributes the larger number of finds in his table that are from Egypt to the facts that it has been better studied and the soil conditions lead to better preservation. Therefore, there appears to be no area where there is gradual series of developments in the production of faience that seem to logically lead onto the first glass and the realisation of its true potential. As has often been pointed out, glass-making appears suddenly, and while faience-making supplies a general background, Peltenburg believes that it does not provide the direct impetus.

The development of early glass

Peltenburg characterises two stages in the development of early glass-making. The first stage accounts for the pre 1600BC glass which is produced apparently accidentally. The use is sporadic and infrequent, and the particular properties of glass may not have been evident, or known to the maker. Glass artefacts occur in isolation, which suggests to Peltenburg that the technology was still not firmly established or well known. The second stage in the development occurs around 1600BC in several areas and involves the development and use of several specialised techniques for the working of glass. Glass is produced over a sustained period rather than in isolated incidents, but its production is still restricted to a limited number of sites. This final stage glass is typified by objects that illustrate that the particular properties of glass have been fully realised and the pieces therefore exhibit characteristics unique to glass-working. Frequently the artefacts produced are luxury items, either core formed vessels or jewellery, often linked to temples or palaces. The various techniques needed for the production of this type of glass are suddenly developed, including fusing, trailing and the incorporation of new metal oxides for the production of new colours. Peltenburg draws analogies between this sudden development of a whole glass technology with the explosion in the manufacture and use of iron. He sees that after a long period of experimentation, the "particular historical circumstances are the mainspring" for the development. The historical situation in the 17th to 16th centuries BC in the Near East is not well understood, but seem to centre on an alliance between the Hurrian and intrusive and politically dominant Mitanni groups. The mixture seems to have developed into a heterogeneous society, founded on novel political institutions. Peltenburg sees this historical background where there is an environment conducive to the modification of historical practices as being the significant factor concerning the sudden development of glass. It also accounts for the distribution of the early core-formed glass vessels, and gives a reason why after about 1000 years of sporadic and on occasion accidental production, the discovery was not made and exploited earlier.

There has been some suggestion (Peltenburg 1987, Kaczmarczyk and Hedges 1983) that the metal oxides used for colorants often approximate to the ratios found in contemporary bronzes. Glass is sometimes found in association with gold appliqué and with bronze which suggests that the glass and metal-workers at least sometimes worked together closely. This is also apparent from the glassmakers' knowledge of the properties of metal oxides used in colorants including copper, cobalt, antimony, and lead. Peltenburg argues that the techniques of glass production are also close to those of metal-working, since while faience is produced in many cases essentially by a cold process, glass involves a hot technology, and the working of plastic glass is similar to the working of metal in an equivalent state. Core-forming is a common metallurgical technique, and very similar in both crafts. As Peltenburg warns, however, the theory of the inter-relationship between glass and metal- workers "at present rests on all too slender a foundation". Further work in this area will be required.

The earliest glass in Egypt

The most securely dated examples of glass do not appear in Egypt until the 18th Dynasty. There are examples of glass earlier than this, but the attributions of many of these are still controversial (Lilyquist and Brill 1995). The most secure pre-18th Dynasty glass includes two scarabs from the 12th Dynasty, one opaque blue, the other turquoise blue, both inscribed for officials and firmly dated. Many of the other artefacts previously thought to have been of glass are now placed in other categories, but it is nonetheless clear that glass was on occasion produced before the New Kingdom (Lilyquist and Brill 1995).

Much work has been done on the early glass of Egypt, but there is still some doubt as to the nature of the earliest finds. Lilyquist and Brill (1995) have made it clear that there was deliberate production of beads and amulets at the very beginning of the 18th Dynasty. Glass inlays in the 17th Dynasty Ahhotep treasure are associated with objects bearing the name Kamose (last pharaoh of the 17th

Dynasty) and Ahmose (first of the 18th). Beads, plaques and amulets have all been found inscribed for Ahmose and Amenophis I. A preponderance of turquoise blue (copper coloured) glass is noticeable, but colourless glass is known from Deir el-Bahri and inscribed for Hatshepsut and Senenmut. Fragments of a glass vessel found around the tomb of Tuthmosis I seem to indicate that the technical ability to make vessels had been developed. Lilyquist and Brill, however, point out that these fragments may not actually come from the tomb, but may date from the time of Hatshepsut, when a reburial took place, suggesting that the technology required for the manufacture of vessels dates to the reign of Hatshepsut. It is clear though that at least by the time of Hatshepsut/Tuthmosis III vessels were being produced.

The tomb of the foreign wives of Tuthmosis III at Wadi Qirud contains over 1000 glass beads as well as a number of glass vessels. Other glass pieces were found in the tomb of Tuthmosis III himself, and are among the earliest vessels to be found in a securely dated context. Scenes on the walls of the tomb of Rekhmire (Thebes TT100, Davies 1935), who was vizier under Tuthmosis III, may shed light on the origin of this very early glass. In the first hall of the tomb is a representation of foreign peoples making offerings to the King, which are being dutifully recorded by Rekhmire and his assistants. Amongst the offerings from the people of Retenu, established as being North Syria and in this period the Mitanni, are two glass vessels one with a marbleised pattern, and the other predominantly blue with a thick gold rim. The depictions of these two vessels are quite similar to two real vessels found in the Wadi Qirud tomb of the foreign wives of Tuthmosis III (Lilyquist and Brill 1995), and while they may not represent these actual vessels, they do indicate that the giving of such vessels is a reality and that the vessels in the Wadi Qirud tomb are probably foreign imports. Lead isotopic analysis (Lilyquist and Brill 1995) of the marbleised goblet MMA 26.7.1175 also suggests that the lead in this vessel is from a non-Egyptian, probably Mesopotamian source, which supports this interpretation. It is therefore possible to suggest that while some of the glass from these tombs may have been from Egypt, other pieces were probably imported.

Further evidence for glass imports during the reigns of Amenophis III and Akhenaten comes from the diplomatic correspondence known as the "Amarna letters". These have yielded a number of mentions of materials that are thought, by reference to other contexts to be raw glass. Oppenheim (1973) discusses two of them, *eḫlipakku* and *mekku*. The word *eḫlipakku* occurs several times (Amarna letters EA323, EA314 and others), all in the context of demands for this substance by the Pharaoh from his regents and various princelings in various cities in Palestine. A further letter, this time from the royal palace of Ugarit (published in *MRS* 12 as no.6), is one of six dealing with overland trade and lists a series of transactions for which the governor of the town has to pay. Usually the payment is to be in gold or silver, but in the case of no.6, the payment is to be in *mekku* which the scribe "finds for some reason feels in need of a explanatory gloss and so he adds after a "Glossenkeil", the word *eḫlipakku*. Thus we have two words for the same thing" (Oppenheim 1973, p260). The term *mekku* is used in the glass making texts (discussed further in Chapter 5) to mean "the raw material used by craftsmen who fashion the glass containers, beads, etc." of the mid second millennium BC (Oppenheim 1973, p261). Thus Amenophis III and Akhenaten are demanding raw glass from Palestine and Syria, suggesting an import of glass into Egypt even as late as the Amarna period and, as Oppenheim says, "this is in spite of the fact that all ingredients necessary to produce glass objects were abundantly available in Egypt".

GLASS MAKING SITES

Evidence for the local manufacture of glass is rare in the archaeological record and according to Moorey (1994), "often of equivocal significance". Outside Egypt, large and small ingots of glass have been recovered from several sites in Mesopotamia. These include the 'Mitannian Palace' at Tell Brak. At Kar-Tukulti-Ninurta and Nuzi, they have been found in contexts that are dated to the 13th and 14th centuries BC. Moorey correctly states however that this gives little more information than that glass was available in this form at this time.

During the reign of Amenophis III there is evidence of glass-working at Malkata (Lilyquist and Brill 1995, Keller 1983), a palace temple complex at Thebes, and during the reign of his son, Amenophis IV or Akhenaten, there is more extensive evidence from Amarna. The excavated material from Malkata has subsequently been lost. Only preliminary reports were ever published and of the glass manufacturing equipment only a few glass rods can now be found (Keller 1983). Material from Amarna is better preserved and constitutes the best evidence for glass making of the period in the ancient world.

AMARNA - HISTORICAL BACKGROUND

"Amarna" was the capital city of Amenophis IV, or Akhenaten as he was later known. During the reign of his father, Amenophis III, life in Egypt is characterised by a time of virtual peace compared to the vigorous military expansion that had gone before. The political influence of Egypt was spreading, however, Amenophis III's name appearing in such diverse locations as Crete, Mycenae, Aetolia, Anatolia, Yemen, Babylon and Assur (Aldred 1988, Kemp 1989). Amenophis III was one of the greatest builders of the New Kingdom, financing new works throughout Egypt. When he died in the 39th year of his reign, possibly after a short co-regency with his son, Amenophis IV, he left an Egypt of political and religious certainties, a state of almost unprecedented strength and respect both at home and abroad (Kozloff and Bryan 1992). Amenophis IV was the sole ruler from 1378/1352 BC onwards. He was educated at Memphis, but was not trained for the throne from birth, having an elder brother, probably called Tuthmosis, who died at a young age to leave Amenophis IV the heir. At the beginning of his reign Amenophis IV started a conventional programme of building to the god Amun (Kemp 1989) and completed some of the works of his father. He also built a new temple, the Per-Aten, for the god Aten at Karnak, and here there is a feeling of a certain novelty and hastiness in his constructions. For example, the sandstone blocks used in construction were unusually small, which suggests a relatively unspecialised work force (Aldred 1988). The art on the blocks is cruder than that that had gone before and characterised by lively and realistic scenes, which if certain contemporary sources are to be believed, was taught to the sculptors by the king himself. When the Aten cult lost favour and the temples were destroyed, these smaller blocks, known as *talatat* were used as the infill to one of the pylons on Horemheb, and thus 12,000 blocks have been recovered and an attempt made at reconstruction. They provide a valuable insight into the early years of Amenophis IV's reign (Aldred 1988).

Early on in his reign and certainly before the 5th year, Amenophis IV made a bold departure from the traditional careers of the kings of Egypt in that he attempted religious reform (Kemp 1989) seeking to create what appears now to be a new and simpler cult. Much of the complexity of Egyptian religion he simply ignored. The dominant god at Thebes, Amun-Re, was deleted from the record in an organised and systematic period of administered iconoclasm (Kemp 1989) and replaced with a single god, known to the Egyptians as the Aten and represented as the sun's disc. The new temples to the Aten were to be large open courtyards filled with altars, not the traditional darkness and secrecy associated with the old cults and the power of the priests. At least by the 5th year of his reign, the King had gone so far in his adoption of the Aten that he had seen fit to change his names. He changed his Horus name, his two ladies names and his Golden Horus name to remove all references to Karnak and Thebes and replace them with the Aten and Akhetaten. He changed his nomen 'Amenophis' to 'Akhenaten', usually translated as 'Servant of the Aten'.

The founding of Amarna or 'Akhetaten'

In the 4th year of Amenophis IV's reign he and his wife rode out to the site of their new capital, which had been revealed to the king 'by the Aten himself' (Redford 1985) and was to be called 'Akhetaten', translated as 'Horizon of the Aten". The site was roughly halfway between Memphis and Thebes, on a flat stretch of desert on the east bank of the Nile. The site had apparently no previous occupants of any significance and was to all intents a virgin site. The reason it was chosen is unclear, but the stelae set up to record the deed state that Akhenaten was guided by his father the Aten and that it was untainted by the worship of previous gods. Some have speculated that the site may have been chosen in part for its twin hills to the east, between which the sun rises, making it look like a huge representation of the hieroglyph *3ḫt*, meaning "horizon". Perhaps this was what Akhenaten was looking for as he travelled along the Nile looking for a site.

The city was started in the 5th regnal year, and built in great haste allowing Akhenaten and the royal court to move to the site in either the 7th or 8th regnal year. Akhenaten sealed his commitment to the new city by promising that the tombs of himself and his family would be built in the eastern hills in a new Valley of the Kings and his courtiers were expected to do likewise (Kemp 1989). The life of the city was however, brief. The king died in the 17th regnal year and what happened immediately thereafter is unclear, but his eventual successor was Tutankhamen. During Tutankhamen's nine year reign there was a complete return to the old orthodox Egyptian pantheon, with Amun-Re as its head. Akhetaten was progressively abandoned, although the waterfront continued in use for some time. The city became a ghost town, but remained largely complete until the 19th dynasty, when Rameses II carried off most of the stone from the public buildings for use in his new building work at Hermopolis, just across the river. The remains were slowly swallowed by the desert and the site never built on again to any great extent.

Excavation of the site

Amarna is the modern name for the city of Akhetaten. In older papers the name Tell el-Amarna is used, but this is an incorrect combination of the modern village of El-Till in the north and the name of the tribe inhabiting the district, the Beni 'Amran. The nature of Amarna was revealed by excavations in the late 19th and 20th centuries. Flinders Petrie embarked on the first scientific evaluation of the site and the first excavations began in 1891, uncovering the royal palace and making a survey of the site with the aid of a young Howard Carter (Petrie 1894). From 1904 to 1914 the German Oriental society gained the concession. Under the direction of Ludwig von Borchadt, they began the clearance of the eastern residential part of the city, including the studio of Tuthmosis and the numerous royal busts and models that it contained (Borchardt and Ricke 1980). After the Great War the Egypt Exploration Society gained the concession and held it between 1921 and 1937. They made a complete clearance of the main city, including the workmens' village. Since 1977, the EES have resumed a systematic study of the site, making further studies of various of the remains.

Glass-working and glass-making at Amarna

Petrie's excavations at Amarna in 1891-2 revealed evidence of glass-working and glass-making (Petrie 1894). Petrie had a deep knowledge of craft processes which helped him to build up a coherent account of the technology of glass production. Petrie (1894) reports that he found several ("three or four") glass manufactories and two large glazing works, along with extensive collections of clay moulds for the manufacture of faience objects. Subsequent work by Turner (1954) has made a significant contribution to the study of glass working at Amarna and included some of the first modern scientific investigations of the material found. This work has been continued by Nicholson (1995a) who opened new excavations at Amarna in the "moulds" area. Beginning in September 1993, 100 square metres were examined and much of the material recovered related to faience manufacture, leading Nicholson to suggest that the manufacture of glass and faience were carried out on the same site perhaps by the same craftsmen. In addition to faience moulds, some glass waste was recovered, along with a few pieces of glass cane and the fragments of at least one cylindrical vessel. Evidence for a pottery workshop was also found, including a small kiln, typical of the pottery kilns found elsewhere at Amarna along with further kilns full of highly fired clay slag. Further work is necessary to make clearer the function of these kilns, but Nicholson (1995a) believes that "the glassy nature of the slag found in the kilns, their proximity to glass waste and faience production, and to finds connected with the glass-making/working, must make their use in the glass industry a strong possibility". Material from the EES excavations, supported by fragments from museum collections form the principal evidence for this study.

Chapter 2: Analytical Methodology

INTRODUCTION

A number of different analytical techniques were used to determine the microstructure and composition of the vitreous materials and related ceramics obtained for this project. The primary equipment, used for the analysis of the majority of the samples, was the scanning electron microscope (SEM), which has the capability of both analysing the chemical composition of a sample and giving an image of a polished section from which an assessment of the microstructure can be made. Results from the SEM was supported by other analytical techniques which were applied to a smaller subset of the materials, according to the particular information required from the analysis. Each of the these other analytical techniques has particular strengths, such as high accuracy and precision or wide range of possible elements that could be analysed, but also particular weaknesses, such as complicated and lengthy preparation procedures. Each of these techniques therefore had to be carefully employed in order to acquire the desired information.

SCANNING ELECTRON MICROSCOPE

In the scanning electron microscope (SEM), a polished sample is placed in an evacuated chamber and subjected to a stream of electrons from an electron gun which excites the sample and causes it to emit X-rays, which have wavelengths and energies characteristic of the elements present. The SEM is primarily an imaging system. In backscattered electron mode, the phases present can be distinguished on the basis of differences in their atomic number which are revealed by different shades of grey, high atomic numbers giving great concentrations of backscattered electrons and a lighter grey to white image. In addition to this imaging ability, two detectors are available that will give chemical compositions, one which scans the spectrum, measuring each energy/wavelength of the X-rays given off individually (wavelength dispersive system, WDS) and the other which measures all the energy/wavelengths simultaneously (energy dispersive system, EDS). The WDS has the advantage of having high resolving power in distinguishing between the wavelengths of very similar values given off by different elements, but its reliance on wavelengths means that it can only be used at high magnifications where area to be analysed is small. The EDS measures the energy of the X-rays so can be used for large areas and bulk analysis, but unfortunately lacks the resolving power, so is less accurate for elements whose peaks are close together, especially where major peaks lie close enough to each other to overlap, for example the $K\beta$ peak for iron and $K\alpha$ peak for cobalt.

The Cameca SEMprobe at the Research Laboratory for Archaeology and the History of Art, Oxford was used for the majority of the microstructural analysis and all the WDS analysis of the samples. This included all the glass analyses, the glass matrix analyses for the frits and faience and the some of the glaze analyses. The Jeol JSM840 SEM at the British Museum was used to give additional imaging and most of the EDS analysis, specifically the bulk compositions for the ceramics, frits and faience and the remainder of the glaze analyses.

Sample preparation method

The sample was cut to give a cross section through as many of the phases or components of interest as possible and ground on a diamond grinder to give a flat surface. It was then placed, flat surface down, in a 25mm diameter plastic mould, small samples being supported with plastic or metal clips. EPOFIX resin was mixed and poured into the mould under vacuum to expel the air from the resin. The moulds were filled to a depth of about 15mm or enough to cover the sample if this was thicker, and kept under vacuum for 60 minutes to allow the resin to penetrate the sample. The mould and samples were then brought back up to atmospheric pressure and the resin left for 24 hours to harden.

The resin mounted sample blocks were then taken out of the moulds, ground down to expose the flat surfaces of the samples and then polished following the sequence laid out below;

1) TEXMET 1000 cloth, 9 micron diamond polishing compound, 30 minutes at 100rpm
2) TEXMET 1000 cloth, 3 micron diamond polishing compound. 30 minutes at 120rpm
3) MASTERTEX cloth, 1 micron diamond polishing compound. 15 minutes at 150rpm
4) MASTERTEX cloth, ¼ micron diamond polishing compound. 10 minutes at 200rpm

If scratches still persisted after this procedure, stages 3 and 4 were repeated. The block was then carefully cleaned down using an alcohol soaked lens cleaning tissue to remove any grease and placed in the carbon coater and a uniform, thin coat of carbon applied.

Analytical procedure

Standard analytical procedures for each of the two machines were set up, and all samples run on each of the machines followed these conditions to ensure comparability in the results. The choice of

Table 2-1 : Standards for SEM-WDS analysis by Cameca SEMprobe.

Element	Standard	Crystal	Peak	keV	I (c/s/nA)[1]	detection [2]
Na	albite	TAP	Kα	1.041	19.9	0.077
Mg	MgF2	TAP	Kα	1.254	178.5	0.033
Al	Al2O3	TAP	Kα	1.487	453.6	0.027
Si	quartz	TAP	Kα	1.740	469.7	0.042
P	apatite	PET	Kα	2.015	78.0	0.017
Pb	PbTe	PET	Mα	2.346	60.9	0.065
K	orthoclase	PET	Kα	3.312	12.7	0.016
Sn	SnO2	PET	Lα	3.444	316.6	0.039
Sb	metal	PET	Lα	3.605	450.2	0.042
Ca	CaSiO3	PET	Kα	3.609	400.9	0.017
Ti	metal	LIF	Kα	4.508	192.8	0.035
Mn	metal	LIF	Kα	5.895	303.1	0.026
Fe	metal	LIF	Kα	6.400	301.8	0.031
Co	metal	LIF	Kα	6.925	279.2	0.042
Ni	metal	LIF	Kα	7.472	259.3	0.038
Cu	metal	LIF	Kα	8.041	224.5	0.053
Zn	metal	LIF	Kα	8.631	283.1	0.074

Table 2-2 : Standards for SEM-EDS analysis by Jeol JSM840

Element	Standard	Peak	keV
Na	Sheffield glass #1	Kα	1.041
Mg	periclase	Kα	1.254
Al	corundum	Kα	1.487
Si	wollastonite	Kα	1.740
P	apatite	Kα	2.015
Pb	Sheffield glass #1	Mα	2.346
K	orthoclase	Kα	3.312
Sn	metal	Lα	3.444
Sb	metal	Lα	3.605
Ca	wollastonite	Kα	3.609
Ti	metal	Kα	4.508
Mn	metal	Kα	5.895
Fe	metal	Kα	6.400
Co	metal	Kα	6.925
Cu	metal	Kα	8.041

accelerating voltage is determined by the elements present within the specimen. There is a general rule that the accelerating voltage (in kV) should be around twice the highest excitation energy, E_c (in keV) of any significant peak of the elements that it is desired to analyse (Reed 1995). Copper is an important element in the samples and E_c for the Kα peak of copper is 8.980 keV. The intensity of the X-rays given off increases with higher accelerating voltages, but this also leads to greater penetration of the sample and therefore lower spatial resolution and greater absorption, so a balance must be struck. The Oxford Cameca SEMprobe was therefore run at an accelerating voltage of 20 kV, and the Jeol JSM840 at 15kV, following the standard working conditions for ceramics on each of these machines.

The beam current affects the intensity of the X-rays produced, and therefore the time required to collect a statistically significant number of counts. If a very low beam current is selected (<1nA) the analysis time required is prohibitively high, but a high beam current can lead to sample damage. A beam current of 10nA was used on the Cameca SEMprobe and 1.7nA on the Jeol JSM840 once again following the standard operating conditions for each of the machines and giving reasonable analysis times with only small amounts of sample damage. Peaks analysed and standards used for the Cameca WDS are detailed in Table 2-1, and the Jeol EDS in Table 2-2, along with the calculated detection limits for the Cameca WDS for each of the elements.

Magnification

The ceramics and frits are relatively coarse grained and inhomogeneous. This necessitated analysis of their bulk compositions at low magnifications and therefore the use of EDS, since WDS can only be used to analyse very small areas which would give an unrepresentative result. Generally, with large samples, x50 (2mm by 2.4mm area) and x100 magnification (1mm by 1.2mm area) analytical scans were used. With smaller samples x250 was used. Individual grains within the ceramics and frits were identified using point analyses on the EDS.

The glasses were generally homogenous which enabled analysis at higher magnifications and the use of the WDS. In order to determine the optimum operating magnification for the WDS on the Oxford Cameca SEMprobe, the Corning Glass Standard 'A' was run through the WDS at various magnifications and the results compared to the reported analyses for this standard. The results are plotted at Figure 2-1, which shows that at both x1000 and x2000 the total for the analysis is less than 100% due to edge effects in analysing relatively large areas. At x8000 and x16000 the soda composition given by the analysis is much less than the reported value due to the migration of the light sodium ion away from the area being analysed. The best combination occurs at x4000, when all the results are within 2 or 3% of the Corning standard. WDS analysis of glasses was therefore conducted at x4000 magnification, representing an area of approximately 25µm by 30µm.

At the beginning and end of each session on the SEM, the Corning standard was analysed to check that the elemental analysis and the totals still fell within acceptable limits. Each archaeological sample was analysed under the SEM at least three times and an average obtained. Different areas on the sample were analysed, wherever this was possible, to check that the sample was uniform and to give a better average of the sample as a whole.

Bulk analysis of frit samples

An estimate of the bulk composition of the frits was estimated by means of image analysis. In the SEM in backscatter mode as shown in Figure 2-2a, the quartz appears dark grey, whereas the glass is a lighter grey. Some estimate of the proportion of quartz grains was made by image analysis in an attempt to provide information on the proportion of quartz to alkali added as raw material and on the composition of a final glass should all the quartz be dissolved into the melt.

Method

Images for analysis were acquired from the SEM and transferred from the SEM direct into PAINT SHOP PRO 3 in Tagged Image File Format (*.TIF). The images selected contained only the two phases of interest (quartz and glass), plus void space. Where images showed other phases or regions (for example a boundary with a ceramic vessel), these images were either discarded, or, more rarely, the unwanted region was blacked out using image manipulation programs. Low magnification images (all the images used were at x50 magnification) were used to ensure that they were as representative of the whole sample as possible. These images were then manipulated within PAINT SHOP PRO 3 to remove headers and footers and to give the highest brightness contrast possible between the phases.

All the images were then imported in *.TIF into VISILOG for processing. VISILOG has the capability of counting the number of pixels in an image that have a brightness between two thresholds set by the user. If the correct thresholds are set, the number of pixels that makes up void space (Figure 2-2b) and quartz of the image (Figure 2-2c) can be calculated and displayed as a binary black/white image. These threshold settings varied from sample to sample according to the brightness of the image. The HOLE_FILL algorithm within the VISILOG program was then applied to the thresholded images of the quartz grains. This algorithm fills in those pixels that are completely surrounded by pixels of the image and therefore removes any 'holes' in the quartz (Figure 2-2d). Examination of the images obtained with and without using the HOLE_FILL algorithm shows that while the algorithm is not completely accurate in its targeting of quartz, the areas obtained using it are closer to those seen in the SEM image (Figure 2-2a) than the unfiltered images (Figure 2-2c).

Calculations

VISILOG area analysis gives the number of pixels that make up the phase analysed. Since all the images analysed contained void spaces and many included the edge of the specimens, the void space had to be taken into account in order to give a true %Quartz value. The proportion of the image that represents quartz Q_V, is therefore expressed by the equation;

$$Q_V = \frac{Q_A}{T - V_A}$$

where Q_A is the number of pixels in the image thresholded for the brightness of the quartz with the HOLE_FILL algorithm(Figure 2-2d), T is 272384, the total number of pixels in the image, and V_A is the number of pixels of void space (Figure 2-2b). The proportion of quartz in the frit by weight (Q_W) is therefore given by;

$$Q_W = \frac{Q_V \rho_Q}{\rho_G}$$

where ρ_Q is the density of quartz and ρ_Q is the density of the glass. The value of ρ_Q was estimated as 2.55, the density of a 10% CaO Roman glass which has a similar composition to the frit glass matrices.

INDUCTIVELY COUPLED PLASMA MASS SPECTROMETRY

Inductively coupled plasma mass spectrometry (ICP-MS) is a technique used in for the measurement of both elemental and isotopic compositions of a sample and its use in archaeology is relatively new. The sample is introduced to a high velocity argon plasma torch in a liquid form where it is instantly atomised and ionised to produce a plasma of single charged ions. This mixture is then introduced to a quadrupole mass spectrometer where the entire mass range from 4.5 to 239 atomic mass units may be scanned in about one second (Young et al, 1997). The technique has the capability of analysing many elements individually by reference to standards. In theory, all elements except H, He, C, N, O, F, Ne and Ar can be measured, but in practise isotopic effects can limit this further. For example, $^{40}Ar^{16}O$ interferes with the ^{56}Fe isotope which makes up 92% of natural iron, meaning that iron composition has to be calculated based on the abundance of the minor iron isotopes which leads to significantly reduced precision. However, apart from these rare interferences, ICP-MS is particularly useful for the analysis of the transition metals and rare earth elements, where detection limits as low as 50 parts per trillion are typical (Young et al, 1997) and precision of the order of +/-1 to 3%.

The ICP-MS in the Department of Earth Sciences at University College, Cardiff was used in this project for the accurate characterisation of the composition of glasses and glazes with particular reference to the transition metals associated with their colorants, where its low detection limits for these elements make it particularly appropriate. For inhomogeneous rocks and ceramics quite large samples are required (ideally 200mg, but definitely >50mg), but very small samples can be used for homogeneous glass. Some elements of interest could not be measured, including nickel, since the Cardiff machine contains internal componentry made of nickel which can contaminate the plasma and lead to a meaningless result.

Sample preparation and analytical procedure

The samples were run in batches of between ten and thirty. Each sample was crushed, dried, weighed and placed in a savillex vial and a spare empty vial was added to the batch for the total procedural blank. The vials were then transferred to a fume cupboard and their contents wetted with a few drops of primar concentrated HNO_3. 3-4mls of romil HF was then added, the lids placed on the vials and the vials heated overnight to digest the sample. The lids were then removed and the HF evaporated until nearly dry, when 3-4mls of primar concentrated HNO_3 was added and the samples left to evaporate again, this time to complete dryness. 5-6mls of 5M primar HNO_3 was then added and heated on a hotplate for several hours. This usually resulted in crystal clear solutions, but in the rare case that there was still a residue, the sample was evaporated to dryness again and 5-6mls more 5M primar HNO_3 added and the procedure repeated until there was no residue. The samples were then transferred to individual 100ml flasks and diluted with distilled water to make a 100ml volume in total. They were then ready for running on the ICP-MS.

Each sample was run twice, each time comparing it to different suites of standards. Since it was desired to analyse both major elements like magnesium and calcium, and trace elements like tin and zinc, the sample peak intensity had to be carefully monitored to ensure that the detector was not overloaded. Often the concentrated sample derived from the sample preparation above had to be diluted 10, 50 or even 100 times and re-run in order to allow the detector to measure the more abundant elements. The results derived from the ICP-MS were measured in parts per million by weight and since all the significant elements in a sample were not measured by ICP-MS, the results derived could not be expressed as weight percents normalised to 100%. These results are therefore not directly comparable with the weight percents derived from the SEM-WDS unless ratios of the elements are used.

THERMAL IONISATION MASS SPECTROMETRY

Thermal ionisation mass spectrometry (TIMS) was used for the measurement of lead isotopic ratios of the glasses and glazes in this project that were coloured by lead antimonate. In the TIMS, the prepared sample is heated to ionisation on a rhenium wire, and the ion beam accelerated by an electric field into the mass spectrometer, where the required atomic masses are selected and funnelled through resolving slits to a series of Faraday collectors where the concentration of each isotope is measured. TIMS is a very accurate method for determining the lead isotope characteristics of a sample, with accuracy and precision typically better than 0.1%.

Sample preparation and analytical procedure

About 20mg of glass was crushed and dissolved in 1ml of 8M (molar) HNO_3 in a teflon beaker, which was then heated and evaporated to dryness. One additional beaker for the total procedural blank was used. The residues were then dissolved again in 0.1ml of 0.1M (molar) HNO_3 and evaporated to dryness. A further in 0.1ml of 0.1M (molar) HNO_3 was added along with one drop of $Cu(NO_3)_2$ in 0.1M (molar) HNO_3 and the solution diluted in 0.5ml of deionised water. The diluted solution was transferred to an anodic deposition cell, a stirrer added and the cell topped up by rinsing the beaker with a further 0.2ml of water which was also placed into the cell. Two platinum electrodes running at 1.8V were added and the cell placed into a machine providing magnetic stirring for 15 minutes. On disassembling the cell lead oxide had coated the anode, which was then gently washed in a drop of deionised water and placed in a clean beaker. The lead oxide was then removed with 0.1ml of 6M HCl which was evaporated to dryness and the residue dissolved again in 0.1ml of 8M (molar) HNO_3 to convert the lead oxide to lead nitrate. After a further evaporation, the sample was mixed with one drop of silica gel and one of phosphoric acid, and evaporated onto a rhenium filament and placed in the source chamber of the TIMS. NBS981 was also regularly run and mass fractionation in the samples compensated for by comparison with the mass fractionation observed in the runs on this standard.

INDUCTIVELY COUPLED PLASMA ATOMIC EMISSION SPECTROSCOPY

Inductively coupled plasma atomic emission spectroscopy (ICP-AES) is a standard technique for the measurement of elemental compositions in geological and archaeological material. The sample is turned into a plasma in the same sort of argon torch as that described under ICP-MS, but in this case, it is thermally excited to the point of atomic emission. This atomic spectra is then detected by a spectrophotometer and used to calculate the elemental composition of the sample. ICP-AES is free of the iron interferences discussed for ICP-MS, so is in this respect better for the analysis of low iron ceramics. It has similar precision to ICP-MS, but has lower sensitivity with detection limits typically in the range of 1 to 10 parts per million for the transition metals.

In this project, the Philips PV8060 ICP-AES at Royal Holloway and Bedford New College was used to give bulk chemical compositions for the ceramics, including both the archaeological samples and the experimentally refired material. These materials are inhomogeneous which meant that once again quite large samples of the order of 100-200mg were required.

Sample preparation and analytical procedure

The analytical procedure for ICP-AES is as described in Hatcher et al (1995). The ceramic sample was crushed to a fine powder and 100mg weighed into a teflon beaker. A 2:1 mixture of HF and $HClO_4$ added and the sample then evaporated to dryness. After cooling, it was redissolved in 10ml of 10% HCl and this solution aspirated directly into the ICP-AES. Silicate standards were run with the unknowns and instrumental drift measured with a drift monitor every 10 samples. The precision and accuracy of the runs on the standards were both better than about 2%.

X-RAY DIFFRACTION

X-ray diffraction (XRD) is a standard technique for the identification of phases in minerals, rocks and ceramics. The powdered sample is subjected to X-rays of a fixed wavelength which are scattered and reflected by the crystal lattices of the phases within the sample. The angle of the sample to the X-rays is progressively changed and the reflected X-rays detected. Constructive interferences leading to high intensities of reflected X-rays occur at certain angles where these angles are characteristic of the spaces between the planes of the lattice of the phases involved. Therefore by measuring the angles at which high intensity reflections occur, the phases can be identified.

In this project, the Philips PW1729 generator with a copper tube with a PW1710 control unit and PW1820 detector at the Department of Materials at the University of Oxford was used. All the experimental ceramics, along with a few archaeological ceramics were analysed and the results discussed in Chapter 4.

Sample preparation and analytical procedure

A 500mg sample of the ceramic was finely powdered and the powder well mixed. The powder was then placed on a glass slide and mixed with acetone, which was allowed to dry so that it gave an area of the slide about 20mm by 15mm in size a smooth, even surface coat. The slide was then placed in the sample chamber of the XRD, and the chamber secured. The sample was then analysed from a 2θ start angle of 5.00° through to an end angle of 70.00° with a step size of 0.05° and a dwell time of one second for each step. Due to high background counts, only phases making up more than about 5% of the composition of the ceramic could be identified with certainty.

Chapter 3: Glass, frits and faience

The site of Amarna has yielded a profusion of artefacts made from a wide range of vitreous materials. They can broadly be split into three groups. The first consists of materials that make up finished objects, in this case glass and faience. As discussed in Chapter 7, objects of faience are very common at Amarna and objects of glass also quite widespread. The second group of materials consists of those such as frit that are intended as an intermediate stage in the production of one of the finished materials, or are the waste products of that production. Third, there is Egyptian blue, which can be used as a pigment or as structural material in the manufacture of vessels and inlays.

This chapter describes in detail representative examples of glass, faience and frit at Amarna. Equal emphasis is given to examples of finished objects and vitreous materials thought by the excavators to be associated with or resulting from production and production debris, especially crucibles and highly fired clay.

DESCRIPTION AND ANALYSIS OF GLASSES

Table 3-1 lists those samples of glass that were acquired for analysis along with their museum location, the number used to refer to them in this study and the original museum number. A short description of the context of the glass is also given. 25 different glass samples were acquired in all. Since several of these were polychrome (for example E5613/1), this led to 29 analyses being obtained.

The typical finished products of glass making from Amarna are amulets, beads and vessels and are discussed in more detail in Chapters 1 and 7. Also included in this group are glasses that are assumed to be chemically finished glasses, but have not yet been incorporated into finished objects. The best examples of this are glass rods, which were intended for incorporation into glass vessels, but for whatever reason have not been thus used. They can be thought of perhaps as the debris of glass working as opposed to that of glass making. Typically the finished glasses in the study are translucent to nearly opaque and are strongly coloured. Often they are partially weathered in that they have a duller finish and a less vibrant colour. Macroscopically all the monochrome glasses and the individual colours of the polychrome glasses appear homogeneous, with no obvious inclusions or structures within them. Of the three polychrome examples (E5613/1, E5613/4 and E5613/6) represented in the group, the first (E5613/1) has yellow and white trails of glass, the next (E5613/4) only yellow and the third (E5613/6) turquoise blue, in each case within the dark blue glass of the body. The trails of coloured glass are typically 1-2mm thick. The glass vessel fragments have white powdery residues associated with the core material adhering to their interior surfaces. The glass rods are typically 2-3mm thick and rounded in cross section.

The second group of glasses, associated with cylindrical vessels and other production debris, are at some intermediate point in, or are waste from, the glass production process. This group also includes drips and spills of glass that have somehow been lost in the workshop and/or discarded. Samples of this group, which were acquired in order to attempt to gain as much information about the production of glass as possible, are typically adhering to the internal or external surface of a cylindrical vessel or fritting pan. Those adhering to the internal surfaces, of which AM22 and AM37 are good examples, are often attached to the white slip that was frequently applied to the interior of cylindrical vessels and appear to be the residues of glass left within the vessel when the remainder of the glass is either poured out hot, or chipped away cold. The external glass appears as drips and spills perhaps accidentally left on the vessels during pouring. The precise placing of the glass on the ceramics, along with analysis of its direction of flow, that is either towards the rim of the vessel or towards the base, and the interpretations that can be drawn from these details are considered in Chapter 4. The glasses adhering to the ceramics are typically darker than those in finished objects, tending towards the black in the case of extreme examples such as 396a. Often they appear close to opaque.

The most obvious macroscopic feature of the glasses is their colour. This is often bright, being especially striking in fresh examples. Most of the glasses examined were blue, but this varies from a very light turquoise blue (e.g. E5613/6) right through to deep blues (e.g. E5613/1 and 68498) while some have a greenish tinge (e.g. AM16). Other colours found in the samples are white, represented in trails on two samples, and mint green, yellow and purple, each with only one representative. Figure 3-1 indicates the range of the coloured glasses sampled.

MICROSCOPIC EXAMINATION

All of the samples were examined in the SEM. Most of the glasses are homogeneous and show no microstructures beyond a few minor vesicles. However, some glasses do show evidence of diagnostic microstructure, which can be important in assessing their method of production. The microstructures fall into three main groups. The first group of structures are associated with the presence of opacifiers within the glass. The second are the results of glass working as opposed to glass making and include marvering of glass rods and striations associated with drawing. The third group

Table 3-1 : Glass Samples acquired for the project

Laboratory No.	Museum	Description
Finished glasses		
E5613/2	Liverpool University	blue glass vessel fragment
E5613/3	Liverpool University	blue glass vessel fragment
E5613/6	Liverpool University	light blue glass vessel fragment
E5613/7	Liverpool University	blue glass vessel fragment
E5613/1	Liverpool University	blue glass vessel fragment, decorated with trails of white and yellow
E5613/4	Liverpool University	blue glass vessel fragment, decorated with a trail of yellow
E5613/5	Liverpool University	blue glass vessel fragment, decorated with opaque turquoise
406	Ashmolean Museum	opaque turquoise glass rod
406a	Ashmolean Museum	translucent blue glass rod
593a	Ashmolean Museum	opaque mint green glass rod
Glasses associated with production debris		
AM16	1993/4 excavations	translucent blue glass [adhering to ceramic]
AM22	1993/4 excavations	translucent blue glass adhering to ceramic with white slip
AM29	1993/4 excavations	translucent blue glass [adhering to fused clay]
AM30	1993/4 excavations	translucent blue glass [adhering to fused clay]
AM35	1993/4 excavations	translucent green-blue glass [adhering to rim]
AM36	1993/4 excavations	translucent green-blue glass adhering to rim of ceramic
AM37	1993/4 excavations	translucent blue glass adhering to ceramic with white slip
AM38	1993/4 excavations	translucent greenish glass adhering to rim of ceramic
396a	Ashmolean Museum	glass [adhering to outside of wall of cylindrical vessel]
396b	Ashmolean Museum	glass [adhering to external base of cylindrical vessel]
396c	Ashmolean Museum	glass [adhering to outside of wall of cylindrical vessel]
68498	British Museum	blue glass adhering to white slip
E5042	Liverpool University	spill of blue glass
E5051	Liverpool University	spill of purple glass
405	Ashmolean Museum	translucent blue glass waste

are associated with the edges of the glass where they contact either the wall of a ceramic or the residue of the core of the core formed vessel.. In addition, other microstructures, including apparent relict grains of the primary batch are occasionally present.

Microstructures associated with opacifiers

Opacifiers are deliberately added by the glass maker to give an opaque glass, the background to their use being discussed in Chapter 1. They are particulate, and in the samples examined in this study, two different types are apparent. The yellow glasses that form the decoration in E5613/1 and E5613/4 and the green glass rod 593a all have particles of lead antimonate ($Pb_2Sb_2O_7$) dispersed throughout. E5613/1 in which a rod of yellow glass has been applied to and marvered into a body of blue glass is illustrated in Figure 3-2. Here, in cross-section, the yellow glass (light grey) appears as a half-ellipse embedded into the blue body glass (mid grey). Throughout the yellow glass are particles of lead antimonate (white), the largest at the bottom of the image being 50μm across, but more typically they are 5-10μm across. Also obvious is the larger number of vesicles in the yellow glass (especially obvious when compared to the body glass), many of which are more linear than the usual spherical vesicles. They are therefore probably formed when the lead antimonate particles are plucked out during sample polishing for examination in the SEM.

The second opacifier observed is calcium antimonate ($Ca_2Sb_2O_7$). This is present in the opaque turquoise glass rod 406 (Figure 3-3), in the white glass trailed decoration of E5613/1 (Figure 3-5) and in the body glass of E5613/6 (Figure 3-4) and E5613/8. Once again, frequent vesicles, are commonly also present, probably caused by the plucking out of particles during polishing and these are most clearly seen in Figure 3-3, a high magnification photomicrograph of glass rod 406. The voids are typically 10-20μm long and often come in clumps. The surviving particles are typically around 10μm across, but many are much smaller, around 1-2μm across and can be seen in the ground mass of the glass. Sometimes, as in the body of glass vessel E5613/6 (Figure 3-4),

Table 3-2 : Oxide compositions for the glasses

Number	SiO_2	Al_2O_3	CaO	MgO	Na_2O	K_2O	FeO	TiO_2	CoO	CuO	P_2O_5	MnO	PbO	SnO_2	Sb_2O_3	SO_3
Blue - cobalt																
E5613/7	62.8	2.5	4.6	3.7	22.1	1.1	1.0	0.2	0.09	0.61	0.2	0.56	0.06	0.03	0.17	0.2
AM37	67.9	2.0	4.6	2.6	21.2	0.6	0.4	0.1	0.16	0.10	0.1	0.16	0.0	0.07	na	na
AM22	64.6	1.8	8.8	2.5	20.2	0.8	0.5	0.1	0.17	0.11	0.0	0.22	0.1	0.05	na	na
E5613/3	59.7	4.1	8.6	4.5	19.8	0.5	0.6	0.2	0.27	0.04	0.2	0.24	0.02	0.01	0.74	0.2
E5613/2	59.4	4.2	8.6	4.6	19.7	0.6	0.7	0.1	0.37	0.06	0.1	0.31	0.01	0.01	0.70	0.2
E5613/1	67.0	1.4	7.4	2.9	19.5	0.9	0.4	0.1	0.05	0.01	0.1	0.07	0.1	0.00	0.03	0.2
E5613/5	64.1	2.1	8.5	3.3	18.6	1.6	0.5	0.1	0.16	0.26	0.1	0.18	0.02	0.01	0.17	0.2
E5613/4	64.2	2.0	8.8	3.3	18.0	1.5	0.5	0.2	0.14	0.32	0.2	0.18	0.03	0.01	0.18	0.2
406a	63.3	1.6	9.1	5.9	17.0	1.9	0.5	0.1	0.20	0.02	0.1	0.09	0.1	0.02	na	na
EA67834	67.0	1.7	9.3	3.6	15.9	1.3	0.4	0.1	0.09	0.36	0.1	0.14	0.0	0.05	na	na
average	*64.0*	*2.3*	*7.8*	*3.7*	*19.2*	*1.1*	*0.5*	*0.1*	*0.2*	*0.2*	*0.1*	*0.21*	*0.0*	*0.0*	*0.3*	*0.2*
Blue - copper																
AM16	63.3	0.6	10.5	3.5	17.7	2.0	0.4	0.0	0.02	1.82	0.1	0.01	0.0	0.02	na	na
405	64.5	0.4	10.4	4.0	15.7	2.3	0.3	0.1	0.01	2.16	0.1	0.02	0.0	0.12	na	na
AM29	70.9	0.9	5.9	2.8	14.5	3.1	0.4	0.2	0.01	1.11	0.1	0.06	0.0	0.06	na	na
AM30	67.4	1.2	8.8	3.5	13.5	2.8	1.0	0.1	0.01	1.37	0.2	0.01	0.1	0.00	na	na
AM38	63.7	2.0	11.2	5.0	12.6	2.6	1.4	0.3	0.01	0.85	0.3	0.08	0.0	0.04	na	na
AM35	67.9	1.6	6.7	4.2	12.6	3.3	0.9	0.2	0.0	1.69	0.42	0.11	0.0	0.19	na	na
AM36	67.0	1.7	9.4	3.3	12.0	3.4	1.1	0.2	0.02	1.18	0.4	0.04	0.0	0.14	na	na
E5613/6	64.3	0.5	7.9	3.5	19.6	2.6	0.2	0.0	0.02	0.52	0.2	0.02	0.03	0.03	0.20	0.3
406	64.1	0.6	5.7	4.9	18.4	2.4	0.3	0.0	0.01	3.17	0.2	0.03	0.2	0.00	na	na
average	*65.9*	*1.1*	*8.5*	*3.8*	*15.2*	*2.7*	*0.7*	*0.1*	*0.0*	*1.5*	*0.2*	*0.04*	*0.0*	*0.1*	*0.2*	*0.3*
Opaque green - copper with Pb_2Sb_2O7																
593a	65.7	1.1	9.1	4.1	14.8	2.4	0.6	0.1	0.00	0.83	0.1	0.0	1.1	0.1	na	na
Opaque white - Ca_2Sb_2O7																
E5613/1	63.8	0.6	9.2	3.7	17.6	2.1	0.3	0.1	0.02	0.01	0.2	0.06	0.05	0.01	2.14	0.2
Opaque yellow - Pb_2Sb_2O7																
E5613/1	60.5	0.5	7.8	3.9	15.5	1.7	0.5	0.1	0.02	0.02	0.2	0.01	7.53	0.03	0.90	0.3
E5613/4	64.4	0.4	7.0	3.5	18.5	1.7	0.4	0.1	0.01	0.18	0.1	0.11	3.07	0.06	0.11	0.2
average	*62.4*	*0.4*	*7.4*	*3.7*	*17.0*	*1.7*	*0.4*	*0.1*	*0.0*	*0.1*	*0.1*	*0.1*	*5.3*	*0.0*	*0.5*	*0.3*
Purple - MnO																
E5051	64.7	0.4	8.7	2.8	19.0	2.2	0.2	0.0	0.00	0.15	0.2	1.22	0.00	0.00	0.07	0.2

the opacifier is present as streaks of particles through the glass, longitudinally quite continuous and 20-30µm wide. This may be evidence for the late mixing of calcium antimonate, where the glass has solidified before completion of mixing and stirring to give a uniform distribution through the glass. The streaks are also visible in the glass in ordinary transmitted light under a low power microscope.

Microstructures derived from glass working

Several microstructures in the glasses are caused by glass working processes. In order to decorate glasses, particularly glass vessels, trails of contrasting coloured glass were sometimes applied to the surface and then smoothed, or marvered. Macroscopically, this results in glasses like those in the top row of Figure 3-1. In cross section, the contrasting coloured rods are semi-circular in shape, the result of marvering an initially cylindrical glass rod. In Figure 3-5, two such semi-circular rods of contrasting thickness are seen, one white (left) and one yellow (right, and in Figure 3-2), which have been applied to a blue glass body.

Another microstructure that results from glass working can be seen in the glass rod 593a, illustrated in Figure 3-6. Striations in the sample that run parallel to the edges of the rod are visible and these are typically very straight and quite long. They appear to be the remains of vesicles that survived in the glass making process and have been stretched out more and more as the glass was drawn into the rod. If the working temperature of the glass is not sufficiently high, and therefore the viscosity remains low, the striations are preserved in the cooled glass rod. However, if the glass rod is strongly heated at some stage during the drawing, then these structures will be lost, which could explain why some rods, for example 406, do not show the same microstructure.

Contact microstructures

Often, where the glass is in contact with a ceramic, a range of structures can be seen. These are usually laterally limited, existing only within a few microns of the glass/ceramic boundary, but in some cases they are much more widespread. Such microstructures tend to be particularly evident where a glass has been in contact with the white slip applied to the ceramic. Figure 3-7 shows the typical limited development of an interaction zone between the white slip and the glass in sample AM22, in which a thin layer of accicular grains of wollastonite, $CaSiO_3$, (light grey) growing into the glass layer (mid grey) can be seen. The white slip is 500-600μm thick and lies immediately to the left of the glass, grading into normal Nile silt ceramic at the left of the picture. In this example, therefore, the microstructure suggests that contamination of the glass by the ceramic wall was very restricted and this is the case for many of the samples.

At the other extreme, as shown in Figure 3-8 for AM30, the contamination is microscopically evident across the sample. In the centre of the image many thin crystals of diopside $CaSiO_3$ (light grey)are preserved in the glass (mid grey), with to the left a large area of partially dissolved lime (white), surrounded by fringing wollastonite laths (very light grey). It is likely that this cluster of crystalline phases, some 1.5mm across, has become detached from the internal lining of the ceramic and incorporated within the body of the glass as a whole, albeit in close proximity to the ceramic.

Relict materials

Relict materials that have survived the glass making process and persist in the finished glasses are rare in the glasses examined, most glasses showing no relict material at all.. Sample E5042, which is a glass drop that was probably spilled during manufacture, is an exception to this, showing rare relict quartz grains and a lime rich material that may be part of the raw materials or may be a fragment of white slip that has come detached from the wall of a ceramic vessel. Very occasional rounded relict quartz grains are present in other glass samples.

COMPOSITION OF THE GLASSES

The composition of the glasses was measured in the SEMprobe using the techniques laid out in Chapter 2. The oxide compositions of the glasses, detailed in Table 3-2, show that the variation in terms of most major elements between the different colours is quite low. The average percentage silica in all of the glasses of each of the different colours is 65%, with an overall range between 59 and 71%. The soda is the dominant alkali at between 14 and 22% with potash secondary at between 0.5 and 3.4%. From the plot of soda against potash for the blue translucent glasses (Figure 3-9) it is noticeable that the cobalt blue glasses have higher soda and lower potash, than the copper blue glasses. The lime content of these blue glasses is typically between 8.0 and 10.0%, although the overall range for the blue glasses is from 4.6 to 11.2% CaO.

As discussed in Chapter 1, it is the transition elements, often present in only small amounts, that are largely responsible for the colouring of the glass and may be added deliberately, or included accidentally with another component of the glass.

The blue glasses split obviously into two groups, those coloured by copper and those coloured by cobalt. The copper glasses which are mid to deep blue contain on average 1.5% CuO, with a range from 0.52 to 3.17% CuO. They have effectively no cobalt, low manganese, tin and antimony and comparatively low alumina. The light opaque turquoise blue glasses, 406 and E5613/6 are coloured by a combination of two of the colorants, copper and calcium antimonate. The predominant colorant is copper, averaging nearly two percent, But the glass also contains calcium antimonate particles, rendering the glass opaque, and possibly also lightening the colour (see Figure 3-3). The deep blue cobalt coloured glasses range from those which have negligible copper, for example 406a and E5613/2, to those that appear to be partly coloured by copper as well as cobalt, for example E5613/7 (0.09% CoO and 0.61% CuO). There is also a clear relationship between the cobalt and certain other elements in the glasses. The cobalt glasses are much higher in alumina than those coloured by other means, averaging 2.3%, twice as much as in the copper coloured glasses. As shown in Figure 3-10, which is a plot of cobalt and alumina for the copper and cobalt coloured glasses, along with the data for the glass matrix of the cobalt coloured frits, the alumina content increases as the amount of cobalt increases. Manganese is also high in the cobalt glasses, averaging 0.21%, compared to 0.04% for the copper glasses. Again there is a link with increasing manganese as cobalt increases. This is plotted in Figure 3-11, with data for the glass matrix of the cobalt coloured frits added to make the relationship clearer. Trace levels of zinc and nickel, not present in the copper glasses are also present in the cobalt glasses that have been examined under EDS.

Opaque white glass is found in the decoration on E5613/1 (Figure 3-5), this being the only example examined. It lacks transition metals in sufficient quantities to impart a strong colour to the glass, but antimony is present at over 2% and it is this as a particulate that renders the glass opaque. The lime content of the white glass is also at the top end of the range for the glasses as a whole.

Opaque yellow glass is present as the decoration on two of the glass vessels, E5613/1 (Figure 3-5) and E5613/4. The compositions of both glasses are very similar, lacking significant amounts of transition elements. They do, however, contain lead in high concentrations. Thus, it is the high lead concentration which is in part present as particulate lead antimonate which imparts the yellow colour. Related to the yellow glass is the single opaque green glass rod 593a. The colour here is imparted by a combination of blue from its copper content of 0.83% and yellow from the presence of lead antimonate particles (see Figure 3-6), the lead in the glass amounting 1.1% PbO.

Purple is an unusual colour in Egyptian glasses and only one example of a purple glass has been examined. It shows that the dominant colorant for this glass is manganese, present at over 1% MnO with no other transition element present in significant quantities and the alumina content also low.

Glass Description Summary

All the glasses examined are of soda-lime-silica type, coloured by cobalt, copper or manganese and opacified by lead antimonate or calcium antimonate. A colorant and a coloured opacifier are sometimes used in combination to impart a third colour, for example, copper blue and lead antimonate yellow to give a green glass. Apart from the opacifiers, the glasses are mostly homogeneous, but some show microstructures that indicate possible techniques of working.

DESCRIPTION AND ANALYSIS OF FRITS

The use of the term "frit" and the previous work done on the subject is discussed at length in Chapter 1. All of the frits obtained for analysis in this project were associated with the ceramic debris from production of vitreous materials, such as fritting pans and cylindrical vessels. Table 3-3 shows the sources of the various frit samples analysed.

Macroscopically the frits consist of two phases: white quartz grains, varying in size from very fine up to 3mm across, in a green or blue glassy matrix, as illustrated in Figure 3-12, together with common vesicles with little else being observable in hand specimen. The relationship of the frits to the ceramic vessels with which they are associated is discussed with the ceramics in Chapter 4.

Microstructure

The frits were examined at a variety of magnifications in the SEM. Away from the junctions with ceramics or lime rich slips, the frits are again seen to contain two, and only two phases: primary quartz grains in a glass matrix. AM32 which is typical of the majority of the frits is shown in Figure 3-13, in which these two phases along with various amounts of void space and vesicles, are visible.

Close to the ceramics, various other phases are evident, specifically primary grains which are ripped out from the wall of the ceramic and secondary grains growing into the glass, most commonly anorthite, devitrite and wollastonite. Both the primary contaminants and the secondary growths are very localised, occurring within the 1.0mm adjacent to the ceramic boundary, the microstructures showing no evidence of any influence of a ceramic outside this zone. The ceramics appear therefore to have no influence on the bulk of the frit, with the possible exception of those frits adjacent to a very lime rich slip which are discussed later. Figure 3-14 shows frit AM31 adhering to the lime rich slip

Table 3-3 : Frit samples acquired for the project and sorted by their colorants.

Laboratory No.	Museum	Description
Cobalt coloured		
AM23	Excavations 1993/4	blue frit adhering to vitrified ceramic
AM28	Excavations 1993/4	blue frit fragment
AM31	Excavations 1993/4	frit adhering to ceramic with white slip
AM32	Excavations 1993/4	blue frit fragment
AM33	Excavations 1993/4	frit adhering to ceramic with white slip
AM34	Excavations 1993/4	blue frit fragment
36457	Petrie Museum	blue frit [adhering to vitrified fritting pan]
36465	Petrie Museum	blue frit in fritting pan
403	Ashmolean	blue frit fragment
67829	British Museum	blue frit [adhering to cylindrical vessel with white slip]
67830	British Museum	blue frit [adhering to white slip (?cylindrical vessel)]
Copper coloured		
25038	Petrie Museum	blue-green frit fragment
Highly contaminated		
67825	British Museum	green-blue coarse frit [adhering to base of fritting pan]

[...] show an association present in hand specimen, but not sampled for further work

Table 3-4 : Qualitative assessment of the microstructure of the quartz grains of the frits.

number	rounded?	bulk fractures?	b to a-quartz	cristobalite?	temperature
25038	no	no	no	no	low
AM31	few	few	?	no	med-low
36457	few	few	yes	no	med-low
67825	some	some	no	no	med-low
67830	few	some	yes	no	med-low
AM33	few	some	yes	no	med-high
AM32	some	some	no	no	med-high
36465	some	some	yes	no	med-high
AM34	some	some	yes	?	med-high
67829	some	some	some	yes	med-high
AM23	some	many	yes	yes	med-high
403	many	some	yes	yes	high
AM28	many	some	yes	yes	high

Table 3-5 : Summary of WDS analyses for the glass matrix of the frit samples.

Number	Associations	SiO_2	Al_2O_3	CaO	MgO	Na_2O	K_2O	FeO	TiO_2	CoO	CuO	MnO
Cobalt coloured												
AM31	white slip + ceramic	71.6	4.1	8.1	1.1	13.5	0.2	0.6	0.2	0.21	0.04	0.27
AM28	white slip + [?fused clay]	76.0	2.7	4.3	2.2	13.2	0.3	0.7	0.1	0.23	0.04	0.20
EA67830	[white slip]	73.8	3.5	4.0	1.6	15.3	0.3	0.8	0.2	0.23	0.04	0.24
403	[white slip + ceramic]	77.2	2.9	3.6	2.3	12.1	0.4	0.7	0.1	0.24	0.06	0.19
EA67829	[white slip + ceramic]	76.4	2.9	3.5	2.3	13.0	0.4	0.8	0.2	0.24	0.04	0.18
AM33	[ceramic + white slip]	76.5	4.8	2.4	3.1	9.4	1.0	1.2	0.5	0.40	0.03	0.63
AM32	[ceramic + white slip]	78.3	5.3	1.2	1.8	10.6	0.8	0.6	0.2	0.38	0.03	0.44
AM34	[white slip]	75.3	7.3	1.2	2.4	11.5	0.7	0.6	0.1	0.43	0.02	0.39
UC36465	[fritting pan]	78.4	5.1	1.1	2.5	10.8	0.4	0.3	0.1	0.28	0.01	0.52
UC36457	[?ceramic]	76.3	5.9	0.9	2.0	12.5	0.5	0.7	0.2	0.37	0.00	0.57
AM23	fused clay	81.0	5.7	0.8	1.0	9.5	1.1	0.2	0.1	0.19	0.01	0.20
average		*76.4*	*4.6*	*2.8*	*2.0*	*12.0*	*0.5*	*0.7*	*0.2*	*0.29*	*0.0*	*0.35*
Copper coloured												
25038	[?white slip]	72.8	1.5	2.1	0.8	13.7	1.5	0.8	0.1	0.01	6.25	0.1
Highly contaminated												
67825	[white slip and fritting pan]	64.8	10.2	4.4	1.9	8.8	1.2	5.9	1.3	0.03	0.88	0.15

of a ceramic, with the typical development of wollastonite close to the boundary in a very similar fashion to that seen in glasses (Figure 3-7). In all the frit samples, the glass matrix is uniform and microstructurally the same, showing no evidence of flow or other structure. No gradient in silica concentration away from the quartz grains is evident.

Table 3-4 shows a qualitative assessment for the microstructure between the quartz grains. Nearly all of the frits show some reaction between the quartz and the glass matrix creating a degree of rounding in the grains and making some of the boundaries diffuse, as illustrated in Figure 3-13 which is AM32, a typical frit with rounded grains in a uniform glass matrix. The degree of rounding, and therefore reaction, varies from the least in 25038 with strongly angular grains (Figure 3-15), to the most in 403 and AM28 (Figure 3-16) with possibly the most. Heat fracturing is present in all the frits other than 25038 and most developed in AM23 with the quartz in AM32, 33 and 34 also being strongly fractured. The typical cracking that runs around the edge of the quartz grains evident in AM32 (Figure 3-13), results from the parting of the grain from the solid glass matrix due to the volume reduction that occurs during the phase transition from β to α-quartz which takes place at around 573°C. Evidence of this transition is present in many of the frits, but since it is not always preserved, its absence cannot be taken as an indication that the grain has not reached this temperature. It is probable that this β to α-quartz transition is the reason behind at least some of the internal cracking as well. A further quartz phase change which is also present in some of samples is the formation of fringes of cristobalite which lie in the glass immediately adjacent to the quartz grains. These are visible in Figure 3-17, a high magnification photomicrograph of frit 403, where it is particularly well developed. Crystobalite is theoretically the stable silica form above 870°C, quartz being the stable state below this.

Discussion

Making generalisations about the microstructure and drawing inferences from them concerning the relative temperatures at which the frits were fired is fraught with difficulty, but it is fairly clear that 25038 shows the least heat alterations and that 403 and AM23 probably show the most. Beyond this, only very tentative assessments can be made and these assessments will vary according to the relative importance one places, for example, on the degree of rounding or the amount of heat fractures and so forth. The final column of Table 3-4 attempts to give some sort of assessment of the relative temperature of formation of the frits by grouping them roughly into four groups on the basis of the microstructure of the quartz grains.

Compositional analysis of the glass matrix

The SEM analysis of the glass matrix of the frits is summarised in Table 3-5 below. It reveals that the frits can be divided into two distinct types, cobalt coloured and copper coloured, by far the most common of which are the cobalt coloured frits, making up 11 of the 13 examined, the copper coloured being the other two samples, 25038 and 67825 of which 67825 is highly contaminated.

The glass matrix of the cobalt coloured frits has silica contents between 71% to 81%, most being in the range 73-78% and averaging 76.4%. AM31, with the lowest silica has the highest lime at 8.1%, which would reduce the proportion of silica. The alkalis are fairly consistent across the samples, soda being dominant and varying from 9.4% to 15.3% averaging 12.0%, potash averaging only 0.5%. The concentrations of most of the elements are reasonably consistent in the frits, only lime shows very wide variation, varying from 0.9% to 8.1%. In Table 3-5 the frits are ordered by decreasing lime content of their glass matrix and as can be seen by comparing the lime levels to the associations, the amount of lime in the glass appears to be linked to the close presence of the lime rich white slip, those samples associated with a white slip being higher in lime.

In terms of minor element composition, all of the frits except 25038 and 67825 contain cobalt as the dominant colorant and only trace amounts of copper. Figure 3-10 is a graph of alumina against cobalt for the frits (and for the cobalt glasses discussed earlier), showing that the alumina content is quite variable, with 25038 being the lowest at 1.5% and AM34 the highest at 7.3%, the alumina apparently being correlated with the cobalt. Figure 3-12 shows a similar graph of manganese against cobalt which again suggests a link between the two, cobalt increasing with increasing manganese.

25038 is similar in many ways in terms of major element composition to the cobalt coloured frits, but contains effectively no cobalt and is coloured instead with by a high level of copper at 6.25% CuO. 25038 has a lime rich slip, but this particular sample was taken at a distance of 20mm or so from the slip and has a lime value of 2.1% with low alumina and manganese reinforcing the suggestion that there is a link between these last two elements and the cobalt content. 67825 is very distinct and again is copper coloured, only 0.88% in this case and lacking cobalt. It contains relatively low soda at 8.8%, but very high FeO, at 5.9% some five times the amount of the next highest frit, over six times the average TiO_2 and 10.2% alumina, twice the average for the cobalt coloured frits and six times that in 25038. The presence in such quantity of FeO, TiO_2 and alumina, along with its association with the base of a highly vitrified fritting pan, strongly suggest that 67825 has been heavily contaminated in these elements during the vitrification of the pan. As a result 67825 is untypical enough to be discounted and not included in the temperature determinations and other work that follows.

Bulk composition of frits

Since the frits contain only two phases and the backscatter contrast of these phases is usually quite distinct, it is possible to use this backscatter contrast along with image analysis programs to determine the percentage quartz in the frits, and therefore obtain some measure of the bulk composition. This is of value because large area analysis is impossible with WDS, and EDS can show errors when dealing with high concentrations of silica and low alumina and soda because of the high degree of overlap in the Kα peaks. The precise methods used in the image analysis are outlined in Chapter II, but since the analysis is only of one cross-section of a small SEM sample, which in turn is taken from a small original sample that might or might not have been representative of the body as a whole, the results in Table 3-6 should be treated as accurate to only ±5%.

The results in Table 3-6 show that AM28 has by far the lowest percentage of quartz at 10%, followed by 403 and rising to 25038, the highest at 69%. This parallels the microscopic textures where AM28 and 403 show the most absorption of quartz into the matrix, the edges of the quartz grains blending into the glass without clear boundaries, and 25038 has the most pristinely preserved quartz, with clearly defined edges and few examples of heat fracturing. Of those samples falling into the middle area, there is a broad trend towards higher temperature microstructures measured in terms of heat fracturing and absorption at the edges of the grains with lower amounts of quartz.

Table 3-6 above gives SiO_2 percentages derived from SEM analyses of the interstitial glasses of the samples. There is no

Table 3-6 : Comparison of calculated total silica from the image analysis if all the quartz were to be absorbed into the glass, with the bulk frit silica analysis by EDS.

Sample	%Quartz by VISILOG analysis	%SiO2 in glass (%S) by WDS	bulk %SiO2 calculated from VISILOG analysis	%SiO2 by EDS of Bulk frit
25038	69	72.8	92	94
36457	49	76.3	88	89
36465	39	78.4	87	90
AM34	36	75.3	84	88
AM31	32	71.6	81	84
AM32	33	78.3	86	87
AM33	29	76.5	83	84
AM23	28	81	87	91
EA67830	25	73.8	80	84
EA67829	21	76.4	81	83
403	14	77.2	81	83
AM28	10	76	78	81

Table 3-7 : Comparison of temperature (°C) determinations by three methods.

sample	viscosity calculations [1]	ternary diagram [2]	ternary $+CaO+K_2O$ [2]	microstructure [3]
AM23	1193	1470	1425	med-high
AM32	1168	1410	1400	med-high
AM34	1166	1360	1350	med-high
36465	1164	1400	1400	med-high
AM33	1163	1450	1425	med-high
36457	1148	1400	1400	med-low
403	1114	1310	1350	high
EA67829	1100	1300	1300	med-high
AM28	1091	1300	1300	high
25038	1076	1260	1225	low
EA67830	1066	1220	1200	med-low
AM31	1065	1220	1175	med-low

inverse agreement between silica in the interstitial glass and the percentage quartz, which is what one would expect if the original amount of silica and alkalis mixed in the batch were constant and the different frits represented different stages in the dissolution of the quartz. Instead it appears that the initial batch composition had some variation. This result is backed up by the results for the bulk silica shown in the third column of Table 3-6 which was calculated using the formula shown below;

$$\%S_T = \%S \times (1 - \%Q/100) + \%Q$$

which shows bulk silica values varying from 78% to 92% indicating a variety of batch recipes. A plot of this value against the value for the silica derived by EDS analysis of the bulk frit (Figure 3-18) shows that there is a constant relationship, the value calculated from image analysis being about 2% lower than the EDS value, perhaps due to the loss of alkalis in the EDS analysis or to or some loss of quartz in the image analysis.

Temperature determinations

Two theoretical techniques for the determination of temperatures have been applied to the frits and the results compared to the tentative qualitative assessment of temperature derived from the microstructure of their quartz grains.

The first value for each sample is derived from the formulae given in Lakatos *et al* (1972), which, based on a large number of experimental runs, calculates the effect of the proportions of the oxides SiO_2, Al_2O_3, Na_2O, Li_2O, BaO, ZnO, PbO, K_2O and MgO on the relationship between viscosity and temperature in a glass, these oxides amounting to typically over 99%

of the composition of the glass matrix of the cobalt coloured frits. The temperatures at which a glass of the composition of the glass matrix would have reached a certain viscosity can therefore be estimated, the viscosity chosen for the calculations in Table 3-7 being 10^4 poise, which is the working point of the glass. From examination of the hand specimens of the frits, it is clear that the viscosity reached this value at least. The second method of estimating temperature for a glass is from the ternary phase diagram, illustrated in Figure 3-19, for silica, soda and lime, which accounts for over 90% of the composition of a typical glass matrix. From this diagram the theoretical temperature at which a glass of that composition would be in equilibrium with the quartz can be determined.

Table 3-7 shows the results derived using the formulae given by Lakatos, *et al* (1972) and from the ternary phase diagram and compares them to those based on the examination of the microstructure (Table 3-4). Two values are determined from the ternary phase diagram, the first temperature being based solely on a normalisation of Na_2O-CaO-SiO_2 and the second, generally lower, by adding the K_2O to the Na_2O and the MgO to the CaO on the assumption that they would act in broadly speaking the same way. Figure 3-20 is a plot of the Lakatos, *et al* (1972) derived temperatures against those from the ternary phase diagram including MgO and CaO in the calculation. These results show that while there is a broad agreement on the relative relationships for the temperatures of the various frits, the absolute temperatures derived by each method differ from each other, the ternary temperature being about 200-250°C higher. This may reflect that the estimated viscosity of the interstitial glass is too low, but is more likely to be related to a lowering of the eutectic temperatures in the ternary due to the presence of elements not taken into account in the simple SiO_2-Na_2O(+K_2O)-CaO(+MgO) system. The temperatures given by the ternary diagram are too high to have been achieved during the Amarna period, and well beyond the failure point of the ceramic crucibles and, indeed the kiln walls themselves. This suggests that the other elements, probably CO_2 and H_2O, are significantly affecting the SiO_2-Na_2O-CaO system, making the ternary diagram of limited use. An attempt to include more elements by plotting on a $NaAlSiO_4$-$CaMgSi_2O_6$-SiO_2 yielded yet higher temperatures, due to the fact that there is too little alumina in the frits to incorporate all the soda into the diagram. It was therefore discounted. More in depth discussion of the temperature data, combined with that derived by experimental refiring of the ceramics, is covered in Chapter 4.

The data of the final column of Table 3-7, the qualitative assessment of the temperatures by examination of the microstructures, shows a broad agreement between the relative calculated values, with those frits assessed as having microstructures typical of relatively low temperatures having relatively low calculated values (by both methods), whilst those with relatively high temperature microstructures have higher calculated values. However, two samples, AM28 and 403 show what have been classified as high temperature microstructures, but have only moderate temperatures relative to those derived for the other frits. This may be due to a variation in initial particle size, these two samples having initially finer grained quartz which therefore reacts quicker and appears to give a higher temperature microstructure.

Summary for frits

All the frits consist of two phases, quartz grains which exhibit a varying degree of alteration due to heat, and a glass matrix which is usually uniform. The matrix is cobalt coloured in all but two examples (which are copper coloured), contains on average 76% silica, 12% soda, low lime (around 1.0%) away from contamination from lime rich slips, and high cobalt (averaging 0.3%) that appears linked to alumina and manganese. The bulk composition shows a variation in silica values from 78-92%, indicating a variety of recipes being used to produce the frit and the temperature of formation of the frit is high, the most reliable estimates putting all the temperatures above 1050ºC and some up to 1200ºC.

DESCRIPTION AND ANALYSIS OF FAIENCE

Faience is the most common of the vitreous materials at Amarna, being far more numerous than glass, frit or Egyptian blue, with most excavated houses having at least one fragment of faience and the majority having many fragments. The distribution and use of faience at Amarna is discussed in Chapter 7.

Acquisition of faience samples

Given the large number of faience objects from Amarna, it was necessary to select quite carefully those to be sampled, with a view to answering those questions posed by the 18th dynasty innovations in the manufacture of vitreous materials and the presence of the frits on the site at Amarna. Lucas Variant D faience, defined as faience with a hard core deliberately coloured blue or blue/green by the presence of cobalt, is one of the innovations of interest. Furthermore it may have a link to the frits and as such, a number of samples were obtained that fell into this category. The use of lead antimonate as a yellow colorant is a second innovation of interest and several examples of faience coloured with this colorant were acquired. In addition to these two innovations in the colorants for faience, faience is used in architectural features, particularly in tiles and inlays, which is again new and is quite widespread. Several examples of tiles, along with a number of inlays, were therefore obtained to examine the processes used in their production. The large number of small clay moulds found at Amarna (see Chapter 7) suggests a large production of moulded faience with ring bezels and shanks being typical of the finished products. A selection of these was acquired to closely tie down the methods of production of such moulded material.

Blue cores - Variant D faience

Macroscopic description

Thirteen faience samples with distinctly blue cores were obtained. In hand specimen the weathered surface of the core often looks only faintly pale blue, but when placed next to ordinary faience with greyish white cores it can be seen to be distinct and in a freshly broken or polished section (Figure 3-22), the cores are deep blue. The contrast is demonstrated in sample 11/674 (Figure 3-21), which is deep blue on a relatively unweathered break at the bottom and pale brown with a hint of blue on the weathered upper surface. This creates a difficulty in sampling, since some ordinary faience produced by efflorescence can have a slightly blue core, and, in a relatively fresh example, can appear to be very similar to

Table 3-8 : Faience samples acquired for the project.

Laboratory No.	Museum	Description
Blue cored faience		
595	Petrie Museum	blue faience kohl tube fragment
23231	Petrie Museum	blue faience Khepresh crown fragment (composite statue)
23419	Petrie Museum	blue faience wig fragment (composite statue)
23420	Petrie Museum	blue faience wig fragment (composite statue)
23505	Petrie Museum	blue faience bowl fragment
391	Ashmolean Museum	blue faience bowl fragment
469	Ashmolean Museum	blue faience vessel
1433*	British Museum	blue faience crown fragment
1764*	British Museum	blue faience wig fragment
176*	British Museum	blue faience cobra inlay fragment
3173*	British Museum	blue faience wig fragment
3556*	British Museum	blue faience inlay
3897*	British Museum	blue faience inlay
Colours		
E5031	Liverpool	white faience bead, inlayed in blue
E5032/3	Liverpool	blue faience disc bead
E5032/4	Liverpool	yellow faience disc bead
E5054/1	Liverpool	yellow faience ring bezel
E5054/2	Liverpool	red faience inlay to yellow ring
770	Petrie Museum	red faience inlay fragment
Tiles		
TA47	Excavations 1993/4	blue faience tile fragment
TA48	Excavations 1993/4	blue faience tile fragment
TA93	Excavations 1993/4	pale green faience tile fragment
E5045	Liverpool	green faience leaf shaped tile
Rings		
TA91	Excavations 1993/4	blue faience ring shank fragment
TA92	Excavations 1993/4	blue faience ring shank fragment
E5053/1	Liverpool	dark blue ring shank
E5053/2	Liverpool	dark blue ring shank
E5053/3	Liverpool	blue faience ring shank
E5053/4	Liverpool	blue faience ring shank
E5053/5	Liverpool	blue faience ring shank
E5053/6	Liverpool	blue faience ring shank
E5057/1	Liverpool	blue faience ring shank
E5057/2	Liverpool	blue faience ring shank

weathered Variant D. It was found on analysis that one of the blue cored specimens, 391, contained no cobalt and was therefore interpreted as ordinary faience produced by efflorescence. It will therefore not be considered further.

The glazes of the samples varied in colour, in fresh examples being deep blue, but tending to weather in some cases to a green colour (Figure 3-22 bottom, right of image). The surfaces of the pieces have all been worked to give the desired pattern, often with inscribed lettering or imitating the hair of a wig, for example. Most of the samples have been worked prior to the formation of the glaze, but some, including the Khepresh crown fragment 23231 which has had small holes drilled through the glaze after it had been fired, have been worked cold afterwards as well. These drilled holes are illustrated by Figure 3-23, which shows a close up of a Khepresh crown fragment 11/12/764 which has identical working

Table 3-9 : The state of preservation of Variant D samples.

Laboratory No.	glaze	interaction zone
Blue cored faience		
3556	100[1]	??, diffuse
469	50[1]	>300, diffuse
* 1764	60[2]	?120, slightly diffuse
3897	missing	missing
23505	30[2]	??, diffuse
* 3173	missing	?300, missing
* 1433	missing	missing
* 176	missing	missing
23420	100[1]	150
23419	50	150
23231	missing	100[2]

[1] some leaching resulting in ilica enrichment;
[2] extensive weathering resulting in partial loss of glaze or inter action layer.

to 23231.

The Variant D fragments sampled include two pieces of crowns and four pieces of wig, all intended as inlays for composite statues, three further pieces of inlays, a bowl and a kohl tube. The method of production of this faience is discussed in Chapter 6, and its archaeological affiliations and importance considered in Chapter 7.

Microstructure of Variant D

The cores of the faience samples have quartz grains cemented together with greater or lesser amounts of glass, but in all cases the amount of glass is significantly greater than that commonly seen in most ordinary faience (see below). The microstructures of various examples of Variant D faience are shown in Figure 3-24, Figure 3-25 and Figure 3-26, ordered by decreasing amount of glass in the cores and, since void space is inversely proportional to this, by increasing porosity. The most highly vitrified samples in Figure 3-24, for example 3556 and 469, have much interstitial glass, the voids in the core tend to be rounded, and the boundary between the interaction layer and the core is diffuse to the point that it is impossible to tell where one starts and the other finishes. Others, such as 3173 and 1433 at the top of Figure 3-25, and in 1764 (Figure 3-24), have more open cores with angular voids and show a sharper demarcation between the interaction layer and the core. Yet others, such as 176 and 23420 at the bottom of Figure 3-25 are coarser grained with less interstitial glass and with a distinctly non-uniform in distribution of the glass, in some areas containing extensive glass whilst in other it is absent entirely. In general, the thickness of the interaction layer somewhat difficult to estimate, especially when the interaction layers or glazes are missing or damaged (see 3897, Figure 3-24, and 23231, Figure 3-26), but the thinnest preserved layer for which a thickness could be given is about 120µm for 23419.

The size of the quartz grains is fairly constant in most of the samples, most having a maximum grain size of between 100-150µm, with the typical grain size being around 20-60µm. However, two, 176 and 23420, are coarser with more large grains up to 200-300µm across and a higher average grain size around 60-100µm. Two samples, 23419 and 23231 (Figure 3-26) show an entirely different pattern. These have coarse grained cores with a maximum grain size of 200-300µm (not shown in Figure 3-26 of 23231, but just off the photomicrograph to the left) and are covered with a thin fine grained layer, 2.0mm thick in 23419 and 2.5mm thick in 23231, with grain size no larger than 100µm and typically <50µm, a structure classified as 'Variant A' by Lucas (1962).

The preservation of the glaze and interaction zones of the Variant D samples is shown in Table 3-9. None of the specimens has a perfect preservation of the glaze coating, all of them being in some way altered or fragmentary. The best preserved glazes (23419, for example) show little alteration but are incomplete and do not cover the whole core, probably due to weathering and loss of the original glaze. In others, 3556 for instance (Figure 3-24), the glaze is apparently preserved, but shows cracking and a darker backscatter contrast than the glass matrix of the interaction layer immediately below it. This indicates leaching of colorants and alkalis to leave a high silica residue which, in extreme cases, can result in a "glaze" analysis with a silica content of over 90%. Better preserved glazes are 50-100µm thick and contain very few or no inclusions of quartz grains picked up from the layers below.

Analysis

The analyses in Table 3-10 show the concentrations of major elements for the glass matrix of the core and for the bulk core of the faience. They have been divided into two groups depending on the nature of the colorant in the cores, the larger group containing cobalt, but only trace levels of copper, averaging 0.07%, in the glass matrix, whereas the numerically smaller group has both cobalt and copper in the core, with copper averaging 1.6% and rising to nearly 3% in the glass matrix of one sample. With the exception of the amount of copper, the glass matrices of the two groups of samples are very similar indeed, particularly in their silica (average 76% for all samples), alumina (7.4%), lime (1.0%), magnesia (1.9%), potash (2.1%), iron (0.6%) and manganese (0.38%)contents, but with the soda and cobalt contents also being close. The bulk silica of the faience is fairly steady at between 86% and 92%, averaging 90%.

Examination of the microstructure of the glazes strongly suggested that they were weathered in many instances and this is borne out by the analyses of their major elements shown in Table 3-11. In several cases the glaze was missing and, in these examples, the analyses were made on the glass matrix of the outermost interaction layer, as close to where the glaze would have been as possible. Silica contents of the analyses varied from 67.7% in UC23505, a virtually unweathered example, to 95.5% in 1764 which shows a high degree of weathering. The analyses of highly weathered glazes were not included in the averages in Table 3-11. The alkali contents are inversely proportional to the silica content, with the high silica examples showing extreme leaching, with 0.0-0.2% soda and 0.7-1.1% potash, while unweathered examples have 17% soda and 2-3% potash. It is possible to construct theoretically the composition of an unweathered glaze from the data for the less weathered of the glazes and this glaze, with 65-70% silica and 15-20% soda, is listed at the bottom of Table 3-11. All the glazes are coloured by both copper, averaging 2.7%, and cobalt, averaging 0.30%, except 1764, which appears to contain only copper, but is

Table 3-10 : Glass matrix and bulk analyses for the blue cored faience

sample	SiO_2	Al_2O_3	CaO	MgO	Na_2O	K_2O	FeO	CoO	CuO	MnO
Glass matrix (Co only in core)										
UC23419	75.9	7.6	0.6	2.3	8.9	2.9	0.6	0.49	0.12	0.45
945	79.1	5.8	1.6	1.5	8.7	1.5	0.4	0.43	0.14	0.24
469	79.2	5.2	1.7	1.5	8.9	1.8	0.4	0.41	0.00	0.44
UC595	75.9	7.2	1.2	1.7	9.6	1.6	1.3	0.38	0.04	0.53
UC23231	75.7	6.1	0.9	1.7	11.8	1.8	0.8	0.48	0.09	0.36
UC23505	74.8	5.8	1.4	2.7	11.1	1.9	0.5	0.54	0.12	0.72
3897	72.6	10.5	0.7	1.7	10.0	2.9	0.5	0.47	0.03	0.25
3556	77.4	8.5	1.3	1.0	8.2	2.3	0.3	0.35	0.08	0.25
3897	72.6	10.5	0.7	1.7	10.0	2.9	0.5	0.47	0.03	0.25
average	*75.9*	*7.5*	*1.1*	*1.8*	*9.7*	*2.2*	*0.6*	*0.45*	*0.07*	*0.39*
Glass matrix (Co+Cu in core)										
1764	78.2	6.8	0.6	1.7	7.1	3.5	0.6	0.30	0.35	0.44
176	70.6	8.1	0.4	3.2	11.8	1.8	0.2	0.33	2.97	0.36
1433	81.4	5.0	1.8	1.0	6.6	1.5	0.4	0.34	1.43	0.27
3173	74.5	9.0	1.0	2.2	9.1	1.0	0.6	0.25	1.63	0.34
average	*76.2*	*7.2*	*1.0*	*2.0*	*8.7*	*2.0*	*0.5*	*0.31*	*1.60*	*0.35*
Bulk Analyses										
UC23419	92.0	2.1	0.4	0.6	2.5	0.5	0.4	0.16	0.08	0.10
UC595	86.2	3.2	1.2	1.1	4.9	1.1	0.3	0.28	0.00	0.39
UC23231	91.2	2.4	0.4	0.4	3.1	0.5	0.3	0.34	0.02	0.07
UC23505	90.5	2.2	0.7	0.8	3.2	0.5	0.2	0.17	0.04	0.22
UC23420	91.1	2.3	0.6	0.6	2.9	0.6	0.3	0.10	0.18	0.03
3556	87.7	3.2	0.8	0.8	4.8	1.1	0.3	0.28	0.09	0.27
average	*89.8*	*2.6*	*0.7*	*0.7*	*3.6*	*0.7*	*0.3*	*0.22*	*0.07*	*0.18*
Glazes										
UC595	71.5	2.1	1.8	1.0	17.1	3.1	0.3	0.29	2.24	0.39
UC23419	70.1	1.2	2.1	0.5	16.0	2.9	0.4	0.15	6.05	0.16
UC23505	67.7	1.4	6.9	0.6	14.6	2.4	0.2	0.13	5.63	0.07
469	73.7	2.0	3.0	1.4	14.6	2.6	0.2	0.37	1.48	0.42
theoretical [4]	65-70	3.0	1.0-2.0	1.0	15-20	2.5-3.0	0.4	0.3	2.0-5.0	0.20
945 [1]	78.1	4.9	0.9	1.1	10.9	2.2	0.3	0.33	1.01	0.14
3556 [2]	92.4	2.8	1.5	0.8	0.0	0.7	0.3	0.22	1.09	0.21
* 1764 [2]	95.5	2.1	2.1	0.6	0.2	0.7	0.6	0.0	3.0	0.12
UC23231 [1, 3]	79.8	4.0	0.4	0.9	9.8	2.3	0.5	0.51	0.79	0.17
* 3173 [1, 3]	79.6	4.1	0.6	1.4	9.1	1.8	0.6	0.22	2.07	0.23
* 1433 [1, 3]	79.8	4.0	0.6	0.8	7.6	3.2	0.4	0.41	2.32	0.18
3897 [2, 3]	93.4	3.1	0.7	0.6	0.2	1.1	0.2	0.09	0.26	0.10

a very weathered example.

A comparison of the compositions of the glazes and glass matrices shows that they are in many ways similar. The original glaze was probably lower in silica than the matrix, although weathering has now made the averages quite similar. However, the glazes are still higher in soda at up to 17.1% and averaging 12.5% compared to an average of 9.3%. Conversely, the alumina in the glaze is less than half that in the matrix, at 3.0% as opposed to 7.4%, and the lime in both is low at 1.0% in the matrix and 2.0% in the glaze (which goes down to 1.3% when the anomalous UC23505 is removed from the figures).

Table 3-12 : Analysis of the white, red, yellow and green faience and blue inlays

sample	colour	SiO_2	Al_2O_3	CaO	MgO	Na_2O	K_2O	FeO	CoO	CuO	MnO	PbO	Sb_2O_3
Glazes													
E5031	white	90.1	0.3	1.7	0.4	0.7	0.3	0.2	0.05	0.16	0.00	4.2	0.0
E5054/2	yellow	56.1	0.8	2.4	1.5	6.1	4.9	4.0	0.0	0.0	0.09	17.3	6.7
E5032/4	yellow	64.9	0.2	2.1	0.5	5.3	2.7	0.7	0.0	0.0	0.06	22.4	0.4
UC770	red	74.8	3.3	2.2	0.5	9.8	2.2	6.7	0.02	0.06	0.00	0.0	0.0
TA93	green	67.7	0.3	1.4	0.2	3.3	1.7	0.2	0.04	7.98	0.02	14.70	1.17
E5045	green	72.9	0.4	2.9	0.5	7.1	1.6	0.8	0.00	2.07	0.14	10.47	0.04
TA47	blue	77.4	0.6	2.5	0.7	7.0	2.4	0.4	0.03	7.60	0.20	0.00	0.00
TA48	blue	75.4	2.0	3.8	1.4	6.6	2.9	0.7	0.01	5.54	0.30	0.00	0.00
Glass matrix													
E5031	blue	75.3	4.1	1.0	1.2	12.2	3.6	0.4	0.36	0.06	0.21	2.6	0.1
E5031	blue	76.7	5.0	0.7	1.1	10.0	3.6	0.3	0.37	0.00	0.18	2.2	0.0
E5031	blue	76.1	4.0	0.8	1.1	12.0	3.3	0.3	0.31	0.04	0.27	2.9	0.0
TA93	green	65.9	0.9	7.7	1.5	4.8	2.2	0.2	0.04	5.08	0.02	10.11	0.38
Bulk analysis													
E5031	white	96.0	0.3	0.9	0.5	0.5	0.3	0.1	0.00	0.00	0.02	0.2	0.0
E5031	white	95.2	0.2	0.8	0.3	0.8	0.5	0.2	0.00	0.00	0.00	0.3	0.2
E5031	blue	86.5	2.0	0.5	0.5	6.0	1.8	0.2	0.13	0.00	0.13	1.2	0.1
E5054/1	yellow	90.8	0.5	0.7	0.4	0.8	0.4	0.6	0.0	0.0	0.08	3.3	1.0
E5032/4	yellow	89.8	0.3	1.2	0.6	1.4	0.6	0.9	0.0	0.0	0.01	2.8	1.4
UC770	red	88.1	2.3	0.9	0.4	1.9	0.5	5.0	0.03	0.04	0.15	0.01	na
E5054/2	red	82.2	2.5	0.7	0.3	1.5	2.2	5.0	0.08	0.00	0.52	0.06	3.64
E5045	green	86.8	0.3	1.8	0.8	1.8	0.6	0.7	0	0.34	0.02	2.57	1.40

Colours

In an attempt to sample as many different faience colours as possible, several samples were acquired, including cobalt blue from small inlays, a white bead, yellow rings and red inlays of two different sizes - the blue, white and yellow colours being of particular interest because these were first used widely in the 18th Dynasty.

White faience bead inlayed blue and red, E5031
The white faience bead, E5031, is flat, roughly triangular and made of white faience inlayed with blue and red in the shape of leaves or petals, the blue inlay only being sampled for the SEM. The top image of Figure 3-27 shows the bead under low magnification with the inlays of blue and red visible on the white background. Below is the cross section of the same bead in the SEM, showing three different areas; the top and bottom of the bead (as marked) are the white faience with a slightly weathered glaze and thick interaction layer onto the medium grained core, whilst the blue inlay in the middle is distinct, being highly vitreous with an almost continuous glass matrix, as shown in close up in Figure 3-28.

The analyses of bead E5031 are given in Table 3-12, which lists all the significant elements. It is clear from the analyses that the white area of the glaze is very low in soda at only 0.7% and high in silica at over 90% which is strongly suggestive of leaching as was suggested from close examination of the microstructure. The glaze is virtually free of almost all the transition metals, having very low levels of iron, copper, cobalt and manganese, but is high in lead, which is present at over 4%. The bulk analysis of the white area is very high in silica at above 95% in the two areas examined and is again almost free of the same transition metals, but contains the lead at 0.2-0.3% PbO. The blue inlay to the bead has a very different composition, with a glass matrix of 76% silica with high soda (11.4%) and potash (3.5%). The glass matrix is coloured by cobalt (0.35%) with only traces of copper and again some lead present (2.6%), perhaps having moved in to the glass matrix of the inlay from the high lead glaze of the white area. The bulk analysis of the blue inlay is lower in silica at 86.5% than the bulk compositions of the white areas.

Yellow faience, E5054/1 and E5032/4
Two samples of yellow faience were obtained; E5054/1, from a ring shank which forms part of an inlayed ring identical to that from which the red inlay E5054/1 was taken, a similar complete ring being illustrated to the left of Figure 3-29, and E5032/4, from a ring bead about 3mm thick. Both samples have a high gloss finish and deep yellow colour as illustrated on the right of Figure

3-29, which is a polished section of E5032/4 in normal light showing the colour of the core and the thin yellow translucent glaze. When viewed in hand specimen, the complete object is very similar in colour to that of the core pictured here. Figure 3-30 is a photomicrograph of the same sample showing the glaze and very distinct interaction layer and occasional particles within the core containing lead (white in the backscattered image in the SEM) and quartz (mid grey) with very little glass matrix.

Table 3-12 gives the analyses of these samples. Analysis of the glass matrix was not possible because it was present only in very small amounts, too small to give meaningful analyses. The bulk analyses of the core show around 90% silica, with only small amounts of alkali (less than 1.5% soda and 1% potash), with the lead content of the core higher than the bulk alkalis at 3.3% compared to 1.2% in the case of E5054/2, and 2.8% compared to 2.0% in E5032/4 and with antimony at 1.0% and 1.4%. Both glazes are very high in lead averaging nearly 20%, with the lead on average twice as abundant as the bulk alkalis and with neither of the glazes containing significant amounts of any colorant other than the lead antimonate yellow. E5054/2 has considerably more iron at 4.0% as compared to 0.7% and this may be linked to the fact that this is a ring with iron coloured red faience inlay and that there may be some contamination from these areas. The antimony content is also quite different in the two glazes, although as discussed with the glasses above, the particulate nature of the antimony makes its proportion in a thin glaze difficult to analyse with any accuracy.

Red architectural inlay 770 and red inlay ring bezel E5054/2
Two examples of red faience were examined, the red architectural inlay 770, which is a fragment taken from one a large number of similarly sized rectangular inlays all approximately 50mm x 20mm, and a small inlay about 5mm long by 3mm wide and 1mm deep that has been placed in a yellow faience ring bezel, E5054/2 (identical in morphology to the complete ring EA22594 shown in Figure 3-29). Both are red brown in hand specimen and 770 is covered, in places incompletely, by an almost colourless translucent glaze, which in has a greenish tinge (illustrated in Figure 3-31, a polished cross section of 770). Figure 3-32 is a photomicrograph of 770 in the SEM showing a layer of fine grained material which has been applied over a coarse core (off picture to the left), known as 'Variant A' by Lucas (1962). This fine grained layer can also be seen in hand specimen and is a frequent feature of these red inlays. The glaze in 770 is incomplete and patchy and the interaction layer quite distinct, whilst E5054/2 has a slightly finer grained core and has no glaze or interaction layer preserved.

Bulk analysis of both samples show that their cores are coloured with iron, containing 5.0% FeO and negligible amounts of any other colorants. In addition, E5054/2 has lead at 3.64% which has apparently diffused into the red inlay from the surrounding yellow ring bezel in which it sits. The glaze is similar, with again high iron, together with 9.8% soda and 2.2% potash.

Faience tiles

Two fragments of blue tile (TA47 and TA48) were obtained from excavation and two fragments of green tile, one from excavation (TA93) and one from a museum collection (E5045). In addition, some twenty other faience tiles were examined and photographed in hand specimen or under low power magnification to give a greater selection of patterns and techniques of working. Many of the tiles from Amarna are painted in bright colours, with flowers, plants, fish and other animals being the most popular motifs, but only E5045, which was leaf-shaped, was large enough in hand specimen to reveal what shape it had been when complete. Some of the tiles have only inlayed decoration (Figure 3-33 left), others have only painted decoration, but many show both techniques used together (Figure 3-33 right and Figure 3-34). The under surface of some of the tiles, for example E5045 (Figure 3-35, top), shows a regular series of ridges and furrows that were suggestive of the piece having been left to dry on a reed or textile mat. The methods of production of these tiles are discussed in detail in Chapter 6.

All the tiles examined in the SEM, and most of those examined in hand specimen have a similar microstructure, in that they have a coarse grained core onto which has been applied a thin fine grained layer varying between 1.0mm and 3.0mm thick (ie. "Variant A" structure, see Figure 3-35, bottom). The core of the blue tiles has little glass matrix, and a grain size up to a maximum of 1.0mm, with a distinct interaction layer and glaze each of the layers having a sharp boundary. In the green tiles, there is more glass matrix and a scattering of particles (white in the backscattered image in the SEM) containing both lead and antimony through both core and glaze.

The analysis of the tiles is shown in Table 3-12. For the blue faience tiles, only the glaze analyses are quoted since they have virtually no glass matrix and therefore no area of glass large enough to get a meaningful result. These glazes are high in copper, at 5-8%, and broadly similar in composition to other copper coloured glazes to be discussed below in the section on faience finger rings.

The green tiles are coloured with a mixture of copper and lead antimonate with a variable amount of copper, at 2-8%, plus between 10 and 15 % PbO and in the case of TA93 significant antimony, the lead being partly absorbed within the glaze and partly combined with antimony in the form of particles. The fact that the antimony is in particulate form makes it difficult to analyse and results, dependent on the number of particles that fell within the area analysed, in some analyses with high antimony contents and some with low. From the bulk analysis of E5045, it can be seen that the amount of copper in the core is low at only 0.34% as opposed to 2.1% in the glaze, whereas the lead is present in the core at significant levels at 2.6%. Throughout the green tiles, the lead is more abundant than the soda, a relationship which is particularly true in the glazes where in the glaze of TA93, for example, there is more than four times the amount of lead than soda.

Faience finger rings

Faience jewellery was common at Amarna, and finger rings were some of the most common types, the moulds used for the manufacture of ring shanks and bezels also being widely found. The production of rings is of interest because it is clear from the presence of moulds that they were moulded and therefore they are conventionally interpreted as being subsequently glazed by efflorescence. Examination of their microstructures and compositions might therefore give an idea of the structures that are formed by moulding followed by efflorescence. These

Table 3-13 : Compositions of the glazes, glass matrix and cores for the blue faience ring shanks.

sample	SiO_2	Al_2O_3	CaO	MgO	Na_2O	K_2O	FeO	CoO	CuO	MnO	SnO_2	Sb_2O_3
Glazes												
E5053/3	76.2	0.2	0.2	0.0	9.0	2.4	0.0	0.20	10.10	0.03	0.3	0.10
E5053/4 *	70.7	0.3	3.1	0.4	8.4	4.5	0.2	0.05	10.60	0.00	0.8	0.19
E5053/5 *	74.9	0.4	0.6	0.0	6.7	5.3	0.2	0.0	11.26	0.00	0.0	0.00
E5053/6 *	78.3	0.9	1.3	0.7	6.9	1.9	0.6	0.07	7.09	0.00	1.0	0.00
E5057/1 *	73.4	0.1	0.5	0.0	7.4	5.4	0.1	0.00	12.53	0.00	0.0	0.00
E5057/2 *	73.9	0.4	1.3	0.5	7.6	5.1	0.1	0.00	9.31	0.00	0.6	0.30
average	*74.6*	*0.4*	*1.2*	*0.3*	*7.7*	*4.1*	*0.2*	*0.05*	*10.15*	*0.00*	*0.5*	*0.1*
Glass matrices												
E5053/3	69.9	0.5	0.6	0.1	5.7	3.2	3.3	0.00	15.84	0.09	0.0	0.02
E5053/4	67.4	0.6	3.4	1.1	6.7	3.9	0.2	0.00	9.39	0.04	6.1	0.20
E5053/5	74.5	1.3	2.6	0.1	5.0	5.1	2.5	0.00	8.55	0.04	0.00	0.14
E5053/6	73.8	0.7	1.6	0.2	7.7	2.0	4.4	0.11	7.33	0.00	0.9	0.13
E5057/1	69.9	1.2	1.9	0.7	6.5	5.9	0.1	0.10	11.93	0.03	0.6	0.11
E5057/2	72.0	0.3	2.8	0.7	6.5	5.3	0.9	0.04	9.12	0.00	0.0	0.20
average	*71.2*	*0.7*	*2.2*	*0.5*	*6.3*	*4.2*	*1.9*	*0.04*	*10.36*	*0.03*	*1.3*	*0.1*
Bulk analysis												
E5053/3	94.7	0.3	0.3	0.1	0.9	0.5	0.3	0.00	1.81	0.05	0.1	0.00
E5053/4	93.9	0.3	0.7	0.2	0.9	0.5	0.2	0.00	1.80	0.00	0.3	0.01
E5053/5	94.9	0.4	0.6	0.1	0.8	0.7	0.2	0.00	1.25	0.00	0.1	0.00
E5053/6	94.6	0.3	0.8	0.3	0.6	0.3	0.2	0.02	1.09	0.00	0.3	0.01
E5057/1	94.8	0.2	0.3	0.1	0.7	0.6	0.1	0.00	1.71	0.01	0.2	0.01
E5057/2	93.9	0.3	0.6	0.2	1.1	0.8	0.2	0.00	1.60	0.00	0.3	0.00
average	*94.5*	*0.3*	*0.5*	*0.2*	*0.8*	*0.6*	*0.2*	*0.00*	*1.54*	*0.01*	*0.2*	*0.0*
Co in core and glaze												
Glazes												
E5053/1	81.3	4.7	1.2	1.4	7.9	2.2	0.3	0.44	0.06	0.21	0.0	0.00
E5053/2	72.8	14.2	0.5	1.2	7.0	3.4	0.2	0.15	0.0	0.01	0.0	0.00
average	*77.1*	*9.4*	*0.8*	*1.3*	*7.4*	*2.8*	*0.2*	*0.30*	*0.03*	*0.11*	*0.0*	*0.00*
Glass Matrix												
E5053/1	81.3	5.1	1.7	1.5	7.2	1.6	0.3	0.43	0.03	0.17	0.1	0.00
E5053/2	79.1	6.4	2.0	1.6	7.4	1.9	0.3	0.33	0.0	0.30	0.0	0.00
average	*80.2*	*5.7*	*1.9*	*1.5*	*7.3*	*1.7*	*0.3*	*0.38*	*0.01*	*0.23*	*0.0*	*0.00*
Bulk analysis												
E5053/1	88.4	2.6	0.8	0.7	5.4	0.8	0.2	0.21	0.00	0.11	0.1	0.00
E5053/2	88.0	2.6	0.8	0.8	5.3	0.9	0.2	0.22	0.05	0.15	0.2	0.02
average	*88.2*	*2.6*	*0.8*	*0.7*	*5.3*	*0.9*	*0.2*	*0.22*	*0.03*	*0.13*	*0.1*	*0.01*

observations can then be compared to other objects in which the production method and glazing technique used is less clear.

Figure 3-36 shows two examples of complete faience finger rings, one dark blue and coloured with cobalt (right), and one light blue and coloured with copper (left). Ten fragments of ring shanks were obtained from the Liverpool University Museum and the recent 1993/4 excavations, two, E5053/1 and E5053/2, dark blue and similar to the right ring in Figure 3-36, the other eight were a lighter blue to turquoise, similar to the left ring. Two fragments of rings in yellow were also sampled, but these have been discussed under section 3.6.3.2 above. Figure 3-37 shows a polished cross section of the ring shanks, revealing their internal colouring, the colour of the glazes, and some idea of their cross sectional shape, which in most cases is three to five sided with very rounded corners and between 3.0-4.0mm in thickness.

Figure 3-38 shows the microstructures of the finger ring shanks, the left half of the figure revealing that many of them are remarkably consistent, with a 40-60μm thick glaze overlying an 300-400μm thick interaction layer which has a sharp boundary to the core. E5053/6 and E5057/2 are slight deviations from this, showing thicker glazes at around 100μm and, in the case of E5057/2, a thicker interaction layer at 500μm. With all these samples that are light blue, the amount of glass matrix in the core is moderate and less than most of the examples of Variant D samples discussed above. However, the dark blue samples 5053/1 and 5053/2 are distinct and more similar in microstructure to the Variant D, with thick glazes at 100μm and very broad interaction layers with no easily definable boundary with the core, which themselves are characterised by abundant glass. They are distinct from the Variant D samples in section 3.6.2 in that unlike the Variant D, they have no copper colorant in either their cores or glaze.

The major element compositions of glaze and glass matrix of the core for the light blue rings are quite similar and show little variability in most elements, with silica contents of 70-75%. The potash values are an exception to this, however, and vary both in the glazes and in the glass matrices. The colorant in both the core and glaze of the light blue examples is copper, and the amount of copper is quite high in some samples, rising to over 15% in the glass matrix of E5053/3, and 12.5% in the glaze of 5057/1 and averaging just over 10% for both matrix and glaze. The bulk cores are very high in silica at 94.3% and contain 1.5% copper on average, all samples having greater than 1.0%.

The two dark blue beads' compositions are distinctly different and similar to the Variant D discussed earlier. Their glazes are quite badly leached, necessitating analysis close to the outermost interaction layer and resulting in a variation between samples in silica (81.3% and 72.8%) and alumina (4.7% and 14.2%). Conversely, the unweathered glass matrices are much more similar in composition. Both the glass matrix and glaze show that the colorant is cobalt, which may be linked to higher manganese and alumina, and that copper is not present. This is a similar pattern to that seen in the glasses and frits and is most probably the result of the cobalt colorant also being rich in alumina and manganese.

Summary for faience

The faience examined exhibits a wide range of different colorants, structures, compositions and methods of working. The Variant D essentially shows the use of two colorants, cobalt alone in the cores of the majority of the samples, and copper and cobalt in the cores of some and the glazes of all. The Variant D faiences are quite highly vitrified and as such have a higher bulk alkali content than 'Ordinary' faience. This former type of faience is often used in inlays, both large in size for use in composite statues and architectural features, and smaller scale such as the blue inlay for the white bead. The colours examined showed the use of iron colorant for red, lead antimonate for yellow and lead alone for white. The yellow faience, in particular, is very high in lead, the lead being far more abundant than the alkalis. The tiles show evidence of both inlaying and painting onto a body that is typically Variant A in structure. The finger rings show a very consistent microstructure with moderate, but significant glass matrix in the body and predominantly copper colouring.

Chapter 4: Ceramics associated with vitreous materials

INTRODUCTION

The earliest Egyptian ceramics are wares made primarily of clay which were moulded and shaped while wet, left to dry and then fired to harden (Lucas and Harris 1962). The principal use of such material in Egypt is in vessels and as such it is known from the Neolithic period onwards. While the first efforts were crude, by the Badarian period finely produced wares were common, showing evidence of some skill in manipulation of the material while wet and control of conditions of firing. Clay based ceramics found in Egypt show a range of firing conditions from very low fired, porous, baked clay, to high fired vitrified vessels. Arnold and Bourriau (1993) amongst others state that there are two major groups of Egyptian clay based ceramics which can distinguished by the type of clays used, the first type consisting of pottery made of alluvial clays primarily from the Nile silt and rich in iron oxides and organics, and the second consisting of high lime marl clays which are of restricted distribution, Qena and Ballas in Upper Egypt being two of the most important sites (Lucas and Harris 1962). When fired these two clays produce pottery with distinctive characteristics, the Nile silt producing a range of colours from reddish brown to dark browns when fired between 1000°C and 1100°C, whilst the marls tend to stay light grey to grey in colour until in excess of 1100ºC.

CERAMICS ASSOCIATED WITH VITREOUS MATERIALS

Petrie's excavations at Amarna in 1891-2 uncovered a significant number of ceramic vessels with glass and frit associated that, along with nearby kilns and other debris, Petrie believed was conclusive evidence of glass-working and glass-making. He published the account of this excavation in 1893, but regrettably as Nicholson (in press) points out, the 1894 account is confused, particularly where it relates to the ceramics and the confusion is compounded in later accounts. Petrie initially distinguished three types of vessel that he at least believed were involved in the glass-making process, the first being a large flattish vessel roughly 10 inches across and some 3 inches deep of which Petrie uncovered a broken example of filled with frit and therefore interpreted the group as "fritting pans". The second type of vessel of which many fragments were found was a "cylindrical jar", which was reconstructed to be about "7 inches across and 5 inches high" and frequently had "glaze run down the outside of them, from the closed end to the open end". Turner (1954) also mentions these jars , stating that "a glassy layer or broad stripe, light green in one case, dark blue in another, runs from the bottom of the cylindrical crucible towards the rim". Petrie interpreted this by suggesting that the cylindrical jars were stood open end down on the floor of the kiln and the fritting pans were stood on top of them, the spills from the pans therefore running down the jars from the closed end to the open end. From the shape of pieces of glass recovered, many of which have ceramic adhering to them, Petrie reconstructs a third class of vessel, a "crucible" which he states are "2 or 3 inches in depth and diameter". These original pieces of these vessels have not been found by any later workers and further excavation has not led to new finds.

Subsequent work by Turner (1954) has made a significant contribution to the study of glass-working at Amarna. Turner points out that the cylindrical jars were definitely used for glass melting, since there is glass adhering to the inside of some of the fragments and he suggests that the glass apparently running from the base to the rim on the outside of the vessels might have resulted from their use as stands during which time they were given a preliminary firing. Nicholson disagrees, believing that the vessels were fired before they had any contact with glass-making, but concedes that the vessels may have had a dual purpose. He observes that a preliminary examination of the surface finds appears to suggest that the exterior of the base was tapered to fit the rim of similar vessels making them stackable and allowing them to be used as saggars if this was desired. However, Nicholson agrees with Turner that the vessels were used for the melting of glass. Moreover, he goes further than this to suggest that the vessels were used as moulds for the making of ingots, believing that the glass ingots that would be produced in using these vessels as moulds would be similar to those ingots found in the Ulu Burun shipwreck off the Turkish coast which is probably contemporary with the late 18th or early 19th Dynasties. Nicholson supports this statement by pointing out that some of the glass ingots from the wreck have tapered sides, as do some of the vessels, the walls of which are markedly thickened toward the base, and that the diameter of the ingots is approximately 6 inches, similar to the reconstructed diameter of some of the vessels. Nicholson also states that the composition of the glass ingots matches those of Egyptian core formed vessels and the glass adhering to the cylindrical vessels very well.

Source of specimens

Table 4-1 lists the Amarna glass working specimens in the Petrie Museum UCL, Ashmolean Museum, British Museum, Ägyptisches Museums Berlin and Nationalmuseet Copenhagen that were examined in this study. In addition to the specimens listed above, some details of four other vessels were obtained from an excavation report by Weatherhead and Buckley

Table 4-1 : List of the ceramics associated with glass working at Amarna.

No.	Museum	Type	Preservation	Association	Vitrification
UC8988	Petrie	cyl. vess.	wall + part base	frit on underside of base	low
UC25248	Petrie	cyl. vess.	wall	glass on inside	moderate
UC36458	Petrie	cyl. vess.	part base + wall	powdery frit inside, glass outside	low
UC36462A	Petrie	cyl. vess.	wall	glass inside wall + outside base	moderate
UC36462B	Petrie	cyl. vess.	rim + wall	glass inside + outside	low
1893.1-41(396b)	Ashmolean	cyl. vess.	fragment of wall	glass on bottom of base	moderate
1893.1-41(396c)	Ashmolean	cyl. vess.	wall + base	glass on inside + outside, frit on bottom of base	low
1893.1-41(396a)	Ashmolean	cyl. vess.	¼ base + wall	glass on outside of wall	low
UC8989	Petrie	cyl. vess.	wall + part base	glass on outside of wall	low
7407	Copenhagen	cyl. vess.	base+part wall	glass on bottom of base	low
7408a	Copenhagen	cyl. vess.	part base + wall	glass on bottom of base	low
7408b	Copenhagen	cyl. vess.	part base + wall	glass outside wall + bottom base	low
7408c	Copenhagen	cyl. vess.	rim + wall	glass on outside of wall+rim	low
EA67827	BM	cyl. vess.	fragment of base	glass on inside and outside	high
12267a	Berlin	cyl. vess.	part base + wall	glass outside wall + bottom base	moderate
12267b	Berlin	cyl. vess.	wall + part base	glass outside wall + bottom base	moderate
UC36457	Petrie	fritting pan	fragment of base	frit inside + underside	high
UC40568	Petrie	fritting pan	fragment of base	frit on underside of base	high
EA67825	BM	fritting pan	fragment of base	frit on underside of base	high
21977	Berlin	fritting pan	fragment of base	frit inside	moderate
1893.1-41(398)	Ashmolean	fritting pan?	fragment of base	Egyptian blue inside	low

"cyl. vess." cylindrical vessel, "BM" British Museum, "powder" refers to a turquoise powdery, friable frit quite different from the other frits discussed.

Table 4-2 : Analysis by SEM EDS of ceramics associated with vitreous materials.

sample	type	SiO_2	Al_2O_3	Fe_2O_3	MgO	CaO	Na_2O	K_2O	TiO_2
AM22	cylindrical vessel	56.0	15.5	9.1	0.3	10.5	3.4	2.7	1.8
AM29	cylindrical vessel	58.2	15.6	10.3	0.9	5.7	1.2	3.5	4.1
AM31	cylindrical vessel	59.5	17.1	11.0	1.0	2.8	5.1	0.8	1.0
AM33	cylindrical vessel	57.1	12.5	8.8	1.7	8.8	6.8	1.4	2.0
AM23	fused clay	58.8	15.8	10.7	1.5	8.5	0.0	1.7	2.4
AM24	fused clay	59.3	16.3	12.8	2.0	4.9	0.0	1.7	2.5
AM25	fused clay	63.8	15.0	10.3	1.5	5.0	0.0	1.7	2.1
AM26	fused clay	60.2	17.6	10.8	1.6	5.6	0.0	1.7	2.3

(1989) which contains drawings of vessels that were associated with glass and discovered in the surface surveys of 1987 and 1988 and are included because they expand the amount of data available and reveal the height of the vessels. Only the general shape of the fragment, the vessel's reconstructed size and the position of associated glass is recorded in the paper, other details, for example the nature of the glass and the presence or absence of white slip not being mentioned. The vessels are referred to here by their figure number in that paper (10.6.1, 10.6.3, 10.6.5 and 10.6.8) since no other number or information is given about them.

Composition

In addition to the large pieces detailed in Table 4-1, several small fragments of ceramics, either associated with vitreous materials or themselves highly vitrified, were obtained from the 1993/4 excavations at Amarna for analysis and these are listed in Table 4-2 along with their associations and their bulk analysis by SEM EDS. The full list of these samples is given in Table 4-7, where their mineralogy is discussed in connection with firing temperatures.

The analyses from Table 4-2 are plotted on Figure 4-1, along with ICP analyses of Nile silts and kiln wall fragments from Table 4-8 (which are discussed later) and the composition fields for Nile silt and marl clays from Noll (1981). As can be seen, all of the analyses for the ceramics fall in the field for Nile silt, well away from the lime rich marls, suggesting that the ceramics associated with vitreous materials from Amarna are made of Nile silt.

Morphology and associations

Examination of the specimens revealed that there are only two designs of vessel associated with glass working, "cylindrical vessels" and "fritting pans", no trace of Petrie's third class of vessel, the smaller "crucibles" being found in the collections. None of examples are complete, the largest fragments of cylindrical vessel being UC8988, which is about a third of the vessel with both wall and base, and 7407, an almost complete base with a large proportion of the wall preserved. However, due to the consistency in the size of the cylindrical vessels, evidenced by the curvature of the walls of the surviving pieces, a good idea of the size of a complete example of this type of vessel can be obtained. This is more difficult with the fritting pans, the remains of which are more fragmentary.

Cylindrical vessels

The cylindrical vessels, illustrated in Figure 4-2a and b and Figure 4-3 to Figure 4-6, are generally large solid vessels and fall morphologically into two slightly different sub-groups. The first group consist of vessels that have a flat, commonly 20mm thick base and vertical or very near vertical walls around 12mm thick thinning slightly to the rim, the wall joining the base at close to a right angle. Of the 20 cylindrical vessels in the study, 16 fit into this group and are quite consistent in size and shape, the reconstructions varying between 230mm and 180mm in diameter, averaging at about 200mm. Only one specimen, 10.6.8, preserves the full height of the vessel from base to rim, being 130mm high, but the thickness of the walls and general rate of thinning of the other specimens (for example UC36462A and UC36462B) suggests that the others would have been similar. The internal surface of 11 of these vessels has a 1.0mm thick layer of white slip adhering to it, whilst three vessels have no slip (7407, UC36458 and UC8988) and no information is available on the other two (10.6.5 and 10.6.8).

In the second sub-group, the walls of the vessels are convex and thicken up to 25mm at the base which is much more pronounced than the first sub-group, with a reconstructed diameter varying between 220mm and 240mm, so they are therefore, on average, slightly larger too. No rim is evident on any of the fragments, but UC8989 preserves the highest fragment of wall, being preserved to 90mm high and 9mm thick at the thinnest part, but from the rate of thinning of the wall, it is unlikely that the walls of UC8989 were greater than 110mm high. None of this second group have an internal white slip. UC8988, a typical example of the first group of vessels with vertical walls is shown in Figure 4-3b, and UC8989, an example of the second subgroup are both in Figure 4-3a and Figure 4-4. In both groups, the impressions of potter's hand and finger marks are frequently found internally and externally, these being particularly evident in 12267a, Figure 4-2. The cylindrical vessels are in most cases much less vitrified than the fritting pans and hold their shape well, with the exception of EA67827, which is probably a fragment of a cylindrical vessel which has melted and flowed, even showing evidence of bubbling in places.

Of the 20 cylindrical vessels, all except one, 10.6.3, has one or more of the vitreous materials adhering to one or more of its surfaces, 16 being associated with glass alone, one with frit alone (UC8988), one with both glass and frit (1893.1-41(396c)) and one with both glass and Egyptian blue (UC36458). Of the 16 cylindrical vessels associated with glass, 11 have glass only on the external surface, two only on the internal surface (10.6.5 and UC25248), and three on both internal and external (UC36462A, UC36462B and EA67827), with all vessels with internal glass (possibly except 10.6.5, where the presence of white slip is not recorded in the diagram from which the paper is drawn) having an internal white slip as well. Two vessels (1893.1-41(396b) and 7408c) have an internal white slip, but no internal glass. The glass adhering to most of the vessels is dark in colour often almost black when it is present in a thin layer or dark blue when it is thicker. The glass often covers quite a large proportion of the external surface of the vessel, for example in 8989 a large part of the wall of the surviving fragment being covered with glass (Figure 4-4). Most of the external glass shows no "way-up structures" to indicate which way the glass flowed, but a great deal of the glass is on the underside of the base of the vessels, which in itself suggests that they may have been inverted. Adhering to 1893.1-41(396a), however, there is a teardrop shaped ribbon of glass that makes it clear that in this case at least, the glass flowed from the rim to the base, as illustrated in Figure 4-5. In addition to having glass on the internal and external surfaces, 1893.1-41(396c) and EA67827 have glass preserved on the fractured surface of the wall of the vessel, suggesting that, in these cases, at least some of the glass adhered to the vessel after it had been broken. A further complication is 12267a, which has the rim of a second vessel of similar diameter to itself adhering to its base, Figure 4-6. None of the four cylindrical vessels of the second subgroup with the convex walls has internal glass, or indeed any evidence of use internally, three of them have thin glass deposits externally and one (10.6.3) appears unused.

Two cylindrical vessels have frit adhering to them (1893.1-41(396c) and UC8988). The frit being mid blue and coarse grained with clearly visible quartz grains lies on the underside of the base of the vessel and on the lower part of the external surface of the wall. In 1893.1-41(396c), where both are present, the frit seems to overlay the glass. UC36458, has Egyptian blue on the internal surface of the base and glass on the external wall with no overlap between the two.

Fritting pans

"Fritting pans" is a term which was used originally by Petrie to describe a shallow saucer-like vessel approximately 250mm across and 70mm deep at its deepest, and which has been preserved in the literature. Only six vessels that come close to fitting this description were found in the collections, four of them highly vitrified fragments of ceramic associated with frit and two relatively unvitrified, one of which was associated with Egyptian blue (Figure 4-7). Two of the vitrified examples (UC40568 and EA67825) have been taken to such a temperature that they have flowed around a quartz pebble, which is now preserved stuck within the base of the vessels. A example (UC36457) shows evidence of warping and what is left of the pan has broken into two fragments and become embedded in frit (Figure 4-8).The fourth example (21977) is also blackened and heat damaged. The fifth pan-like vessel (1893.1-41(398)) is very different in that it has not been vitrified to the same extent, looking very similar to the cylindrical vessels in its degree of vitrification. This vessel which has Egyptian blue adhering to it, is similar to a sixth uncatalogued pan in the Petrie Museum in terms of both morphology and degree of vitrification, but the latter has no evidence of use in terms of adhering vitreous materials.

The intense vitrification and subsequent warping of the vessels, along with the relatively small number and size of fragments preserved, makes it very difficult to reconstruct the original size and shape of the pans. However, it is probable, as Petrie states, that they were no more than 70mm high with a width probably anything from 200mm to 300mm and about 10-12mm thick.

None of the six possible examples of fritting pans have glass associations. One, the uncatalogued UC example is unused and another, 1893.1-41(398) has Egyptian blue on the internal surface, but no frit. This leaves four pans associated with frit. In all of these the frit is mid blue, coarse grained and hard, lying only on the underside of the base of UC40568 and EA67825 and only on the internal surface in 21977. In UC36457, large quantities of frit are found adhering to the internal surface of the vessel which is fractured, allowing the frit to leak through the fractures onto the external surface. While UC40568 and EA67825 both have internal white slips, 1893.1-41(398) and 21977 have none and in UC36457 it is not possible to say, since all the internal surfaces are covered with frit.

Interpretation

It is clear from the different vessel types and the presence of three distinct vitreous materials that the interpretation of the vessels associated with glass and frit at Amarna must attempt to explain the presence of glass and frit only on the external surfaces of some of the vessels, the fact that both frit and glass can be found adhering to a single vessel and the presence of glass on fracture surfaces. Some cylindrical vessels, including UC36462A, UC36462B, UC25248 and 1893.1-41(396c), have internal glass remains adhering to the interior white slips and it is possible that these vessels were used as containers for the melting of glass. Similarly, glass adhering to the external surfaces of these vessels might be explained as spills running down the outside of the vessel that would inevitably result when the molten glass was poured. However, no "way up structures" are preserved on these particular samples to definitely indicate the direction of flow of the glass, which makes this interpretation speculative. Nicholson (1993, 1995b) hypothesised that some of the vessels may be moulds and may have had internal slips to prevent the glass sticking. The mould would be broken when the glass had solidified, the ingot removed and the broken mould discarded.

In contrast, this explanation cannot account for the presence of vessels with external glass, but no internal glass, which is the commonest mode of occurrence accounting for 11 of the cylindrical vessels. Petrie's interpretation for these vessels was that they were used upside down as stands upon which the fritting pans were balanced when they were in the kiln (Figure 4-9) and that spills from the pans ran down the external surfaces of the vessels, but not the internal. While this would account for spills on the outside of the vessels, the fritting pans are in general much more vitrified than the cylindrical vessels, which would not be expected if both were right next to each other in the kiln. With the fritting pans and resting on four point contacts on the edges of the cylindrical vessels, both would be very prone to failure. The fact that the vessels are in many cases not highly vitrified, argues that some of them were not in the kiln at all, but may instead have been used upside down as stands on which glass could be poured into moulds or shaped. If the pouring was done with the cylindrical vessel mould supported on other upturned vessels, then the occasional breakage would lead to molten glass pouring all over the outside of the upturned base of the vessel leading to exactly the sort glass coverage seen in many of the vessels (for example, 12267a, in Figure 4-3 and UC8989, Figure 4-4).

The fritting pans are divided into two types. Four of them are highly vitrified and associated with coarse frit, one (1893.1-41(398)) is not vitrified and contains powdery Egyptian blue and one is morphologically identical to 1893.1-41(398), but empty. Therefore it is possible that the fritting pans were used for two different purposes, one involving high temperatures and coarse frit, and the other lower temperatures and Egyptian blue. For the reasons already given, it is unlikely that the fritting pans stood on the cylindrical vessels in the kiln as Petrie speculated. UC40568 and EA67825 (which may be fragments of the same vessel) both have a quartz pebble enclosed in their base where the ceramic has flowed down and covered it during firing, positive evidence to suggest that the fritting pans stood on top of the quartz pebbles which lay on the floor of the kiln (Petrie 1894). Some of these pebbles are now stored in the British Museum and Petrie Museum UCL (for example, British Museum EA67824) and show splashes of frit on them, indicating that the floor of the kiln was spattered with frit. This frit would have been transferred to the base of any vessel subsequently placed in the kiln, which may explain why cylindrical vessels 1893.1-41(396c) and UC8988 have frit on their undersides. The shattered and warped state of the highly vitreous fritting pans suggests that the floor of the kiln could have received further frit from failures in these vessels during firing.

The use of fritting pans for Egyptian blue, and the presence of external glass and internal Egyptian blue on cylindrical vessel UC36458 suggests that perhaps both types of vessels were used or reused for two separate purposes, one involving glass and a second involving Egyptian blue. That frit and glass were heated in the same kiln is supported by the presence of both frit and glass on the bottom of cylindrical vessel 1893.1-41(396c). This association between glass and Egyptian blue may indicate that the same specialists were producing these two products, or, at least, there was a use of common vessels between two closely connected industries.

Conclusions

There are two principal vessel morphologies associated with vitreous materials at Amarna. Cylindrical vessels which were probably used for melting, pouring and moulding glass and as supports for this process, and fritting pans which were fired at high temperature with coarse grained frit, or used as containers for Egyptian blue. Quartz pebbles incorporated into the base of two of the fritting pans along with the relatively low degree of vitrification of most of the cylindrical vessels, suggests that the pans were placed on the floor of the kiln, and not on the upturned cylindrical vessels as suggested by Petrie. The presence of powdery frit and glass associated with the same cylindrical vessel, along with coarse frit and glass in the same kiln, suggests that the industries for the production or manipulation of all three vitreous materials were closely allied and may have been operated by the same workers.

EXPERIMENTAL DETERMINATION OF FIRING TEMPERATURES FOR CERAMICS

The ceramics used in the manufacture of vitreous materials have been subjected to high temperatures during their use, an estimation of which would give an indication of the capabilities of the workers and the kilns that were used, and may provide information on the range of fuels, weather and wind conditions that were necessary for a successful firing.

As the firing temperature and duration of firing increases for a ceramic, the microstructure and mineralogy changes in a predictable way. This predictable variation in the microstructure and mineralogy for a known starting composition allows an estimation of the conditions of firing to be made for ancient ceramics where the nature of the original clay is known. The Amarna ceramics that are associated with the manufacture of vitreous materials are made from Nile silt, the chemical and mineralogical nature of which is well known. Therefore, if modern Nile silt is experimentally fired and the microstructures and mineralogies subsequently obtained are close to those observed in the ancient ceramic, it is (theoretically) possible to deduce that the ancient ceramic was subjected to the same conditions.

In order for a useful method to be drawn up for the estimation of firing temperatures in ancient ceramics from experimentally produced samples, a number of different factors have to be considered. Firstly, the raw material should be as close as possible to that used in antiquity which was not a problem in this case since Nile silt from Amarna was available and proved to have a very similar chemical and mineralogical composition to those ancient ceramics examined. Secondly, the method of preparation of the raw silt and subsequent firing should replicate that used in antiquity. Since, at Amarna, the silt appears to have been used straight from the ground without grinding or elaborate sorting and the ceramics show that they were fired under oxidising conditions, this condition again can be met.

Turner (1954) tried to estimate the temperatures that the Amarna ceramics associated with vitreous materials would stand by heating 2.0-3.0g of ancient samples. His results showed that the "unused" vessels (ie. those not with glass making frit adhering) would have become fluid when subjected to prolonged heating at temperatures of 1150-1200°C, whereas those with glass making frit adhering became fluid at a lower temperature, below 1100°C. 1100ºC is close to what Turner considers to be the maximum attainable temperature in Egypt at this period, but he states that even if temperatures of higher than 1100°C could be obtained, the ceramics used would not withstand them, concluding, "All the evidence therefore suggests that the upper temperature limit in the glass-melting operations ... did not exceed 1100°C".

Experimental Method

In order to further explore the temperatures reached by the ceramics involved in glass making, samples of Nile silt were fired at various temperatures (Table 4-3). The silts were packed wet into blocks 30mm square and 9mm deep, this depth being selected because it is close to the thickness of the wall of the cylindrical vessel, and dried at 150ºC for 30 minutes. The temperature then was increased by 15ºC per minute up to the firing temperature which was held for 60 minutes. From Turner's work, the probable range of interest for the ceramics was between 950°C and 1150°C, so samples were therefore fired first at these extremes for each of three durations, 60 minutes, 300 minutes and 600 minutes. The results showed that while there was a very slight change in the mineralogy of the resultant ceramic (as observed under the SEM) between 60 and 300 minutes, there was no observed change at either 950°C or 1150°C between 300 and 600 minutes. Therefore, the mineral assemblage in the ceramic therefore appears to require some period between 60 and 300 minutes in order to equilibrate. Further firing beyond this time did not further affect the system, so it is very probable that the changes that are going to occur in the mineralogy and microstructure within a reasonable time frame have all occurred by the 300 minute point. It was decided therefore to fire all the blocks for this duration to allow for comparison, and to consider only these blocks when it came to interpreting the ancient samples where the firing

Table 4-3 : Range of temperatures and durations selected for experimental refiring., an 'X' showing that these conditions were simulated.

	duration (mins)			
temperature (deg C)	60	300	600	techniques
unfired				ICP,XRD
650		X		mic,ICP
950	X	X	X	mic,SEM,ICP
1000		X		mic,SEM
1050		X		mic,SEM,ICP,XRD
1100		X		mic,SEM
1150	X	X	X	mic,SEM,ICP,XRD
1200		X		mic,SEM
1250	X			mic,SEM,ICP,XRD

Table 4-4 : ICP AAS analysis of various refired Nile silts from Amarna.

sample	SiO_2	Al_2O_3	Fe_2O_3	MgO	CaO	Na_2O	K_2O	TiO_2
PN13								
unfired	62.55	15.70	11.10	3.10	3.33	1.05	0.97	1.73
650	58.74	17.28	11.99	3.36	3.95	1.17	1.08	1.92
AM45								
950	57.21	16.31	11.34	3.23	6.91	1.12	1.35	1.91
1050	57.22	16.62	11.66	3.30	6.07	1.15	1.37	1.98
1150	56.81	16.75	11.82	3.28	6.13	1.15	1.40	2.00
1250	55.63	16.67	12.36	3.33	6.85	1.06	1.35	2.04

temperature is unknown. The exception to this was the 1250°C block which had already completely melted by the 60 minute mark and was therefore not fired further. Table 4-3 shows a summary of the different temperatures to which the experimental ceramics were fired and the analytical techniques subsequently used on them.

The experimentally refired blocks were examined by a number of different analytical techniques in order to characterise the changes in the ceramic caused by the variations in temperature. The bulk composition of the material was determined by SEM with attached EDS and also by ICP AAS. The mineralogy was determined by SEM and XRD. The microstructure was examined by eye, under the binocular microscope and under the SEM. All the methodology used is discussed below.

Results

The analysis of the refired ceramics by SEM with attached EDS and by ICP AAS shows a broad agreement between the two techniques for most of the oxide percentages considered. While the silica, lime, iron and titanium agree well and the potash reasonably so, the alumina is consistently higher by about 1% absolute in the SEM, the soda (just over 1% by ICP) is not detected and the magnesia content is about twice that measured by SEM. This loss of soda, which is a well known weakness of the SEM with EDS, is due to migration of light elements away from the electron beam and is discussed elsewhere (see Chapter 2). The variation in the alumina and magnesia is probably due to the difficulty in accurately distinguishing the silicon, aluminium and magnesium Kα peaks which overlap using the EDS and make accurate determination of the alumina difficult. For these reasons, the ICP analysis will be taken as more likely to be accurate than the SEM and will be used in the discussions that follow.

Table 4-4 shows a summary of the ICP analyses for the major elements of the refired Nile silts. Silts from two different areas at Amarna were analysed, PN13 and AM45, both of which are very similar in their compositions except in their lime content which is 6.5% in AM45, but only about 3.2% in PN13. Data for the refired samples shows that, within the error of measurement, there is little variation in most of the oxide percentages of the Nile silt as the temperature rises from 950°C to 1250°C. There is, however, a systematic increase in the iron content in AM45 from 11.3 to 12.3% and in titanium across this temperature range.

Mineralogy - SEM

The analytical data for several examples of the ceramics associated with vitreous materials is discussed above and is shown in Table and Figure 4-1, whilst the mineralogy of the Nile silt making up the individual ceramic samples recovered from Amarna is discussed below. Here the changes observed under the SEM in each of the minerals that make up the ceramics as the temperature of firing is increased is discussed, the results being summarised in Figure 4-10. They can be divided into the primary minerals, principally quartz, felspars, pyroxenes and iron oxides, and the secondary minerals, iron oxides, felspars, and pyroxenes.

Primary mineralogy

The most abundant mineral observed in Nile silt for all the firing temperatures is primary quartz, which ranges up to 500μm across and shows rare weak evidence of euhedral shape, tentatively suggesting a vein mineral source for some of the grains. At 950°C, the quartz is relatively unaltered and cohesive with rare examples of heat cracks in a few grains, but by 1100°C the grains are starting to react at the edges to give a more rounded shape, with heat shattering common and present in nearly all of the large grains. The reaction of the grain boundaries continues giving more and more rounded grains as the temperature increases until at 1150°C most of the grains are rounded with sharp grain boundaries to the surrounding glass. By 1200°C the boundaries are becoming more diffuse and graded and at 1250°C there are many examples of grains being almost completely dissolved leaving only a "ghost" of the shape of the grain with extensive resorbtion along the heat cracks.

The primary felspars can be divided into a calcic plagioclase close to anorthite which is common, and a potash rich alkali felspar (although an apparently unique grain of soda-rich felspar was detected) which is less common, both being present as grains up to 150μm in length. The plagioclase grains occasionally show the characteristic lath shape, and many show the plagioclase multiple lamella twinning. At 950°C, the grains are largely unaltered, with slight development of alteration along the boundaries between the twins, but by 1050°C extensive heat cracking is common and by 1150ºC most of the large grains have extensive reaction along the cleavage planes. By 1200°C, the smaller grains have dissolved completely leaving the rare large grains alone. While at 1250°C, these grains are still evident, they now have highly diffuse boundaries, their shape being demarcated by the absence of secondary iron oxides within the grains as compared to the surrounding glass. The cleavage planes are occasionally still visible under high magnification. At low temperatures the K-felspar is unaltered and shows good cleavage which with increasing temperature becomes more and more altered until by 1100°C there are distinct linear reaction areas along the cleavage and the edges of the grain are beginning to appear ragged. No K-felspar was detected above the firing temperature of 1100°C.

Ilmenite is present in abundance as primary grains at 650°C, in the form of small opaque grains rarely up to 70μm across, but more typically smaller, the great majority of the grains that are opaque in thin section and have a bright backscatter contrast in the SEM being ilmenite, with rarer hematite. The grains are often euhedral and remain largely unchanged at high temperatures, but above 1150°C they occasionally show exsolution into Fe-rich and Ti-rich phases. A primary calcic pyroxene (with lesser amounts of iron and magnesium) is another common mineral within the silt, with fresh grains largely unaltered up to 1150°C, and frequently showing acicular or elongated shape up to 250μm long. Around 1150°C, alteration rims around the grains are evident, and the grains show more evidence of heat cracks. Above 1150°C, no primary pyroxene is present.

The other primary phases which are rare, never being represented by more than one or two grains per section, include amphibole (hornblende), biotite, apatite, olivine, epidote and garnet. These rare grains give valuable information as to the nature of the source material for the silt, but their rarity makes them of very limited use in comparing ceramics with differing firing temperatures, since no inference can be drawn from their absence in a particular sample. They are therefore not considered further in the refiring experiments.

Secondary mineralogy

At 1100-1150°C and above, small secondary opaques, iron rich with traces of titanium and typically 1-4μm across, begin to form, occurring throughout the glassy matrix of the samples but often clustering around iron rich phases such as ilmenite grains. By 1250°C they are very common occurring throughout the extensive glass phase. In some of the more vitreous ancient samples, secondary grains of anorthite and diopside are also very common, often forming extensive lattice works of elongate secondary grains, but these were not observed in the experimentally refired samples.

Mineralogy - XRD

The XRD methodology is described in Chapter 2 and the results displayed in Table 4-5. Despite the complex nature of the silt and the many phases present, leading to a low peak to background ratio and a high degree of interference in the samples, enough information can be extracted from the data to support the conclusions drawn from the study of the development of the mineralogy using the SEM. The unfired silt contains clay minerals (smectites and illites) with their characteristic high d spacings, along with phases such as micas and calcite that generally will not stand high temperatures. K felspar is also present in the unfired silt, but by 1050°C, only a suggestion of possible K felspar is left, along with the stable and ubiquitous quartz and plagioclase. At this temperature is the first significant development of hematite with its strongest 2.69Å and 2.51Å peaks. These are much stronger at 1150°C, with the secondary hematite peaks also evident at 2.20Å and 1.69Å suggesting that more hematite is present, and stronger again at 1250ºC as can be seen in Figure 4-11, which shows the original traces of the XRD for 2θ values of 30-40. Quartz and plagioclase continue and are still present at 1250°C. The cristobalite 4.05Å peak is first present in strength at 1150°C and continues almost equally strong at 1250°C, giving at 1250°C a mineral assemblage of primary quartz and plagioclase, with secondary hematite and cristobalite.

Microstructure

Preliminary macroscopic examination of the refired Nile silt blocks macroscopically revealed that from 650ºC up until 1050ºC, the block was a reddish colour, the surfaces were rough and granular and the block remained coherent not losing its shape. At 1100°C, there is a slight darkening of the colour of the block which becomes definite at 1150ºC, developing a shiny surface suggesting that it is highly vitrified, and at this later temperature, the block is just beginning to sag under its own weight and is therefore on the verge of failure, but does not yet adhere to the surface upon which it was fired. At 1200°C, the block is a very dark brown with shiny rounded surfaces where it has started to flow, showing that a vessel of the same material would have failed before this temperature. The block also adheres to whatever it is sitting on in the furnace. At 1250ºC the block is black in colour, melts and flows freely and cannot be freed from the base of the kiln. Away from contact with glass or frit, therefore, the failure temperature is around 1150ºC or slightly higher, but before 1200°C is reached.

Figure 4-12a-d illustrate polished sections, as seen under the SEM in backscattered mode, of the Nile silt fired at six temperatures between 950ºC and 1250ºC, showing the changes in the microstructure through this temperature range. The changes with temperature in the microstructures of the individual phases are discussed under the mineralogy, but perhaps of more interest and use are the overall changes in the microstructure in the ceramic as a whole which centre around the nature and extent of the glass phase and the characteristics of the void spaces which are intrinsically linked to this. Maniatis and Tite (1979) give a scheme for the classification of the nature and extent of the vitrification in a ceramic, identifying stages through which a ceramic goes on progressively heating to higher temperatures, and assigning labels to these from partial vitrification (PV) through extensive vitrification (EV) to continuous vitrification (CV). As a final stage at the very top end of the temperature range, continuous vitrification develops coarse spherical bloating pores (CV(BP)) (Tite et al 1982). At 950°C (Figure 4-12a), in the experimental samples, there is already extensive vitrification (EV), with the edges of the voids still rough, tending to linearity and concentrated around the mineral grains as parting spaces. By 1050°C (Figure 4-12b), the glass phase is more extensive and the voids are rounder at the edges and

Table 4-5 : Summary of the data derived from the XRD analysis of the unfired Nile silt and that fired to 1050 °C, 1150°C and 1250°C.

2 theta	peak str.	d spacing	unfired	1050	1150	1250
6.02		14.69	smectite (1)			
19.81-19.95	4	4.48-4.45	illite and micas (1)			
20.85	3	4.26	quartz (low) (1)	quartz (low) (4)	quartz (low) (4)	quartz (low) (2)
20.90-21.62	6	4.25-4.11	K-felspar (<1)	K-felspar (2)		
21.94	10	4.05			cristobalite (4)	cristobalite (3)
22.05	0	4.03	plagioclase (<1)			plagioclase (3)
26.68	10	3.34	quartz (low) (10)	quartz (low) (10)	quartz (low) (10)	quartz (low) (10)
26.68-27.70	10	3.34-3.22	K-felspar (<1)	? K-felspar (<1)		
28.14-27.79	10	3.17-3.21	plagioclase (<1)	plagioclase (7)	plagioclase (5)	plagioclase (5)
29.37	10	3.04	calcite (<1)			
30.61-30.29	7	2.92-2.95	plagioclase (<1)	plagioclase (2)	plagioclase (2)	plagioclase (1)
33.30	10	2.69		hematite (1)	hematite (3)	hematite (3)
33.82	5	2.65	plagioclase (<1)			
35.33	8	2.54	ilmenite (<1)			
35.77	5	2.51	hematite (<1)	hematite (1)	hematite (3)	hematite (3)
36.52	1	2.46			quartz (low) (<1)	
39.52	1	2.28			quartz (low) (<1)	quartz (low) (<1)
40.25	3	2.24	ilmenite (<1)			
41.02	2	2.20			hematite (<1)	hematite (<1)
50.20	1	1.81	quartz (low) (<1)	quartz (low) (<1)	quartz (low) (<1)	quartz (low) (1)
54.20	6	1.69			hematite (<1)	hematite (<1)
53.05	10to6	1.73	ilmenite (<1)			
60.01		1.54			(<1)	(<1)
62.54		1.48			(<1)	(<1)
64.07		1.45			(<1)	(<1)

"peak str." gives the theoretical relative strength of that peak for the relevant mineral compared to the largest peak, which is given the value 10. Numbers in brackets give the relative strength of the peak compare to the largest peak in the analysis, in these cases always the quartz 3.34Å peak.

less ragged. By 1150°C (Figure 4-12d), continuous vitrification (CV) is evident where the mineral grains look as though they are "floating" in the glass matrix and the extensive pores are more rounded and beginning to join up. This process continues and by 1200°C and above the pores are becoming spherical with none of the elongate features still evident at 1150ºC. These pores grow in size and decrease in number through to 1250ºC (Figure 4-12f) where many of the smaller ones have joined to give coarse bloating pores (CV(BP)).

Comparison of analytical techniques

The analytical techniques of XRD and SEM are at least in theory capable of distinguishing the different firing temperatures of the ceramics, each having its strengths and weaknesses. Mineralogy in both SEM and XRD shows a clear change in the samples from the low temperature end to the higher temperatures. Of the two, XRD appears less sensitive, and there is always the possibility of missing minor phases that are present in only small amounts in the high background scatter of the trace which can obscure minor peaks. XRD also has the disadvantage of requiring quite a powder sample, which cannot subsequently be used for other optical or SEM microscopy, not always possible in the current study, where very few of the pieces of ancient ceramics are large enough to permit a sample for XRD to be taken. Examination of the mineralogy under the SEM has fewer disadvantages, being capable of showing small and rare grains although their presence in any given section however, cannot be guaranteed, and therefore their *absence* cannot be taken as absolutely indicative of their disappearance from the system, only the *presence* of relict primary grains (showing lower temperatures) or developing secondary grains (showing higher temperatures), can be used reliably as indicators of temperature.

Examination of the microstructure, since it involves examining the whole of the ceramic and does not rely on the unpredictable presence of possible rare grains, is perhaps the greatest use in the determination of temperature and is especially valuable because it shows the greatest variation through the temperature range of greatest interest, that is 1050-1200°C as the ceramic approaches

Table 4-6 : Results derived from SEM, for primary and secondary phases and extent of vitrification for the experimentally refired Nile silt AM45.

Temperature	Vitrification	primary phases	secondary phases
unfired	none	qz,pl,Kf,il,py,*am,*mi	none
650	none	qz,pl,Kf,il,py,*am,*mi	none
950	PV	qz,pl,Kf,il,py,*am,?*mi	none
1050	PV	qz,pl,Kf,il,py,*am	none
1100	EV	qz,pl,*Kf,il,py,?*am	*io
1150	EV-CV	qz,pl,il,py	io
1200	CV	qz,pl,il,py	io
1250	CV(BP)	qz,*pl,il	io

Table 4-7 : Associations of ceramic samples of this study, mineralogy derived from the SEM.

Sample	Type	Vitrification	primary phases	secondary phases	temperature
AM37	cylindrical vessel	EV	qz,py,il,pl,*am	*io,wo,py	1150
AM38	cylindrical vessel	EV	qz,pl,Kf,il	*io	1100-1150
AM22	cylindrical vessel	EV	qz,py,il,pl,*am	py	1100-1150
AM31	cylindrical vessel	EV	qz,pl,py,il,*am,*ap	*io	1100-1150
AM33	cylindrical vessel	EV	qz,py,il,pl,?gy,*ap	pl	1150
AM36	? vessel	EV	qz,py,Kf,il,*am	none	1100
AM40	? vessel	PV-CV(BP) [1]	qz,pl,Kf,*il	io,pl	1050-1250
AM39	? vessel	CV	qz,pl,il	io,*py	1200
AM29	clay fragment	PV	qz,pl,Kf,py,il	none	950-1000
AM24	fused clay	CV	qz,pl,il	io,py	1200
AM26	fused clay	CV	qz,pl,*il	io,py	1200
AM23	fused clay	CV(BP)	qz,il	io,pl	1200
AM25	fused clay	CV(BP)	qz,pl,*il	io	1200-1250

"PV" partial vitrification, "EV" extensive vitrification, "CV" continuous vitrification (after Maniatis and Tite, 1979), "CV(BP)" presence of bloating pores Tite et al (1982). Phases listed are; qz=quartz, pl=plagioclase, py=pyroxene, il=ilmenite, am=amphibole, Kf=K felspar, ap=apatite, wo=wollastonite, gy=gypsum, io=iron oxides. "" indicates that the phase is rare. Temperature is derived by the comparison of the ancient samples to the refired reference collection with the emphasis on the microstructure supported by evidence from the mineralogy. [1] different degrees of vitrification seen on different parts of the vessel.*

failure. The degree of vitrification and development and nature of the pore structure of the body, therefore, give the clearest indication of temperature progression. As a result, it is the microstructure as seen in the SEM, supported by the evidence from the mineralogy drawn primarily from the SEM, but wherever possible from XRD as well, that will be used as a reference scale against which the ancient examples can be compared and some indication of their firing temperature hopefully derived. Microstructure and mineralogy associated with the different temperatures are summarised in Table 4-6.

COMPARISON OF ANCIENT EXAMPLES TO REFIRED REFERENCE COLLECTION

Table 4-7 gives a summary of the information on the Amarna ceramics associated with vitreous materials detailed in Chapter 3. The second column of the table gives the interpretation of the ceramic type derived from the morphology of the fragment, the third the level of vitrification of the samples (after Maniatis and Tite, 1979) from partial vitrification to continuous vitrification with coarse spherical bloating pores and the fourth/fifth the mineralogy taken straight from the sample descriptions from work on the SEM. The estimated temperatures in Table 4-7 are derived primarily from a characterisation of the microstructure of the ancient ceramic, particularly the degree of vitrification and the shape of the pores, supported by the mineralogy and compared to the reference characteristics for the various temperatures found in Table 4-6. It should be noted that secondary development of pyroxenes and calcic plagioclases was observed in the ancient samples (Figure 4-13), but not in the refired reference samples. The reason for this difference is not clear, but is probably linked to the rate of cooling of the material. Secondary minerals such as these form as the ceramic cools through a critical nucleation point and continue to grow below this point until either the appropriate elements are no longer available for their growth, or the temperature drops below the level at which these elements are able to freely

Table 4-8 : Analysis by ICP AAS of the kiln wall samples compared to an average for the experimentally refired Nile silt (AM45) and two other Nile silts (PN6A and PN13A).

sample	SiO_2	Al_2O_3	Fe_2O_3	MgO	CaO	Na_2O	K_2O	TiO_2
AM42A	62.55	14.19	10.14	3.01	4.59	1.53	1.55	1.88
AM42B	62.80	13.54	9.49	2.85	6.12	1.51	1.35	1.81
AM42C	59.67	15.58	11.21	3.16	4.90	1.58	1.28	2.03
AM45	56.72	16.59	11.80	3.29	6.49	1.12	1.37	1.98
PN6A	59.29	14.07	11.93	3.41	5.14	1.58	1.20	2.77
PN13A	62.55	15.70	11.10	3.10	3.33	1.05	0.97	1.73

Table 4-9 : Summary of microstructure and mineralogy for the kiln wall fragments at the surface and that at distance X into the wall.

	at the kiln wall surface			distance	atdistance X from surface		
sample	vitrification	mineralogy	temperature	(X)cm	vitrification	mineralogy	temperature
AM42A	CV	qz,il,2io,*2pl	1150	7	PV	qz,pl, py,il,*Kf,?2io	1050
AM42B	EV	qz,pl, py,il,2io	1100	5	PV	qz,pl, py,il,am	1050
AM42C	CV(BP)	qz,il,2io,2pl	1250	20	EV	qz,pl, py,il,2io	1100

Phases listed are; qz=quartz, pl=plagioclase, py=pyroxene, il=ilmenite, am=amphibole, Kf=K felspar, ap=apatite, io=iron oxides. "" indicates that the phase is rare ,"2" that it is secondary. Estimated temperatures are given for the different areas by comparison with the experimentally refired Nile silt references.*

move. This is shown schematically in Figure 4-14 in which crystal size and number of crystals are plotted against the temperature. If the temperature falls slowly, small number of large crystals are formed, a little faster and large numbers of small crystals form, whilst if the temperature falls very rapidly through both these zones, no crystals are formed at all. The refired samples used in the comparison study were cooled relatively rapidly, certainly faster than they would have done in an ancient kiln, the size of the one excavated at Amarna O45.1 and discussed in the following section. Therefore it seems probable that these samples have not developed secondary pyroxenes or plagioclase because they have rapidly passed through the temperature range where nucleation is possible. Experiments in firing further experimental pieces from 1200ºC and 1150ºC and cooling them slowly a period of up to five hours as they drop the first 100ºC was attempted, but again no secondary pyroxenes or plagioclase was observed. It seems possible that the temperature range where these minerals grow may be quite restricted and therefore very difficult to replicate in the laboratory. Alternatively, although less likely, several cycles of firing and cooling may be required.

All of the ceramics except one (AM40) showed a uniform microstructure and mineralogy across the whole of the sample, suggesting that the entire sample was subjected to the same temperature and given time to equilibrate. The temperatures derived vary from about 950°C up to 1250°C, with the samples classed as "cylindrical vessels" having, in general, the lowest temperatures of between 1050°C and 1150° and the "fused clay" having the highest temperatures, the majority being right up at 1200-1250°C, above the failure temperature of the ceramic. Exceptions to this are AM29, a clay fragment which appears to have been subjected to temperatures of only 950-1000°C and AM40, a sample with an anomalously high tin content perhaps reflecting its involvement in metal working rather than glass working, in which vitrification varies across the sample from partial vitrified (1050°C) at one end through to continuous vitrification at the other (1250°C). Of the other samples that were possibly from vessels (marked "? vessel" in Table 4-7), AM36 was subjected to a temperature of 1050-1100°C which is within the same temperature range as the cylindrical vessels, tentatively suggesting that it may be such a vessel, whilst AM39 was subjected to a higher temperatures of 1150-1200°C, closer to the range for the fused clay.

KILN WALL FRAGMENTS

As discussed below in Chapter 7, examples of kilns were uncovered at site O45.1 in the 1993 and 1994 excavations at Amarna (Nicholson 1995), two of which are large and two smaller. At least one of these latter was associated with faience and/or pottery production. From one of the large kilns, known as Kiln 3, three samples of the kiln wall were recovered for examination.

Each of the three kiln wall samples was taken from the inner wall of the kiln and shows a marked graduation in microstructure from the high temperature internal surface of the kiln wall preserved in the sample, to the lower temperature of the core of the wall itself. A summary of the major element analyses (from a sample taken at their less vitrified ends) by ICP AAS is given in Table 4-8 along with various examples of the fired Nile silt for comparison. This shows that, while there is some variation in the compositions, the AM42 kiln wall samples all fall within the range of composition of the Nile silt samples. This result, along with evidence from the SEM showing that the mineralogies are also the same, indicates that the kiln walls were made of the local Nile silt, the same material as used in the ceramics employed in the production of vitreous materials.

Analyses along a run of 27 points across the largest sample, AM42c, from the high temperature kiln wall to a depth of 33mm into the wall at the low temperature end of the sample, were set up on the SEM (illustrated in Figure 4-16). This revealed that the composition of the sample was consistent throughout, any variation present being accounted for by the chance presence in the analytical field of large grains. This result effectively rules out the possibility of a thin lime rich lining being applied at least to this part of the kiln. A plot of the analytical totals for the SEM with EDS before normalising is shown in Figure 4-17, revealing that there is a plateau of around 90% at the high temperature end falling rapidly at a depth between 20 and 25mm into the wall to a second plateau at 60% at the relatively cool outer end. The total is linked to the number and extent of void spaces, the more the voids the lower the total (since the SEM spends more of its time analysing resin or empty space). Therefore, the higher totals at the high temperature end represent a marked drop in the amount of void space as can be seen clearly in Figure 4-15b, which is taken at a depth of 20-23mm below the kiln wall surface and marks the onset of continuous vitrification at about 1100°C, and a marked increase in the amount of glass. The inner surface itself (Figure 4-15a) shows continuous vitrification with large bloating pores and therefore, from comparison to the experimentally fired samples, this surface must have reached a temperature of around 1250°C (Figure 4-15c). The other two kiln wall samples, AM42a and AM42b show a similar pattern of high temperature inner surfaces and cool internal zones, although they are less clear and the maximum temperature is lower. Table 4-9 shows a summary of the various microstructures and mineralogies preserved on all three of the kiln fragments both at the inner surface of the kiln wall and at a depth within the fragment. From this it is possible to derive two temperatures (columns 4 and 8 of Table 4-9) in exactly the same way as was done for the ceramic vessel fragments.

Firing duration

When a kiln is fired, the surface of the kiln wall will be at the temperature of the air in the kiln close to the wall. Therefore, the higher the temperature of the kiln firing, or the longer the firing duration, the further the penetration of the heat into the wall until at a distance from the surface of the kiln wall, the temperature will fall to be that of the surroundings. Since the temperature of the air in the kiln can be estimated from an examination of the microstructure of the surface of the kiln wall, it is theoretically possible to calculate the heat flow, as modelled in Tite et al (1982) and thus estimate of the duration of the firing from the temperature that the ceramic has reached at some distance into the kiln wall. If the wall is brought rapidly up to the working temperature (T_s) and held constant for a time (t) the temperature (T) at depth (X) into the wall, assuming that there is a linear heat flow perpendicular to the wall surface, is given by:

$$\frac{T-T_0}{T_S-T_0}=1-\mathrm{erf}\left(\frac{X}{2(\alpha t)^{1/2}}\right)$$

where T_0 is the initial temperature away from the surface of the kiln wall (in this case 20°C), α is the thermal diffusivity and erf(y) is the error integral;

$$\frac{2}{\sqrt{\pi_0}}\int^{\zeta} e^{-\xi^2}d\xi$$

Table 4-10 : Firing durations for the Amarna kiln 3 derived from Table 4-9 and using the methodology of Tite et al (1982).

sample	surface Ts (C)	distance (X)cm	temperature Ti (C)	duration t
AM42A	1150	7	1050	65min
AM42B	1100	5	1050	120min
AM42C	1250	20	1100	280min

Using the thermal diffusivity of 0.005cm²/s assumed in Tite et al (1982), then for the three kiln wall samples the operating conditions (maximum temperature, duration of firing) for the Amarna kiln 3 from the three wall samples can be calculated (Table 4-10). Thus, the durations of firing appear to vary from a little over an hour to over four hours. It is clear from replication studies of kiln operation (Nicholson 1998), that different parts of a kiln can be at very different temperatures at the same stage in the firing, hotspots being a common feature. Thus the temperature of the surface of the kiln wall at one point may not be representative of the temperature of the kiln as a whole, or the temperature of the working area where the crucibles are placed. It is, however, consistent with the very high temperatures seen in the fused clay samples and the fritting pans, but higher than that seen in the cylindrical vessels, where the other two samples, showing lower temperatures and durations, are a better fit.

SUMMARY FOR CERAMICS ASSOCIATED WITH VITREOUS MATERIALS

The ceramics are all made from Nile silt and have been subjected to high temperatures to varying degrees. They fall into two types: cylindrical vessels, which are generally lower fired and associated most often with glass teared externally, and fritting pans, which are higher fired and most often associated with frit. Rarer examples of cylindrical vessels containing frit and both types containing Egyptian blue have been observed. The theory that both types were used together in the same kiln in the production of frit is disputed due to the variation in the temperatures that the vessels have been subjected to and the high probability of failure in the fritting pan if it was balanced on the sharp edges of the cylindrical vessels as Petrie hypothesised. The possibility that the cylindrical vessels are used more frequently in glass *working* is put forward to explain the external glass and relatively slightly lower temperatures (1100-1150°C). Reuse of vessels and the close relationship of frit, glass and Egyptian blue is also evident.

The kiln fragments show that the internal surface of the kiln was made of Nile silt, apparently without any lime rich lining in the areas sampled and that this has been subjected to temperatures of up to 1250ºC for periods of over 4 hours. Other areas of the kiln wall are at lower temperatures, and the limited number of samples makes it difficult to interpret exactly the conditions in the kiln through a firing.

Chapter 5: Raw materials of ancient production

INTRODUCTION

Which raw materials were used for the manufacture of the vitreous materials is of great importance in the study of the production of the materials, since only by accurately ascertaining the starting materials is it possible to determine the full production process and assess the amount of labour, degree of selection of raw materials and the various processes involved. In addition, raw materials can give valuable indications of the social setting under which the production was carried out. For example, the use of rare non-locally sourced materials could, in the presence of other supporting evidence, argue for a high degree of organisation in supply, which in turn might suggest either a strong merchant class or high degree of centralisation of control of the industry, either of which could be influencing trade or exchange in materials, workers and manufacturing information. The evidence for raw materials comes from two principal sources; the ancient textural evidence for glass making and the close analysis of the microstructure and chemical composition of the vitreous materials. There is virtually no early textural evidence concerning the production of faience, but a close link between the glass and faience industries, might suggest that some of the raw materials were common to both, so evidence from glass making texts could be significant for faience too.

EVIDENCE ANCIENT TEXTS

A search through ancient texts reveals that in several periods texts are preserved that are interpreted as describing the making of glass. These texts range in date from Babylonian texts of the mid second millennium BC, through Roman texts and ultimately right up to those of Europe in the 16th century AD. Only the earliest are of interest in this study and these are the Assyrian texts from the library of King Assurbanipal (669-627 BC) at Nineveh and the writings of Pliny the Elder in the first century AD. Of the 20,000 fragments recovered at Nineveh, a few dozen fragments contain information concerning the manufacture of glasses and glazes. Following early work by Campbell Thompson (1925), a major re-examination of these texts was carried out by Oppenheim (1970), assisted by Brill, who provided advice on technical matters. Oppenheim states that the most of the texts are Assyrian in date, but refer in part to earlier traditions dating back to the mid-second millennium, the same age as rarer Middle Babylonian texts found in the same place. The texts must, however, be treated with great caution. As Moorey (1994, page 210) and others point out, ancient texts such as these "served administrative or literary purposes... [and] were far removed from practical knowledge". Moorey goes on "current understanding of the texts is, at the most basic level, faulty and incomplete. They employ a large number of rare and unique words, refer obscurely to chemical procedures, and lack a scientific mode of presentation". In addition, the later texts were copied from texts sometimes 1000 years older, perhaps as a routine part of the training of scribes, and these scribes were almost certainly somewhat divorced from the craftsmen who were actually preparing the glass. Even given these reservations, Brill identifies the words for two raw materials that were used by the Assyrians, the first being *immanaku* which Brill interprets as quartzite pebbles gathered from a river bed and the second is *ahussu* a plant ash almost certainly from a desert plant, and most probably from *Salsola kali*. He convincingly argues that the plant ash could have added all the calcium, iron and magnesium oxides needed to make up the composition of Assyrian glass and specifically demonstrates that lime did not need to be added separately, a clear difference from the interpretation of Campbell Thompson (1925). Brill also uncovers the probable Assyrian words for copper oxide (for making red glass) and copper or a copper alloy (for making blue glass).

Pliny the Elder, writing in the first century AD, gave information about the manufacture of glass before and during his time. He states that, before his times, "soda" and "magnesium" were added to the sand, along with "brilliant stones", "shells and quarried sand". During his time, however, this procedure was simplified and very white sand from the mouth of the River Volturnus was used, pounded in a pestle and mortar and "three quarters of a part soda" added, the result being fused in a furnace. The use of finely powdered siliceous rock or sand plus plant ash remained the basis for the manufacture of glass right up to the early 19th century and in many areas of the world, up to the present day.

EVIDENCE FROM THE CHEMISTRY OF GLASSES

As has already been stated in Chapter 1, as a general rule ancient glasses are compositionally more complex than modern glasses. This may be due to a number of factors. Firstly, the raw materials are generally impure, and it was difficult for glass makers in antiquity to realise that they were impure and that there was something to be gained by purifying them. In this way, certain elements were unintentionally added to the glass with the raw components, elements that would not necessarily have had a beneficial effect on the characteristics of the glass, and sometimes could have been counter productive. Secondly, elements may have been leached from the crucibles and associated equipment while the constituents were being processed. Thirdly, there is clear evidence from ancient and medieval texts that a good deal of ritual was associated with the production of glass, and therefore components that now appear rather peculiar may have been added as a part of long standing tradition or superstition (Oppenheim 1970). These additions again may not have had any beneficial effect in terms of improving the finished product.

From the evidence of the texts, as well as from the chemical composition of ancient glasses and modern knowledge of the properties of silica rich melts, it is clear that there are three key components in the production of ancient glass; silica, alkali and colorants and/or opacifiers, plus more minor components including the possible separate addition of lime.

Silica

Silica is the dominant element in the vast majority of ancient vitreous materials in all the vitreous materials from Egypt during the Dynastic period. Turner (1956a) has suggested two sources of silica that may have been used for the manufacture of vitreous materials and specifically glass in ancient periods. The first was natural sand derived from rivers or coasts and the second white quartzite pebbles from river beds. Brill's interpretation in Oppenheim (1970) of Mesopotamian texts, suggests that quartzite pebbles may have been used in the manufacture of the early Mesopotamian glasses, whilst the use of sand has been widely attested in the manufacture of glass by several ancient authors in the Roman period and later, including Pliny and Agricola. The primary requisite for a silica source is that it should be as free as possible from any impurities, in particular the transition metals. Iron is especially troublesome, since even in relatively minor amounts it can impart a strong colour to the resultant glass, which is undesirable if the glass maker is attempting to use the more subtle colouring of one of the other transition metals. The secondary consideration is the ease of crushing the material. Sand requires less effort to crush and can even be used in the raw state, whereas pebbles require some effort to reduce the grain size until it is fine enough to be able to be used in the batch. However, this factor should not be emphasised strongly, since, as discussed in Chapter 7, archaeological evidence suggests that Egypt was not short of manpower, and that the modern concepts of efficiency are not necessarily applicable.

Despite careful selection, iron is often brought into the glass with the silica, due to the fact that it is finely mixed in with sands to the extent where it cannot be extracted. Table 5-1 shows analyses of two Egyptian sands from Amarna, demonstrating that iron is present in Egyptian sands to a concentration that would be more than sufficient to give the levels seen in the vitreous materials, averaging 0.6% in the glasses and 0.7% in the frits (Chapter 3). It is, however, possible that the iron was unintentionally added along with ores of copper and other elements used for colouring purposes, and a purer source of silica used. What is clear from Table 5-1 is that the sands are very variable in composition and that a vitreous material resulting from identical processing of the two different sands in Table 5-1 would have very different compositions. The resultant vitreous materials do not vary in composition by such an extent, which may indicate that there was a very careful selection of sand indeed, or that quartzite pebbles were used, a source that would be less compositionally variable.

Glass preserves virtually nothing in the way of the microstructure of the silica from which it was produced, with the rare grains that are preserved being highly altered and of questionable source, so that may not be of assistance. With vitreous materials such as faience, frit and Egyptian blue, however, this is not the case. The quartz grains that make up these materials are frequently only slightly altered from their primary state. In the case of frits, for example UC36465 shown in Figure 5-1, the grains are in some cases up to 6mm long, angular in shape and show none of the rounding that is typical of aeolian grains (see Figure 5-2), all of which strongly suggests that they are crushed quartzite pebbles.

The geology of the site of Amarna does not help in determining the source of the silica, since both sand and quartzite pebbles (which are brought down the Nile and are commonly found scattered over the Nile flood plain) are both common. Petrie (1894) working on the site stated that he found kiln waste in the form of white quartz pebbles. From the occasional drips of glaze on them, he deduced that the pebbles formed the floor of the kiln. Here, he hypothesised, they would be fired many times and would subsequently crack and shatter, thus making them easier to crush to produce the raw materials for future glass-making purposes. Petrie therefore thought that pebbles were the source. Two of the fragments of glass making vessels, UC40568 and EA67825, have the remains of quartzite pebbles adhering to them, which suggests that the pebbles were, at very least, involved in the processes somewhere.

To summarise, in terms of chemistry, the two sources of silica are not easily distinguishable. The microstructure of the frits suggests that their source was crushed quartzite pebbles, which may also be true of faience, where grains in the core are often suggestive of these pebbles. For glass, on the other hand the situation is unclear, but the low iron and the apparent lack of variability in glasses

Table 5-1 : Sand analyses by ICP-AAS

sample	SiO_2	Al_2O_3	CaO	MgO	Na_2O	K_2O	FeO	TiO_2	MnO
AM44 surface sand	87.36	2.79	5.50	0.55	0.55	0.53	1.76	0.76	0.04
AM43 sub-surface sand	73.56	3.29	18.11	0.90	0.83	0.61	2.02	0.52	0.04

points very tentatively once again to the use of quartzite pebbles.

Alkali

Turner (1956c) lists three sources of alkaline salts that were available to the glass-maker; large scale natural evaporitic deposits from former land locked seas and lakes, the ashes of vegetable matter, and local artificial evaporites resulting from the evaporation from pans of sea or river water, perhaps with enrichment of salts due to efflorescence of the soils. A number of large inland lakes and seas that produce evaporitic deposits are known, Wadi Natrun and El Kab being the most significant in Egypt. These sources produce a range of compounds including sodium sulphate, carbonates and bicarbonates. Specimens of natron from the Wadi Natrun have been found in Egyptian tombs and it was widely used in ancient times as a detergent, for medicines and for embalming among other things. It is therefore reasonable to look towards natron as a possible source of alkali for vitreous materials. However, as Turner points out, natron is compositionally very variable, both in ancient and modern samples. Therefore, if natron was used in vitreous materials, it must have been very carefully selected on some physical grounds, the selection being a lot more careful than for the natron used for other applications in Egypt, which shows a much wider compositional range than the alkalis seen in glass.

Alkali obtained by artificial evaporation would obviously vary according to the source, whether it be sea or river water. Pliny mentions the use of Nile water to obtain "nitrum", giving two places where this was carried out, Naucratis in the Delta, and Memphis. Turner points out that obtaining salts by this process would be very labour intensive, requiring about 7000 parts of Nile water for every one part of salt obtained. However, the annual flooding of the Nile would, after the subsequent evaporation, leave the sands beneath it slightly enriched in salts and it has been proposed by Ian Freestone that this enrichment would be increased by efflorescence by the soil during drying. There is, however, no evidence that salts were obtained in this way during the New Kingdom.

Plant ash has through the centuries been a major component in the production of glass. Ancient writers have frequently mentioned ash from both coastal and inland plants as being used in glass-making, including *salicornia* (Campbell Thompson 1925) and *Salsola kali* (Brill 1970) at 7th century BC Nineveh. Analyses carried out by Turner have revealed the complex and variable nature of ashes derived from coastal and marine plants, about which relatively little work has been done. Since the chemical make up of the plant is dependent on the mineralogical and chemical makeup of the soil in which it is grown, the plant's chemical analysis will vary according to the region in which it is grown. It is even possible that the chemical make up will vary according to the weather. For example in cases of extreme drought the soil will become leached of certain nutrients and thus the chemical make up of the soil, and thus the plant, will change. Turner rightly points out that the great variability in the alkali from all sources, and the variability in particular of the ash from different plants and even within one source of plant natron, would make the production of "clear, bright, homogeneous glass" extremely problematical.

The total alkali content of glass has a very wide range. Some alkali-lime-magnesia-silica glasses from 18th and 19th Dynasty Egypt have soda contents of over 28% (Caley 1962). Occasional Amarna glasses have been analysed with soda in excess of 22% and since these glasses often also contain potash, total alkalis in some rare instances exceeded 30%. In the glasses of this study, the soda varied from 12-22% and potash from 0.5-3.5%. As commented on in Chapter 3, there is a clear difference in the proportions of soda and potash in the different coloured glasses, with the cobalt coloured glasses being higher in soda and lower in potash than the copper coloured ones.

In attempting to determine the source of the alkali used in the glass, various authors have cited the magnesia content as being important in addition to the soda and potash. Magnesium oxide is of almost universal occurrence in the glasses, most ancient glasses contain between 2 and 5% magnesia, and the Amarna glasses of Chapter 3 tend to have relatively high magnesia contents, averaging 3.7%. Sayre and Smith (1961, 1967 and 1974) noted that in the some 400 glasses that they examined (from Europe to India and the second millennia BC to the 12th century AD), there was a correlation between the concentration of magnesia and some of the alkalis, high magnesia correlating with a relatively high potash level, but with no correlation between the amounts of lime and magnesia. This slightly higher level of potash is compensated for by a reduction in the amount of soda, so that in the samples that he examined the total alkalis remained approximately the same. Sayre goes further to speculate that it might be possible to divide the magnesia concentrations into two almost discrete ranges, with virtually no overlap, which correlate with two discrete ranges in the potash concentrations. From this he draws the conclusion that there are two statistically different types of glass produced in the period and area that the study covered. Glass from 1500 to 800 BC in the Eastern Mediterranean, Egypt and Mesopotamian was of the high magnesia (3-7%), high potash (1-4%) type, whilst in the following period up to the latter part of the first millennium AD, low magnesia (0.5-2.5%), low potash (0.1-1.0%) glass predominates. By analysing different plant ashes from desert plants and others and comparing them to natron analyses, Brill (1970) and Lilyquist and Brill (1995) link this variation in the magnesia and potash with a change in the nature of the alkali used in the glass batch, the earlier high magnesia and potash being use of plant ash, whereas the later low magnesia and potash represents use of natron.

Figure 5-3 shows a graph of the percentage concentration of potash against magnesia for the glasses of this study with areas marked where glasses produced by plant ash based alkali and natron based alkali should plot (after Lilyquist and Brill, 1995). As can be seen, the glasses coloured by copper and cobalt plot in distinct fields, the copper coloured glasses falling in the plant ash zone, and the cobalt coloured glasses with similar magnesia to these, but lower potash. This difference is unlikely to be caused by the mixing of plant ash and natron in the cobalt glasses, since in this case these glasses would plot on a mixing line between the two fields. Instead, they seem to have similar potash to the natron samples, but higher magnesia. This extra magnesia could come from the addition of a dolomitic limestone, but there is apparently no relationship between the lime and the magnesia in the cobalt glasses, which tends to suggest this is not the case. Alternatively, the magnesia could be coming in with the cobalt colorant. The source of the cobalt is thought to be the alums of the Western Oases (as discussed in detail below) and these alums contain 6.0-8.5% MgO (Table 5-3, Kaczmarczyk 1986) which would greatly increase the amount of MgO in the final glass. Segnit (1983) states clearly that it was the magnesium bearing sulphates that contained the minor elements manganese, cobalt and nickel. The CoO:MgO ratio in the alums averages 0.054 in the four analyses of Kaczmarczyk (1986), which would mean that if the alums added to the glasses had the same ratio, then given the average CoO content of the cobalt glasses of the study was 0.16% (Chapter 3), the amount of MgO accounted for in the glass by the addition of alums would be 3.0%. Given the average MgO of the cobalt coloured glass was 3.6%, the remainder that is interpreted as coming from the alkali, would be 0.6%, which would fall in the field for glass produced using natron. Figure 5-4 shows the potash and magnesia graph replotted with the magnesia reduced in the cobalt coloured glasses by 3.0%, revealing that, in this case, all the cobalt glasses, except one, plot in or close to the natron field. The exception to this pattern is the cobalt coloured glass 403a, which has a very high MgO content at 5.9% (the highest of all the glasses analysed) and K_2O content of 1.9% which, when the 3.0% of MgO is removed that is interpreted as the addition from the alum is removed, plots very close to the plant ash field and that for the copper glasses.

The source of the alkali for the other vitreous materials is less clear cut. The glass matrix of the cobalt coloured frits have very low potash, averaging 0.5%, and moderate MgO at an average of 2.0%, which looks similar to the cobalt glasses with a natron based alkali enriched in magnesia from the alums, but with the enrichment not as pronounced as in the glasses. A natron source is therefore suggested for these. In contrast, the copper coloured blue faience rings and other copper blue faience has a very low in MgO content at only 0.3%, and very high potash content at 4.1%, which suggests that the alkali may be a plant ash of very different composition to that used in the glasses. The Variant D faience plots between the plant ash and natron fields when silica enrichment is taken into account, and is discussed further in Chapter 6.

It seems that broadly speaking, therefore, there are two source of alkali for the glasses at Amarna, plant ash used for all the copper coloured glasses and possibly one of the cobalt ones, and natron used for the majority of the cobalt coloured glasses, whilst the other materials show more complicated patterns perhaps derived from the mixing of sources as discussed in the production of these materials in Chapter 6.

Colorants and opacifiers

The deliberate addition of compounds containing transition metals in order to alter the colour or appearance is a widespread and common technique in the manufacture of ancient vitreous materials. Chapter 1 details how the chemistry of the colorants affects the colours and the effects of the presence of more than one of the transition elements. The most common colorant of Egyptian vitreous materials is copper, but at Amarna this is followed and maybe superseded in popularity by cobalt, an 18th dynasty innovation. These elements are commonly found associated with other metals which are present in their ores and which are often incorporated, presumably accidentally, into the vitreous materials. Those associated elements can often be identified within the material, and this can, in some circumstances, lead to a tentative provenancing of the source of the colorant. With this in mind, 28 samples of Amarna glasses and frits were taken analysed by ICP-MS following procedures laid down in Chapter 2, since this method has better detection limits than SEM-WDS and the levels at which some of the transition metals are present in many of the glasses is close to the SEM-WDS detection limits. The results of the analyses for the elements of particular interest are displayed in Table 5-2.

Copper blue

Copper oxide is frequently added as a colouring agent. Which, depending on the conditions, gives a blue, green, or opaque red colour. Copper is used from at least the Nagada I period (*c.* 4000BC) in faience glazes and is the commonest form of colorant in the early Egyptian glasses. Sayre and Smith (1974) found that the copper contents of those Egyptian glasses that were deliberately coloured with copper was in the range of 0.59-1.45% CuO, with an average of 0.93%. This is lower than the copper coloured Amarna glasses analysed by SEM-WDS in this study (Chapter 3) which have copper contents of between 0.85% and 3.12%, averaging 1.45% CuO. In faience glazes the copper composition can be much higher, up to 12.5% CuO in the case of the ring shank E5057/1, and averaging over 10% in the faience rings, despite the fact that some of the glazes are slightly weathered. Kaczmarczyk and Hedges (1983) speculate that since copper is very mobile within a silicate matrix, copper contents below 0.1% CuO are probably caused by accidental contamination from bronze or copper vessels and tools.

Geilmann (1955) observed in his analysis of a wide range of ancient glasses that several of them had copper, tin and lead present in proportions similar to those found in bronze of the same period as the glass. Oxidised or corroded bronze could therefore have been included in the glass melts in order to introduce copper as a colorant. Sayre found a similar pattern amongst some of his glasses as did Kaczmarczyk and Hedges (1983), who showed that the presence of tin in the glazes of faience corresponds to the beginning of the 18th dynasty when there was a significant rise in the amount of bronze being used in Egypt. While known in the Middle Kingdom, bronze only really becomes common in the New Kingdom (Lucas and Harris 1962).

The ICP-MS analyses in Table 5-2 for copper and tin are plotted in Figure 5-5. They show that the data falls into four groups; low copper (<150ppm) with significant cobalt (432-4714ppm), significant copper (614-4498ppm) and cobalt (589-1969ppm), high copper (>9900ppm) with very low cobalt (<40ppm), and those with some lead colorant. The first two groups are closely related, one sample, E5613B, being able to be placed in either group, and together these two groups lie on a line, the slope of which gives a Sn:Cu ratio of 0.069 (SnO_2:CuO of 0.071). Kaczmarczyk and Hedges (1983) give the average Sn:Cu ratio for 18th dynasty bronze as 0.070, which strongly suggests that in these first two groups of glasses where cobalt and copper are both present, bronze could have been used as the way the copper was introduced. A second explanation for these groups could be that these are cobalt glasses which are very contaminated from bronze tools that have been used in the processing. However, if this was the case, some of them would have 0.56% CuO contamination, the norm being 0.25% CuO, and Kaczmarczyk and Hedges (1983) speculate that

*Table 5-2 : ICP-MS results for glass and frits (marked *), all results given in % or ppm as indicated.)*

sample	colour	^{25}Mg	^{27}Al	^{43}Ca	^{55}Mn	^{59}Co	^{63}Cu	^{66}Zn	^{75}As	^{208}Pb	^{118}Sn	^{121}Sb
TA69	blue	1.82%	0.33%	8.85%	140	9	1.48%	42	21	290	160	59
AM16	blue	1.69%	0.31%	6.87%	210	39	1.20%	59	69	43	240	3
TA98	blue	1.82%	0.56%	7.37%	710	33	0.99%	52	62	63	210	8
UC22920B	blue	2.30%	?	6.31%	?	17	?	110	25	57	800	13
UC22922	blue	0.21%	0.18%	6.32%	46	14	?	39	250	70	?	17
MCR2639	blue	0.98%	0.85%	2.37%	930	800	0.45%	780	5	150	290	11
E5613C	blue	1.37%	0.78%	5.12%	0.30%	570	0.27%	670	8	33	210	0.11%
UC22096	blue	1.62%	0.97%	6.60%	0.17%	590	0.23%	560	20	110	110	0.11%
E5613D	blue	1.68%	0.99%	6.06%	0.14%	930	0.22%	0.12%	9	72	140	0.11%
MCR2625	blue	2.14%	1.81%	4.96%	900	590	0.22%	580	19	300	170	656
E5613E	blue	1.61%	0.87%	5.67%	0.14%	950	0.21%	0.13%	10	72	160	0.13%
E5613F	blue	1.22%	0.60%	3.96%	770	680	0.21%	470	7	74	8	720
E5042	blue	1.53%	0.82%	5.31%	0.12%	620	0.17%	880	7	200	78	?
UC22909A	blue	0.93%	0.55%	3.66%	660	730	0.14%	770	16	51	110	?
AM37	blue	1.17%	1.59%	3.80%	0.12%	931	928	520	3	120	87	72
E5613B	blue	1.86%	1.52%	5.63%	0.17%	0.20%	614	0.21%	8	68	40	1.05%
UC22922A	blue	0.95%	3.39%	0.79%	0.47%	0.47%	96	0.15%	1	23	18	12
AM28 frit *	blue	0.92%	0.99%	2.63%	0.10%	0.14%	138	0.23%	1	110	10	450
AM32 frit *	blue	0.48%	1.43%	0.47%	0.16%	0.16%	54	0.21%	1	8	2	2
AM34 frit *	blue	0.48%	1.45%	0.47%	0.16%	0.16%	46	0.21%	0	10	3	1
MCR1967H	blue	2.72%	1.25%	6.58%	0.19%	980	141	0.11%	1	31	12	33
MCR1967A	blue	0.70%	0.96%	2.10%	750	950	149	0.12%	9	110	13	270
E5053	blue	0.28%	0.95%	0.45%	760	750	183	0.14%	1	23	14	5
E5613A	blue	1.11%	0.60%	4.69%	710	430	120	490	1	9	3	6
UC22918BB	green	1.52%	0.32%	4.93%	190	9	0.63%	62	11	0.28%	590	620
UC22918BA	green	1.50%	0.41%	5.55%	190	8	0.63%	770	20	1.28%	620	?
UC29166	yellow	1.81%	0.20%	4.56%	130	3	371	620	19	2.06%	18	0.34%
E5051	purple	0.84%	0.12%	4.01%	0.76%	19	123	250	3	36	?	?
alum 6A	white	2.63%	5.70%	0.19%	0.46%	0.21%	14	0.37%	0	0	na	na
alum 3A	white	3.15%	4.62%	0.06%	0.46%	0.27%	68	0.48%	0	5	na	na

only copper contents below 0.1% CuO are caused by accidental contamination from bronze or copper vessels and tools. Contamination on this scale would therefore be a less likely explanation than deliberate addition. The third group with high copper and no cobalt does not fall on the same line, having a similar level of tin to the significant copper and cobalt group, but two or three times the copper, giving a Sn:Cu ratio of 0.02. This perhaps suggests that this group has the same bronze colorant as the significant copper and cobalt group, but with additional copper being added, perhaps, since these samples have virtually no cobalt, to give a deeper blue colour than the level of bronze colorant alone would create. The lead coloured glasses have one with little copper (yellow in colour) and two with around 6300ppm Cu which are green and plot above the line that would indicate bronze perhaps suggesting that the lead colorant has brought in extra tin to a copper or bronze colorant, but without further examples it is not possible to say more than this.

Cobalt blue

Farnsworth and Ritchie (1938) proved Cobalt was deliberately added to Egyptian glass for the first time during the New Kingdom. Cobalt is common in the production of deep blue, violet and indigo glasses during this period, although after the New Kingdom it virtually disappears through the Third Intermediate Period until the Late Period, where it is again introduced to the production of blue glasses and then for black. In the right conditions cobalt can produce colours from pink and purple to green and blue. Cobalt in Egyptian New Kingdom glasses where it is employed, falls in the range 0.075 to 0.55%. In the cobalt blue Amarna glasses analysed in Chapter 3, the cobalt content varies between 0.03% and 0.37%, averaging 0.20%. The frits are slightly more cobalt rich, averaging 0.29% and extending up to 0.42% in the glass matrix of the frits, a level which is very similar to that found in the cobalt coloured faience.

The cobalt in all the vitreous materials in Chapter 3 is strongly associated with other elements, the most abundant of which is alumina. The alumina content of the finished vitreous material could partly be derived from the presence of alumina bearing clay minerals as impurities if sands were used for raw materials and variations in the proportion of clay minerals within the sands could lead to some of the variation seen within the glasses. However, consulting the SEM-WDS data from Chapter 3, it can be seen that cobalt coloured vitreous materials have a significantly higher alumina content than copper coloured glasses which average 1.1% alumina, compared to an average 2.2% alumina extending up to 4.2% for cobalt glasses. Of the frits, the lowest alumina content at 1.5% is the copper coloured one, whereas the lowest cobalt coloured frit is 2.7% and these average 4.6%. This relationship between cobalt and alumina, and indeed cobalt and manganese, nickel and zinc has been commented on in Chapter 3 (see Figures 3-10 and 3-11). These relationships are supported by the ICP-MS data which is plotted in Figure 5-6, Figure 5-7 and Figure 5-8 for Co-Al, Co-Mn and Co-Zn respectively. The slopes of these graphs lead to the ratios shown in Table 5-4 which show good agreement between ICP-MS and SEM-WDS for the two ratios where both techniques could be used. Nickel could not be measured in the ICP-MS since it had internal componentry made of nickel which would contaminate the plasma, and it was present at too low a level to be reliably measured in the SEM-WDS.

In addition to its use in vitreous materials, cobalt is also used as a colorant in the New Kingdom in the paint applied to some ceramics, known as 'palace ware' or 'blue-painted pottery', so called because it always has blue in its colour scheme (Figure 5-9). The ware appears in the middle of the 18th Dynasty, seems to have been manufactured until the 20th and is especially common during the reigns of Amenophis III and Akhetaten (Boston 1982). It is usually made of Nile silt, covered with a lime slip and painted often in linear designs, animals and flowers in blue and black. The blue paint applied to the surface of these ceramics looks remarkably consistent both in terms of colour and in the way the paint takes to the slip. Riederer (1974) and Noll (1981), amongst others, have identified this pigment as cobalt aluminate, an artificial spinel. One sample of the blue painted ceramic numbered AM21, which was acquired from the 1993/4 excavation, a cross section of which under the SEM is shown in Figure 5-10, reveals a Nile silt ceramic with a lime rich slip 60µm thick above it, overlain by the blue painted layer 30µm thick and quite uniform in thickness. Occasional grains of pigment can be seen in this layer and analysis was carried out on one of these (marked in the figure) and the resultant EDS spectrum is shown in Figure 5-11. As can be seen, the grain is rich in alumina, with lesser concentrations of silica, lime and potash, plus significant amounts of the transition metals cobalt, manganese, iron, nickel and zinc, which is the same pattern of transition metals as in the vitreous materials, which strongly suggests that the colorants have a similar source.

The source of the cobalt in Egyptian vitreous materials has been the object of a great deal of debate. A variety of sources have been suggested, two in Egypt, the nickel ore site of St John's Island

Table 5-3 : Analyses of alums from the Western Oases (Kaczmarczyk, 1986).

samples	MgO	Al_2O_3	MnO	Fe_2O_3	CoO	NiO	ZnO	SO_3
alum 1	6.1	14.0	0.88	0.37	0.40	0.33	0.80	46.9
alum 2	8.5	10.7	0.93	0.25	0.43	0.58	0.99	44.7
alum 3	6.5	16.0	0.84	0.32	0.32	0.18	0.60	-
alum 4	6.0	16.1	1.13	0.35	0.31	0.16	0.62	-
average	*6.8*	*14.2*	*0.95*	*0.32*	*0.37*	*0.31*	*0.75*	*45.8*

The bulk of the rest of the analyses is water, Na_2O, K_2O, CaO and CuO combined being <0.2%.

Table 5-4 : Comparison of Co ratios derived from graphs for ICP-MS and SEM-WDS.

	glass + frits		alums	
technique	ICP-MS	SEM-WDS	AAS [1]	ICP-MS
Co:Al	7%	-	-	8%
$CoO:Al_2O_3$	5%	7%	3%	5%
Co:Mn	73%	-	-	46%
CoO:MnO	69%	110%	39%	43%
Co:Zn	77%	-	-	57%
CoO:ZnO	84%	na	49%	62%

[1] *from Kaczmarczyk (1986)*

and a source in the Western Oases, and several further afield, a source in what is now Iran being the only one conventionally taken as being likely (Kaczmarczyk and Hedges (1983) and Kaczmarczyk (1986)). St John's Island is not known to have been worked in antiquity, and is therefore ruled out by Kaczmarczyk. The Iranian source is thought to be the source of the cobalt used in the production of Islamic blue and white wares and is a manganese-free arseno-sulphide deposit. It therefore has a chemistry very unlike the chemistry of the Egyptian materials, which appear to show a significant manganese concentration and contain little arsenic.

This leaves the sources in the Oases of the Western Desert, particularly the Dakhla and Kharga Oases. There are no modern economically viable sources of cobalt in these Oases (Andrews 1962), but there is however, abundant evidence (Lucas and Harris 1962) that the Oases have been worked in antiquity for their deposits of alum. Beadnell (1901) mentions that while there were reports of cobalt bearing minerals at the Dakhla Oasis, he could not find any there, a result that is supported by the work of Segnit (1983) at Dakhla, who finds very low concentrations of cobalt in the alums from this site. The Kharga Oasis, however is a different matter. Beadnell (1909) drew up the geological survey of this site describing a series of gently dipping beds, slightly folded and faulted in some areas, with basal Upper Cretaceous sandstones, shales, and limestones, topped by a white chalk and with Eocene and more recent beds above. Within the Upper Cretaceous series, forming the bottom of the majority of the Kharga depression, is the Surface water sandstones containing sulphates of aluminium and magnesium, occurring as thin bands within the beds "and seldom exceeding a few centimetres in thickness" (Beadnell 1909). The alum is "frequently of a delicate rose tint" and analyses (Richmond and Orff 1892, Kaczmarczyk 1986) show that they contain cobalt at the level of 0.3-0.7% CoO (Kaczmarczyk 1986) or 1.02% CoO (Richmond and Orff 1892). Kaczmarczyk (1986) shows that this cobalt is accompanied by significant amounts of alumina, manganese, iron, nickel, zinc and sulphur, his analyses being reproduced in Table 5-3. Two of the alum samples analysed by Kaczmarczyk (although it is unclear which two) were reanalysed by ICP-MS, the ratios of cobalt to the other significant transition metals being calculated and compared to the other materials and techniques (Table 5-4). The results show that this is a close correlation between the ratios of cobalt and alumina, manganese and zinc in the glasses and in the alums, which is also seen on the graphs of the ICP-MS data shown in Figure 5-6, Figure 5-7 and Figure 5-8.

There is a strong resemblance, therefore, between the chemistry of the cobalt used in the vitreous materials and ceramics and that of the cobalt bearing alums of the Kharga Oasis. Given the certainty that the Oasis was worked in antiquity, it is reasonable to assume that the pink colour of some of the alums would have come to the notice of the miners and that this alum was subsequently used as the cobalt colorant in the vitreous materials at Amarna. The method for concentrating the cobalt into a useable form has been investigated by Noll (1981). He has speculated that it is formed by precipitation from a solution of the cobalt bearing alum, aided by the addition of a strong alkali, perhaps ammonia or sodium carbonate. The resulting gel was then heated to 800-1000°C to form the cobalt aluminate spinel which is used as the pigment. However, it is clear from the analyses of the alums and the glasses that there would not be a need to concentrate the cobalt in the alums greatly and it is possible that it was used after only preliminary preparatory heating to drive off the sulphur.

A possible confirmation from texts?

A large number of texts survive from the New Kingdom detailing tax lists and lists of gifts and supplies delivered by governors of various outlying areas to the King. It is possible therefore that if the Kharga was the source of the cobalt, these documents contain evidence as to the source of the cobalt colorant. It should be noted, however, that many of the papyri are incomplete and thus do not give all the details that we would like and a number of the areas with which the lists deal have names that we can no longer assign to a geographical area with absolute certainty, both of which cause problems in interpretation. However, by far the biggest problem is that we have no idea what the Egyptian term for the cobalt colorant was, other than the fact that since it is a mineral in its raw state it might have the characteristic determinative for "mineral" in Egyptian. Giddy (1987) has completed a detailed study of papyri concerning the Oases of the Western Desert. She agrees with the work of Sethe (1920) in isolating the term *wḥ3t* for "oasis" and "of the oases" and she goes on to differentiate *wḥ3t rsyt* "Southern Oasis", first mentioned in New Kingdom texts, as the term for the Kharga-Dakhla group.

Giddy analyses one text in particular that is of special interest to the question of where the cobalt mineral may have come from. This is the Papyrus Chester Beatty IV (BM 10684). It is thought to be from Deir el-Medina, and dates from the late 19th or 20th dynasty. Giddy (1987) provides the first translation and analysis of the text and the quotes which follow are from this work. The document lists those things from *wḥ3t* which were likely to be of interest to the Nile Valley, being "one of those letters... the purpose of which is to familiarise the writer or reader with a number of rare words and names of things". The lines of text of interest are lines 9,10 to 9,13, since in line 9,14 it introduces a different category of "things", fish from *w3ḏ-wr*, indicating that the *wḥ3t* items are over by then. Since the papyrus is damaged, it is not clear whether all the items listed in lines 9,10 to 9,13 are from *wḥ3t*, or whether another introductory line is missing. Giddy argues convincingly, however, that the layout of the text, the groupings of the "things" and the break in the presentation with the word "another" all suggest that the list is cohesive and all deal with *wḥ3t*.

A list of things from *wḥ3t* starts with line 9,10 with the introduction "Come (let me tell thee) concerning the things of which are numberless" and then goes on to list plant produce (9,11) starting with (?)grapes, then a number of unusual plants perhaps with details of their containers or amounts ending with *b(n)ìt rš3* "many dates". This is followed by different types of baskets or other woven produce (line 9,12). Line 9,13 deals with stone and mineral produce from *wḥ3t*, beginning with *inr dš(r)t n ḳd* "red stone for building" and *ibnw ḥḏ(t)* "white alum". A piece of the text is then lost before the line ends with *p3yw m dnìwt,* "*p3y* stones in baskets". The red stone is a reference to the characteristic red sandstone that is found in all the Western Oases. As discussed above, alum is widespread in the Western Oases, but this is the earliest reference known to alum from *wḥ3t* in Egyptian texts. The *p3y* stones are of the greatest interest and are difficult to attribute, but they are probably some sort of mineral from their determinative. Giddy lists two other minerals listed also in Harris (1961), *p3ḳt*(?) and *p3g3w*, but thinks that *p3y* is neither of these. She speculates from the context and her knowledge of many other similar lists that *p3y* are "perhaps small semi-precious stones or some sort of 'earthy' mineral".

So the evidence from the text suggests that the unknown *p3y* stones are rare and of value to the people of the Nile Valley. The word used is unusual to the average Egyptian scribe, hence it features on this list of rare words and names, comes from the Oases of the Western Desert, and is associated in the script with red sandstones and alums. The translator speculates that they may be semi-precious and/or earthy. This is a very close description of the likely characters and source of the cobalt ore minerals as deduced from the chemistry of the vitreous materials described above. No other mineral would have such a close fit to the description. It is certainly possible that this one line from the papyrus details the shipment of a certain number of baskets of the ore to the Nile Valley from either the Dakhla or Kharga Oases. If this is the case, it represents a very rare tie up between the chemistry and the contemporary written accounts.

Manganese purple

As discussed in Chapter 1, manganese can be used in different glasses both as a colorant and decolourant and as such a whole range of colours can be influenced by the presence of manganese. Examples of such colours occur from 1500BC and could include dark blue, violet, red, 'colourless', greenish yellow and light and dark green. However, its deliberate use in New Kingdom Egypt, however, seems to be restricted to violets and blacks.

As discussed above, manganese is strongly associated with the presence of cobalt in vitreous materials, and as such is present at up to 0.3% MnO in the glasses discussed in Chapter 3, and up to 0.6% in the glass matrix of the frits, where the cobalt is at a higher concentration. In the glasses and faience coloured solely by copper, manganese is never present at concentrations greater than 0.11%, and usually at a lot lower concentration than this. Only one example of a glass apparently deliberately coloured by manganese was analysed. This is E5051 which contains 1.35% MnO, with only 0.3% FeO, and no other transition metals at measurable levels. Manganese minerals are common in Egypt and are one of the compounds used for eye paint in Egypt, although galena is more common (Lucas and Harris 1962). The source of the Amarna manganese is not known.

Lead antimonate yellow

Lead based colorants are used in Egypt for the production of yellow opaque glasses and yellow faience (Lucas and Harris 1962, Kaczmarczyk and Hedges 1983 and many others). Most sources state that the most common form of the lead in these materials is lead antimonate, $Pb_2Sb_2O_7$.

The coloration and opacity in Amarna glasses and faience are caused by small particles that appear white in backscatter mode on the SEM, indicating the presence of high atomic number elements (described in Chapter 3). To assess what these particles consisted of, 21 point analyses were made of the bright particles in three different samples, two of faience (E5032/4 and E5054/1) and one of glass (E5613/1). The results are plotted in Figure 5-12, which shows the lead and antimony weight percents and the trend line for the analyses. The theoretical trend line for stochiometric lead antimonate $Pb_2Sb_2O_7$ is also illustrated. As can be seen the data are quite closely grouped and the slope of the trend line gives a PbO : Sb_2O_5 ratio of 1:1.61 for the weight percents, which translates to 1:2.10 in atomic percent, very close to the stochiometric formula of 1:2. The weight percent trend line for stochiometric lead antimonate has a PbO : Sb_2O_5 ratio of 1:1.53 and is plotted on Figure 5-12 for comparison. From this data it is clear that the samples tested plot close to the stochiometric formula in all cases, but that the average is slightly enriched in antimony over the lead. This may reflect the fact that the lead is more mobile and at high temperature leaches out into the surrounding glass. Analyses for the compositions of the yellow glass, plus these for the faience glaze and glass matrix of the faience cores are also plotted on Figure 5-12 for comparison and indicate that the glaze and the interstitial glass are very enriched in lead.

Minerals containing the elements lead and antimony are known in the natural world. Nickel and Nichols (1991) list 57 minerals containing these elements, of which only nine, most of them sulphides, do not contain significant amounts of either transition metals or arsenic, the most common of these being boulangerite $Pb_5Sb_4S_{11}$, but most of them being very rare. None of the references for these nine minerals list Egypt or Western Asia as sources and although the sulphide bodies of the Eastern Desert would be the right environment for the presence of such minerals, there is no evidence that they are there in any quantity or that they were worked. It seems unlikely, therefore that the lead and antimony were added to the glass as one naturally occurring mineral. A more likely solution, is that the antimony and lead sources were mined separately and combined later, perhaps as a result of other metalworking processes. The primary antimony ore is stibnite (Sb_2S_3), which has quite wide occurrence in sulphide ore deposits and is frequently associated with ores of lead, silver and gold, all of which were used widely in antiquity. Stibnite is often oxidised to yellow antimony oxides such as cervantite (Sb_2O_4), which would be noticeable and may have been the reason for its early selection as a colorant. The precise provenance of antimony in Egyptian glasses remains to be solved, however. Possible sources include the ancient gold mines at Wadi Ballit in the Eastern Desert (Hume 1937) and Fawakir, quartz veins at Gebel el Ineigi, East of Edfu (Azer 1966) and further afield in Turkey and Morocco, some of which were mined in antiquity. The use of stibnite in Egyptian kohls has also been speculated on (Lucas and Harris, 1962), but this seems to make up only a small fraction of the total kohls (generally black compounds used for eye paint) and more work using modern analytical techniques is required to further elucidate this area.

Figure 5-13 shows a phase diagram of the $PbO\text{-}Sb_2O_5$ system which may cast light on the reasons for the excess of PbO in the vitreous materials. As can be seen the melting point of $2PbO.Sb_2O_5$ is around 1300°C, which is above that thought attainable in antiquity. With increasing PbO, however, the melting temperature is reduced until when PbO concentration is about 92% the eutectic at 820°C is reached. The bulk compositions from the Amarna vitreous materials lie along this liquidus between the peritectic and eutectic. The additional PbO therefore lowers the temperature at which the $Pb_2Sb_2O_7$ will melt, making it more easily absorbed. Only when this has at least partially occurred will the bright yellow colour typical of $Pb_2Sb_2O_7$ be produced, Chambers and Rigg (1925) finding that the best colours for the system were produced at a PbO:Sb_2O_5 ratio of between 1:1 and 3:1, although this could be

Table 5-5 : Pb isotope results. Analysis of glass and faience by TIMS giving the three major ratios used for the provenancing of lead. Also given is the type of material analysed and its source. All samples are from Amarna.

sample	type	source	207/206	208/206	206/204
5032/4	yellow faience bead	Liverpool	0.8072	2.0152	19.3925
5045	green faience tile	Liverpool	0.8124	2.0282	19.2285
5053	yellow faience ring shank	Liverpool	0.8073	2.0157	19.3958
5054a	yellow faience ring shank	Liverpool	0.8074	2.0166	19.4000
5054b	yellow faience ring bezel	Liverpool	0.8078	2.0185	19.4193
5613a	yellow decoration to glass vessel	Liverpool	0.8102	2.0200	19.3084
5613b	yellow decoration to glass vessel	Liverpool	0.8118	2.0200	19.2890
2664	glass rod	1993/4 excavation	0.8110	2.0235	19.3228
2647	glass rod	1993/4 excavation	0.8088	2.0178	19.3410
UC22916	glass rod	Petrie Museum	0.8094	2.0193	19.3307
UC22937A	glass rod	Petrie Museum	0.8261	2.0543	19.0036

higher and at the higher end, these would fall in the range of the liquidus of Figure 5-13 and match the values obtained in Egyptian material.

The source of the lead used in the colouring of Egyptian vitreous materials has been the subject of great debate. It has been assumed that it came from a similar source to the lead metal (used occasionally in Egyptian objects, such as weights) or to the galena (PbS used widely in cosmetics, particularly eye paints). Lucas and Harris (1962) stated that before the 18th Dynasty the source of the lead was local, but after that they admit the possibility of importation from outside. The source of these leads and galenas has subsequently been worked on by numerous groups (Brill *et al* 1974, Stos-Gale and Gale 1981, Lilyquist and Brill 1995, El Goresy *et al* 1997) who have concentrated on the use of lead isotopic techniques to identify the ore bodies from where the various materials have been obtained. Figure 5-14 shows results from these various studies plotted on a Pb isotope graph of Pb^{208}/Pb^{206} against Pb^{207}/Pb^{206}. The ore sources plot in fairly discrete areas, but the number of analyses from each body is low, ranging from two to six, and the kohls and glasses plotted are all 18th Dynasty. Table 5-5 shows the new results of the present work which are also plotted on Figure 5-14.

As can be seen, the results of the new Amarna glasses and faience all plot between 0.807-0.813 (Pb^{207}/Pb^{206}) except one outlier, the glass rod (UC22937A), which has a Pb^{207}/Pb^{206} of 0.8261. This mirrors 18th Dynasty glasses analysed by Lilyquist and Brill (1995), where the majority of glasses plot between 0.808 and 0.815 with the Wadi Qirud glasses falling as two outliers at 0.8253 and 0.8226. Lilyquist and Brill (1995) also have two early glasses from the tomb of Tuthmosis III plotting much higher at 0.8380. The 18th Dynasty kohls form a tight group averaging below 0.805, but on the general trend line of the glasses. All the ores except those from Gabal Zeit plot below this line, and higher in Pb^{207}/Pb^{206} than the majority of the glasses. The kohl results fall close to the area of the two analyses for the Gabal Zeit ores, and some of the Amarna vitreous materials also fall close to the same ores, other trailing off on the same line towards the Wadi Qirud glasses.

Interpreting the data is difficult, since there are insufficient points for one to be sure of the size of the fields concerned and the amount of scatter of the points. The clearest pattern is that the Gabal Zeit galenas, known to have been worked in the 18th Dynasty, are consistent in Pb isotope composition with the 18th Dynasty kohls analysed by Brill (1974), which suggests that Gabal Zeit could be the source for these kohls, whereas other ore bodies are demonstrably different. Less clear is the pattern for the vitreous materials. There appears to be no distinction between the faience and the glasses, with the majority plotting together. What is clear is that the earlier glasses from the Tomb of Tuthmosis III and Wadi Qirud, plus one of the Amarna glass rods plot well away from this area. The Wadi Qirud samples fall in the area characterised as "Mesopotamian" by Lilyquist and Brill (1995), since this is the area where Mesopotamian glazes plot. The Wadi Qirud samples and the Amarna glass rod UC22937A are therefore possibly consistent with a Mesopotamian source, as yet unidentified. The other glasses and faiences plot between the Gabal Zeit galenas and the Wadi Qirud samples, but closer to the Gabal Zeit end. The presence of a glass rod of possible Mesopotamian isotopic characteristics at Amarna, although it is only one sample and so should be treated with care, suggests that even by the time of Amarna, there is still some Mesopotamian characteristics glass in the form of rods or ingots being incorporated in the melts. This may be because the glass is still being imported to Egypt during the Amarna period, or it may be in the form of "old stock" that has been moved around with the glassworkers as they moved. Either way, this material would be contaminating any Egyptian Pb-characteristics glass if it is added in to a melt or reworked with new glass. Is tempting, therefore, to see the pattern of glass and faience lead isotope characteristics from Figure 5-14 as a mixing line with locally made lead coloured glass with Gabal Zeit type Pb-characteristics being mixed with Lilyquist and Brill's "Mesopotamian" type, exemplified by the Wadi Qirud samples and glass rod UC22937A. This should be taken only as a working hypothesis from this preliminary data, further analyses, particularly of ore bodies, being required to confirm or deny its possibility.

Calcium antimonate
Only one example of white opaque glass coloured with calcium antimonate was sampled. The source of this colorant/opacifier is currently unknown.

Deliberate addition of lime?

Whether, in ancient glasses, there has been the deliberate addition of a lime bearing compound is hotly debated (Turner (1956a), Campbell Thompson (1925), Brill (1970) and many others). Lime is present in plant ash, and most sands, although not in significant quantities in natron. The lime content of some sands or plant ashes, if they were used, is enough to account for lime in some glasses where the lowest lime contents are of the order of 2%, but in other glasses their lime appears too high to be dismissed in this way, many early Egyptian glasses lying in the range 8 to 11%, the highest from Amarna in this study (AM38) has a lime content of 11.2%. Of the ancient authors, Pliny mentions the deliberate addition of lime in the form of "shells and quarried sand", and Campbell Thompson (1925) has argued that the Assyrian recipes indicate the deliberate addition of lime, but this is disputed by Brill (1970). The arguments for the addition of lime in this study are circumstantial and largely negative. Some of the vitreous materials, particularly the glass matrices of the Variant D faience and the frits (away from contamination), show low lime contents, averaging about 1.0%, which suggests that there was a source of silica free of lime, ie. quartz pebbles, that was available to the faience workers. Although a different lime rich silica source could have been used for the glass, if quartz pebbles are the source of the silica, an addition of secondary lime, probably in a form free from magnesia, would be necessary. Limestone is freely and widely available in Egypt and would be an obvious source for this lime.

SUMMARY OF RAW MATERIAL SOURCES

A summary for the sources of the raw materials used for the production of ancient vitrified materials in diagrammatic is shown in Figure 5-15, which divides the various materials into those available within the confines of Amarna as laid out by the boundary stelae (see Chapter 7), those available within Egypt of the 18th Dynasty and those from foreign sources. Quartz pebbles are the probable source of silica for the frits and perhaps the most likely source for the glass as well and it appears both plant ash and natron were used for alkali. Most of the materials are local, sand, quartz pebbles and plant ash all being available within a few miles of the centre of Amarna. Other materials are available further afield, but still in Egypt, including natron, the most likely source of which is the Wadi Natrun, and cobalt colorants which appear from their chemical composition to be sourced to alums of the Oases of the Western Desert. The copper colorants are partly bronze, whilst others are coloured by copper or copper ore and it is speculated that this comes from waste from the metalworking areas, with the most likely ultimate source for the original copper ores lying outside Egypt. Lead isotope provenancing of lead based colorants has indicated a complex pattern, but may include lead from both Gabal Zeit on the Red Sea coast of Egypt and a second lead possibly of Mesopotamian origin.

Chapter 6:
Methods of production

INTRODUCTION

Chapter 5 identified the raw materials used in the production of glass, frit and faience from their microstructure and composition. On the basis of these raw materials, this chapter goes on to discuss their processing that leads ultimately to the finished product. Petrie's hypothesis that the frits represent an intermediate stage in the production of glass is discussed in detail and the other possibilities laid out. The evidence for the stages involved in the processing of Ordinary faience, Variant A faience and especially Variant D faience is then deduced and verified by experimental reproduction. The similarities in the production routines between the different vitreous materials and between the vitreous materials and other industries are pointed out. These similarities have implications for theory that the discovery of glass and its initial development in the 18th Dynasty was intimately linked with developments in one or more of the other industries. Information on these links and new developments allows speculation on what are the key factors in the discovery of glass making and the exploitation of this discovery on such a large scale at Amarna.

THE PRODUCTION OF FRITS

Chapter 5 demonstrated that the raw materials of the frits were quartz pebbles, an alkali which was probably natron and a colorant, usually cobalt, but occasionally copper. The method of production of the frits is discussed briefly in Petrie (1894) and is not in much doubt. The large grains of quartz visible in the frits means that the pebbles were coarsely crushed, leaving occasional fragments 4-5mm across mixed in with much finer powdered quartz. This variable grain size material was well mixed with the other components and the resulting batch placed in a fritting pan and fired. The results of Chapter 3 and Chapter 4 both suggested that the firing temperature was high, perhaps up to 1150-1200°C.

Experimental reproduction

To verify the production method deduced from the ancient material and detailed above, the frits were experimentally reproduced in the laboratory. Each batch from F0 to F5 was made up to slightly different compositions as shown in Table 6-1, one of which (F2) lies close to the average soda to silica ratios for the bulk frits, two (F0 and F1) slightly higher in soda to silica ratio and two (F3 and F5) slightly lower The batches were prepared from sodium carbonate and quartz laboratory reagents, coloured with a little cobalt nitrate and thoroughly mixed. Each batch was divided into three crucibles, which were then fired to 950°C, 1050°C and 1150°C for three hours. The resulting experimental frits were examined under the optical microscope and SEM.

Frits of different compositions (F0 to F5) showed a variety of different degrees of melting over the firing temperature range. When fired at 950°C, the low silica experimental frit F0 was very vitreous and had melted completely to form a homogeneous glass. At the other extreme, the high silica F5 showed very little reaction when fired at 950°C, and only slight vitrification at 1050°C and 1150°C. The microstructures of both of these extreme experimental frits lie outside the range of those observed in the ancient frits. F1, F2 and F3 show little reaction when fired to 950°C and only slightly more at 1050°C. When fired to 1150°C, they develop a good vitreous matrix with unreacted quartz grains and, where mixing of the raw materials has been thorough, no other phases are

Table 6-1 : Compositions of the experimental frits

Number	SiO_2	Al_2O_3	CaO	MgO	Na_2O	FeO	CoO	CuO
F0	76	0	0	0	23	0	0.3	0
F1	81	0	0	0	18	0	0.3	0
F2	85	0	0	0	14	0	0.3	0
F3	90	0	0	0	9	0	0.3	0
F5	95	0	0	0	4	0	0.3	0
F6	80	3	2	2	12	1	0.2	0

Table 6-2 : Comparison of average glass matrix analysis for the ancient cobalt frits with the experimental frit F6

Number	SiO_2	Al_2O_3	CaO	MgO	Na_2O	K_2O	FeO	TiO_2	CoO	CuO	MnO
average cobalt frit	76.4	4.6	2.8	2.0	12.0	0.5	0.7	0.2	0.29	0.03	0.35
AM28	76.0	2.7	4.3	2.2	13.2	0.3	0.7	0.1	0.23	0.04	0.20
experimental frit F6	75.9	3.0	3.2	2.3	13.4	0.0	1.3	0.0	0.45	0.00	0.08

observed. Therefore when fired to 1150°C, frits F1, F2 and F3 show microstructures that are closest to the ancient frits, and this observation confirms the estimates for the firing temperature for the ancient frits derived in Chapters 3 and 4.

However, one significant difference between these three experimental frits and the ancient frits relates to the backscatter contrast of the glass and quartz phases. In the ancient frit these are distinct, with the quartz dark grey and the glass light grey, as shown in Figure 6-1, but the experimental frits F1 to F3 showed almost no backscatter contrast since the glass contained low concentrations of elements with high atomic number. A further frit, F6 was therefore produced with alumina, lime, magnesium and iron contents close to that of the ancient examples (Table 6-1) and following the results of the earlier work, this was fired at 1150°C for 5 hours. The addition of these other components raised the backscatter contrast and made the two phases more easily visible as can be seen in Figure 6-2. Figure 6-2 also shows that frit F6 also has many of the microstructural peculiarities seen in the frits, including partial dissolution and rounding of the quartz grains shrinkage cracks both around the edge of the grains and internally as a result of β-quartz to α-quartz transition. Analysis of the glass matrix of frit F6 given in Table 6-2 shows that the composition of the glass matrix of F6 is very similar to that of the ancient frits being, in all important respects nearly identical to that of AM28, the analysis for which is also listed.

THE PRODUCTION OF GLASS

Frits as a glass making intermediate

Petrie (1894) believed that the frits associated with fritting pans were an intermediate stage in the production of glass. However, the exact place that these frits played in this production changes through Petrie's various accounts of the glass making process. In the excavation account, Petrie (1894) seems to confuse the frits with Egyptian Blue and, although the passage is not clear, he does not appear to link them in to the glass making process. The fritting-pans in which they are found are described as being "about 10 inches across and 3 inches deep". In 'The Arts and Crafts of Ancient Egypt' (Petrie, 1909), the frits are involved in the process, being "fairly fused" with colourless glass in a pan "which was about four to five inches across, and held half an inch to an inch deep of the glass". In a lecture to the Newcomen Society (Petrie, 1925), the process of glass making in Egypt has become two staged, first the manufacture of the frit and subsequently its melting in saggars. It is clear that, as Nicholson (in press) says, "as Petrie added to his knowledge of glass and glaze, so his accounts became altered, though without explicitly correcting his earlier work". It is the last stage of this metamorphosis, with frit being a preliminary stage that is subsequently melted to produce glass, that Turner (1954, 1956c) adopts. He states that "the melting process was carried out in two stages, first a comparatively low temperature conversion whereby the sand and the ash (or soda) were converted to frit, the second in which the frit was converted to glass in crucibles in the melting furnace". Turner subsequently defines "low temperature" as below 750°C. It is this sequence of production, based on Petrie's work and adapted by Turner, that has subsequently been repeated throughout the literature (Lucas and Harris 1962, Newton and Davidson 1989, Nicholson 1993, Tait 1991 and many others).

The generally accepted method of glass production at Amarna, drawn from a composite of the papers outlined above is shown in the black sequence of Figure 6-3. The quartz pebbles, plant ash and colorant are crushed and mixed and placed in the fritting pan where they were sintered at a temperature below 750°C. On cooling the frit was chipped off the pan, crushed and placed either in small crucibles (Petrie 1894, these crucibles have not been identified by subsequent workers) or in the cylindrical vessels discussed in Chapter 4 (Turner 1954 and most other accounts). The vessels were then heated to a temperature of probably around 1100°C to form a glass. On cooling, the crucible was chipped off the glass along with areas where mixing had been inadequate and the glass quality poor. The resulting ingot could then have been worked into the desired shape.

Frits and glass compared

If Turner is correct and the frit is a glass making intermediate, then if the whole of the frit is crushed and melted to create the glass and no additional material is added to or taken from the system (in black, Figure 6-3), the bulk compositions of the glass and frit should be very similar. Following this hypothesis, the bulk composition of the frit should equal the bulk composition of the glass.

Table 6-3 shows the average analyses of the frits and glass taken from Chapter 3. Comparison of the bulk frit and the glass compositions shows that the frit has a much higher silica content, and is much lower lime and alkalis. The bulk frit is also richer in the cobalt colorant and its associated alumina. It is clear therefore, that

Table 6-3: Average analyses of cobalt glasses and frits

Number	SiO_2	Al_2O_3	CaO	MgO	Na_2O	K_2O	FeO	TiO_2	CoO	CuO
Cobalt glass [1]	64.1	2.2	7.9	3.6	19.2	1.2	0.5	0.1	0.16	0.17
Cobalt frit - bulk analysis [2]	84.0	3.1	1.9	1.4	8.1	0.3	0.5	0.1	0.20	0.00
Cobalt frit - glass matrix [3]	76	4.6	2.8	2.0	12.0	0.5	0.7	0.2	0.3	0.00

taken from Chapter 3 : [1] Tables 3-2, [2] by combining Tables 3-5 and 3-6, [3] Table 3-5

crushing and melting the frit yields a product with a composition far different from that of the glass, strongly suggesting that the simple currently accepted hypothesis is incorrect.

One of the most significant differences between the compositions of the bulk frits and those of the glass is the silica content. In theory, since the some of the silica is present in the frits in the form of quartz grains, when the frit was crushed it might be possible to pick out some of these grains and hence reduce the silica content of its melted product. This stage in the process is indicated by the red addition to Figure 6-3. The process would be extremely laborious and it would be impossible, in practice, to remove all the remnant quartz. However, assuming that all the unreacted quartz could theoretically be removed, the product of melting would then have the same composition as the glass matrix of the frit, also shown in Table 6-2. While the glass matrix composition is somewhat closer to that of the glass than is the bulk frit in terms of the lime and alkalis contents, it differs even more in terms of the cobalt and alumina contents. Therefore, this modification of the Turner hypothesis to explain the place of the frits in the production of glass is very little more successful than the original hypothesis and can be dismissed.

Given that removing the quartz from the frit does not result in an appropriate composition, the only other way to obtain a final product that chemically resembles glass is to add further material at the post-fritting crushing and mixing stage, as illustrated in blue in Figure 6-3. Both further alkali and lime have to be added in order to lower the proportion of silica and colorants and in this way, when the resulting batch was melted, the product would have the correct glass composition. However, the adding of further alkali and lime at such a late stage would defeat the whole reason for fritting in the first place. As discussed above, the reason for fritting the batch is to convert the carbonates and sulphates into forms that can more easily react with the silica and in the process drive off carbon dioxide and sulphur dioxide and other impurities which would otherwise create bubbles and cloudiness and disrupt the appearance of the glass. The late addition of the carbonates and sulphates contained in the alkali and lime would create exactly these effects again and make the first stage of fritting pointless.

In addition to the problem that the chemical composition of the frits and glasses are significantly different, there is also the problem of the temperature at which the fritting was carried out. As mentioned above, Turner (1954), citing glass making texts mostly from Roman and Medieval periods, sets an upper temperature limit for fritting at 750°C, below the level at which melting would occur. Firing temperature estimates for the frits derived in Chapter 3 (Table 3-7) from their composition and presumed viscosity when at their maximum firing temperature suggest that the frits were subjected to sustained temperatures in the range of 1050-1200°C. These high temperatures are supported by independent temperature estimates for the ceramics associated with vitreous materials discussed in Chapter 4. For example. the microstructure of the cylindrical vessels suggest they have been subjected to temperatures in the range of 1100-1200°C. Although no example of a fritting pan was available for sampling, the hand specimens are generally at least as highly vitrified as the most vitrified of the cylindrical vessels, if not more so, suggesting they were subjected to temperatures of around 1200°C. Therefore temperature data derived from the composition of the frits themselves and from the microstructure of the ceramics with which they are associated both suggest that the Amarna frits were subjected to temperatures far in excess of the 750°C hypothesised for glass frits by Turner.

In summary, the composition of the frits does not match that of the glasses and, without the nonsensical addition of further lime and alkalis, they cannot be remelted to produce a glass of the correct composition. It is therefore most probable that these frits are not associated with the production of glass in the way hypothesised by Turner.

One of Petrie's earlier accounts of the place of the frits in glass making (Petrie 1909) states that they were used as a colorant to be added to colourless glasses. Of all the glass in the collections studied for this work only one possible piece of colourless glass was noted, many other fragments labelled as "glass" in fact being pieces of quartz. This makes colourless glasses extremely rare at Amarna and no analyses of these glasses are available. Lilyquist and Brill (1995) give one example of an analysis of a colourless glass bead inscribed for Hatschepsut (MMA 26.7.746). This has a composition of 64.6% silica, and 20.5% soda and is similar to other 18th Dynasty glasses, but lacks the colorants. Since cobalt contents of frit and cobalt blue glass are similar, in order to produce a cobalt glass with the observed CoO content, the colourless glass would have to be mixed with at least its own weight of frit and in most cases more (Figure 6-4). This would result in a final glass with the correct proportion of colorant, but which once again would be too high in silica and low in both lime and soda. The rarity of the colourless glasses and the discrepancy concerning the compositions therefore again suggest that the hypothesis that frits are used as a colorant for colourless glasses is incorrect.

In summary, neither Turner's remelting hypothesis nor Petrie's colorant hypothesis (Petrie 1909) adequately relates the composition of the frits to the composition of the glasses. Therefore the frits most probably do not represent an intermediate stage in the manufacture of glass as was proposed.

Glass making without fritting

It is possible that a fritting stage was used in the production of glass and that the intermediate product has not been preserved or recognised in the archaeological record. However, in order to investigate whether it was possible to produce a glass of the composition of Amarna glasses without the intermediate fritting stage, several experimental glasses were produced. The Amarna glass composition could be simplified to 65% silica, 20% soda, 10% lime and 5% colorants and impurities, the impurities being ignored. A glass of the required composition was therefore prepared from granular quartz, sodium carbonate and calcium carbonate laboratory reagents with a little cobalt nitrate added for colour. Four separate batches were placed in porcelain crucibles and fired in a laboratory furnace for 10 hours at 950°C, 1000°C, 1050°C and 1100°C. The results were then examined under a binocular microscope. The batches fired to 950°C and 1000°C were not completely fused, the remnant grains of quartz giving them a texture similar to sugar, with the 1000°C sample having slightly fewer remnant grains than that fired at 950°C. However, the batches fired to 1050°C and 1100°C were both good clear glasses with very little unfused material and very few vesicles. Therefore with pure laboratory reagents, glass of the Amarna composition can be produced by firing for ten hours at 1050°C or above, a temperature well within the range possible for ancient kiln technologies and suggesting that, given these pure batch materials a fritting stage is not necessary to produce a good quality, clear glass.

Experimental glass making with more authentic starting materials was attempted by Paul Nicholson at Amarna in 1996 (Nicholson and Jackson 1998). Using local raw materials, he built a replica kiln based as closely as possible on the design of Kiln 3 excavated at O45.1 (Nicholson 1995, discussed in Chapter 4). Local sand was mixed with seaweed ash and a small amount of cobalt colorant and placed in mullite crucibles which were then fired in the kiln. With only a natural draught from the prevailing north wind, Nicholson found it quite easy to attain a temperature of 1000°C. This rose to 1100°C after 5 hours of firing and was held around this temperature for 4 hours, with 1150°C being the maximum temperature reached. While some of the crucibles were damaged in the firing, when one of the survivors was removed on the day after firing in contained "a well formed ingot of deep blue glass... of good quality and relatively free of bubbles and unreacted material" (Nicholson and Jackson 1998).

Summary

Comparison of the chemical composition of the frits and glass show that the frits are most probably not an intermediate stage in the production of glass as has previously been widely suggested. This conclusion is supported by temperature estimates for the formation of the frits, derived from their composition and the microstructure of the vessels associated with them. These estimates suggest that the frits were formed at 1100-1200°C far higher than the suggested maximum of 750°C for a glass making frit. Experimental reproduction of glasses in both the laboratory and the field suggest that it is possible to produce a good glass straight from the raw materials in a one stage process, without the necessity of a fritting stage. It is therefore likely that the Amarna glass was produced in this way and that the frits are an intermediate stage in the production of not glass, but some other vitreous material.

THE PRODUCTION OF ORDINARY FAIENCE

Production of faience, particularly ordinary faience, has been extensively worked on (Tite et al 1983, Vandiver 1982, 1983a and 1998, Tite and Bimson 1986) and consists essentially of three steps, the forming of the shape of the body of the piece, the method of glazing and the firing. The same range of methods of shaping, glazing and firing as used in the production of ordinary faience can also be used to produce many of the faience variants, the production of some of which is subsequently described in this Chapter.

Shaping the body of the faience was accomplished using various techniques that evolved through the many millennia that faience was made. The simplest technique was perhaps modelling by hand and this was still used in the New Kingdom for many pieces, particularly those that were large and needed only in small numbers. Where larger number of small pieces were needed, the New Kingdom saw the introduction of moulding in open face moulds, a technique that enables large numbers of near identical pieces to be produced relatively rapidly compared to individual modelling. It is clear from the examination of New Kingdom faience, that many techniques for shaping the core and subsequent working of the core were being used, including drilling, piercing, incising, inlaying and painting (Vandiver, 1983a), each being employed in a highly competent manner to accomplish the desired effect on an individual pieces.

There are three different glazing techniques that are used for ordinary faience - efflorescence, cementation and direct application (Figure 6-5, adapted from Vandiver 1983a) Each of the three methods of glazing leads to a distinctive combination of macroscopic and microscopic features on the finished vessel. All three methods are in use in the New Kingdom (Vandiver 1983a), although only two, efflorescence and direct application appear to be evident on the faience pieces from this study.

In the efflorescence method, the glazing powder is mixed into the body when moist. As the body dries a proportion of the more soluble of the salts in the powder are carried to the surface of the body by capillary action and precipitate on the surface forming a white crusty layer. Efflorescence is greatest on the areas of the vessel that dry fastest, usually the outer surfaces or rims of a vessel. The inner part of the vessel, particularly the internal surface at the bottom dries only slowly and effloresces less, while a base standing on a surface may not effloresce at all. The thickness of the glaze is directly related to the amount of efflorescence, hence the thickest glaze tends to be on the external surfaces and the thinnest on the base and internal (Figure 6-5a). The thickness and extent of an effloresced layer is dependent on many variables including the composition of the glaze mixture, the type of copper compound, proportion of the glaze mixture to crushed quartz in the body, the grainsize of the body and temperature and air flow during drying.

In the direct application method, the glazing powder, either in a raw state or prefritted, is applied to the surface of the body either by painting or dipping before firing. Direct application of a wet slurry either by painting or dipping tends to result in at least some of the slurry running down the vessel leading to drips and flow lines and an uneven thickness of glaze. There is a tendency for the glaze to pool on the bottom internal surfaces of vessels and at the base of external walls (Figure 6-5b). With prefritted glazes, the minimal penetration of the glaze into the body of the object in direct application leaves a thick glaze and very little interstitial glass in the core.

In the cementation method, the body is buried in the glazing powder which on heating partially melts and volatilises forming a glaze on the surface of the object, perhaps with lime acting as a parting agent. Since in cementation all parts of an object are in equal contact with the glazing powder, this technique results in a glaze of even thickness all over, with no drying or firing marks (Figure 6-5c).

The shaped and glazed object is placed in a kiln for firing. During firing an object with an effloresced or applied glaze, it has to rest on a setter or other surface and this results in irregularities in the glaze where it is contact, called firing marks. These can often be identified in complete well preserved objects and are good indicators of the method of glazing used. No information has survived from the New Kingdom about the shape or number of setters used or how the kiln was stacked so this can only be guessed at. The temperature of firing depends critically on the method of glazing and the composition of the glaze, but all techniques and most compositions require temperatures in the range of 850-1000°C.

The production of blue ordinary faience - finger rings

Lucas and Harris (1962) define ordinary faience as typical Egyptian faience consisting of a white, or slightly tinted, friable core and most usually a blue glaze. In the current work the finger rings are the most obvious examples of ordinary faience.

Chapter 7 discusses the presence of clay moulds at Amarna and points out that clay moulds for both the ring shanks and bezels are common. This strongly suggests that the basic shape for the ring components was formed by pressing the damp faience paste into a mould. In Chapter 3 it was noted that the rings have significant interstitial glass in their cores, which is conventionally taken as evidence for glazing by the efflorescence method, and none of the rings show any drips or runs in the glaze that are indicative of application. Figure 6-6 shows a flow diagram outlining the production sequence for the rings. If the paste was left to dry in the mould, it would be almost impossible to remove, so it would have to be taken out almost immediately while it was still wet. The bezel and shank would then be cemented to each other using a small amount of glaze slurry (Boyce 1995). Thus occasional drips caused when this slurry is applied too thickly can still be seen occasionally in the finished rings. Since handling the object easily damages the effloresced layer, the assembly of the ring would probably have to take place almost immediately it left the mould, again while it is still damp and before this layer has developed. After drying and efflorescence, the ring would then be transferred to the kiln and fired.

The production of other small objects

Small numbers of yellow, red and cobalt blue faience samples were also analysed (Chapter 3), the yellow and cobalt blue samples all being finger rings, and the red faience being inlays. All three colours show evidence that they were produced in the same way as the ordinary faience finger rings. As discussed above, from the presence of so many clay moulds, all the faience rings are thought to be examples of efflorescence glazing. In addition, the cobalt and yellow rings have extensive glass within their cores, and the yellow rings and red inlays have lead antimonate and iron oxide particles respectively in their cores, all caused by the mixing the glazing powder in with the body before efflorescence.

The yellow can therefore be thought of as being produced in the same way as the faience finger rings as described in Figure 6-6 The red faience is probably the same, although is was probably not moulded, but formed by hand. The cobalt blue rings may have been produce in an identical way to Figure 6-6, but it is also possible that since the microstructure is more or less continuous glass, some frit was added to the cores in order to increase the amount of glass. Examination of further samples of these rings would be required to confirm which was more likely.

THE PRODUCTION OF VARIANT A FAIENCE - FAIENCE TILES

As noted in Chapter 3, most of the tiles examined show the presence of fine crushed silica slip applied to a coarser grained base, characteristic of Variant A as defined by Lucas and Harris (1962). Close examination of the bodies of the tiles shows that although most of the grains are quartz, there are rare grains of felspar and mafic minerals, suggesting that sand is used to make the bodies of at least some of these tiles (Chapter 5). The underside of the tiles occasionally shows the impression of a woven fabric (Figure 3-35), the presence of which has been the cause of some speculation. As Nicholson (1998) has proposed, these impressions could be formed by placing the tile on a textile for drying, or alternatively, for keeping it wet while it was worked. Another idea is that the impression could have been transferred from the surface of one of the plaster trays that have been found at Amarna and are of unknown use, but that also frequently show a textile imprint on their flat surface. Perhaps a more likely explanation for the impressions it that they are caused by a cloth that was placed under the tile to prevent it from sticking to the surface on which it was worked, a problem which experimental reproduction of a faience tile has highlighted. The cloth could also be used to aid drying or to keep them moist while they are worked, whatever is required in that part of the processing, but its primary role would still have been to make the tile easier to move around. When the working of the tile was complete and it was dry, the cloth could then easily be peeled off before firing.

Figure 6-7 is a flowchart illustrating the various stages in the production of the tiles from their raw materials and Figure 6-8 illustrates the subsequent inlaying and painting of the tiles in more detail. The process began by laying the base of the tile, consisting of wet sand mixed perhaps with a binder, on a cloth. A wooden frame was perhaps used to give it the correct shape, but such frames have not been identified possibly because they have not survived. The fine grained Variant A layer was then applied, which from its very uniform thickness, was probably applied as a slurry, since other techniques would result in less uniformity. The tile was then left to dry slightly until it was clay-like in consistency, the cloth helping the excess water to be removed from the body. Next the pattern to be inlayed was incised into the surface, often cutting right through to the body. The inlay material, mixed with water to form a wet and clay-like paste, would be gently pushed into the depressions created. This work would take some time, and it would have been necessary to keep the tile damp so that it could be worked. The cloth could easily be damped or the tile moved to place it on a damp pile of cloths to help this. The tile was then left to dry completely, allowing both the Variant A layer and the inlay to effloresce. Then the brown outlines and details were painted on the surface. After allowing it to dry more thoroughly, it was then fired to create the finished object.

THE PRODUCTION OF VARIANT D FAIENCE - HARD, VITREOUS, BLUE CORES

The production of Variant D faience has been discussed in Kaczmarczyk and Hedges (1983) and Vandiver (1983a), both suggesting that ground blue glass was added to the core of the faience to give the typical blue cored vitreous matrix that is observed. The composition and microstructure of the Variant D faience has been described in Chapter 3. This description includes a number of key observations that need to be accounted for in any production sequence. Firstly, the Variant D has a range of microstructures, all with quite vitreous cores, but varying from the very vitreous 3556 down to 23420, where there is less glass and a higher porosity (Figure 3-24). In addition to this, some of the cores have a glass matrix containing both copper and cobalt colorants (1764, 176, 1433, and 3173), whereas the others have only cobalt (Table 3-10). The relationship between the amount of glass in the cores and the colorants of the core and glaze is shown in Table 6-4. As can be seen, there is a slight trend for Variant D faience that is coloured by cobalt alone in the core and cobalt and copper in the glaze to be more vitreous than faience that has cobalt and copper in both the core and glaze, but this is far from pronounced. The exceptions to this pattern are 1764, which is the third most vitreous, but has a unique pattern of a core with a cobalt content similar to the other samples, but low levels of copper (0.35% CuO, compared to the next lowest at 1.43% CuO in 1433) and 23420, which is perhaps the least vitreous, but still has very little copper in the core. In addition, 1764 has a glaze that apparently contains copper and only 0.03% CoO. However, great care should be taken in the interpretation of the glaze data for this sample, since the glaze is mostly weathered completely away and the analysis of that remaining shows a silica content of 95.5% suggesting that this too has been extensively leached. It is therefore possible that this glaze originally contained more cobalt, but that it has subsequently been leached away.

Table 6-4 : Relationship between amount of glass in core of the Variant D faience and the colorants in the core and glaze.

sample	description	colorant in core	colorant in glaze
3556	↑	Co	Co+Cu
469		Co	Co+Cu
1764	increasingly	Co+Cu [1]	Co+Cu [2]
3897	vitreous,	Co	Co+Cu
23505	decreasing	Co	Co+Cu
3173	silica	Co	Co+Cu
1433		Co+Cu	Co+Cu
176		Co+Cu	not preserved
23420		Co	Co+Cu
23419	Variant A	Co	Co+Cu
23231	Variant A	Co	Co+Cu

[1] 1764 has only 0.35% CuO in its core, [2] 1764 has a weathered glaze and only 0.03% CoO, the 'fine grained layer is the equivalent structure to the Variant A.

There are therefore two different types of the Variant D faience, some that have cobalt coloured bodies most of which are highly vitreous cores and some that have cobalt plus copper bodies with perhaps slightly less vitreous cores, both types having copper and cobalt in the glaze. Included in the first type are two samples (23419 and 23231) that have a fine grained quartz layer overlying a coarser core similar to the structure seen in Variant A.

Variant D cores

Table 6-5 shows average analyses for cobalt coloured glass and the glass matrix for the two types of the Variant D faience taken from Chapter 3. If they are compared, it can be seen that the glass is significantly higher in lime and soda and lower in silica, cobalt and alumina. Therefore, if powdered glass was mixed with ground quartz to form the core of Variant D faience, it would result in a very different composition to that seen, with significantly higher lime and alkalis in the cores. Dilution of the lime and soda coming from the powdered glass by partial melting of the quartz grains during firing would make the compositions closer, but this same effect would further lower the cobalt and alumina contents of the cores, giving a composition even less like the actual Variant D glass matrix. Therefore, it seems unlikely that powdered glass was added to the cores of the faience and another reason for the high amounts of glass in the cores of this particular faience as compared to most ordinary faience must be sought. Examination of the glass matrix of the frits given in Table 6-5 shows that they are much closer in composition to the glass matrix of the Variant D, with very similar silica, magnesium and bulk alkalis and where the frits are uncontaminated (see Chapter 3) similar low lime compositions. The average cobalt and alumina compositions of the frit (0.29% CoO and 4.6% Al_2O_3) is between those of the glass (0.10% CoO and 2.2% Al_2O_3) and the glass matrix of the Variant D (0.31% CoO and 7.2% Al_2O_3). However, the range of values for the glass matrices of the frits and the Variant D substantially overlaps. Table 6-5 also shows the analyses of the glass matrices of frit AM32 and Variant D sample 469, which lie in this overlapping composition range and are, to all intents and purposes, identical in composition apart from a slightly different balance of alkalis. Two photomicrographs comparing the microstructures of the frits and Variant D are illustrated in Figure 6-9 which shows that there are some similarities between them. Both have only the two phases, quartz and glass, but the faience tends to have more angular quartz grains which tend to be finer, with a grain size of up to 100μm and averaging about 40-

Table 6-5 : Comparison of average analyses of cobalt glass, frits and Variant D faience along with some analyses for individual samples.

sample	SiO_2	Al_2O_3	CaO	MgO	Na_2O	K_2O	FeO	CoO	CuO	MnO
glass matrix										
Co glass average	64.1	2.2	7.9	3.6	19.2	1.2	0.5	0.10	0.16	0.17
Co frit average	76.4	4.6	2.8 [1]	2.0	12.0	0.5	0.7	0.29	0.00	0.35
Variant D without Cu average	76	7.5	1.1	1.8	9.7	2.2	0.6	0.45	0.07	0.39
Variant D with Cu average	76	7.2	1.0	2.0	9.1	2.0	0.5	0.31	1.60	0.34
Cu frit 25038	73	1.5	2.1 [1]	0.8	13.7	1.5	0.8	0.01	6.25	0.10
frit AM32	78	5.3	1.2	1.8	10.6	0.8	0.6	0.38	0.03	0.44
Variant D 469	79	5.2	1.7	1.5	8.9	1.8	0.4	0.41	0.00	0.44
glaze averages										
Variant D	65-70	3.0	1.0-2.0	1.0	15-20	2.5-3.0	0.4	0.3	2.0-5.0	0.2
faience ring	75	0.4	1.2	0.3	7.7	4.1	0.2	0.05	10.2	0.0
bulk core averages										
Variant D	90	2.6	0.7	0.7	3.6	0.7	0.3	0.22	0.07	0.18
frit	86.5	2.7	1.6 [1]	1.2	7.0	0.3	0.4	0.17	0.00	0.20
frit + 35% quartz	90.0	2.0	1.2 [1]	0.9	5.2	0.2	0.3	0.13	0.00	0.15

[1] *lime in some of the frits artificially high from contamination from lime rich slips (see Chapter 3)*

50µm, whereas the frits can have rare grains up to 3000µm across although more typically the grains are between 300-500µm. Therefore the glass matrix of the frits is compositionally a reasonable match with the glass matrix of the Variant D cores, and both frits and Variant D faience have similar microstructures, the difference in quartz grain size being explained by the need to grind the frit to produce the Variant D.

Table 6-5 further shows that the bulk composition of the frits is lower in silica than the bulk composition of the Variant D faience, so that if the frits were used in the cores of the Variant D faience, further silica in the form of crushed quartz would have to be mixed with them. The composition that would be created if the average frit was mixed with a further 35% by weight ground quartz is also given in Table 6-5. In this mixture, the silica concentration (90%) would be the same as that for the Variant D and the rest of the composition would also be very similar. The ranges of all the oxides of the new mixture would overlap in part with the bulk compositions of the Variant D, although cobalt, alumina and manganese contents of the mixture would still be slightly low, and the alkali content slightly high.

Addition of copper coloured frits

Not all of the frits are cobalt coloured: two of them, 25038 and 68725 are coloured with copper. If these copper frits were mixed in with the cobalt frits to produce the cores of some of the Variant D faience, this could account for the fact that some of the cores have copper in them and some do not. After the cobalt and copper frits were manufactured, they would be difficult to tell apart and if they were used just for their colouring properties, then one would presumably be as good as the other. The glass matrix analyses for the two sorts of Variant D given in Table 6-5 show that the Variant D with both copper and cobalt in the core has a lower cobalt content (and is lower in the associated alumina and manganese) than that with just cobalt as its colorant. This supports the hypothesis that the copper frit might be being used in these samples as a substitute for cobalt, perhaps in times of scarcity of the cobalt colorant.

Shaping the core

To produce the Variant D core, the powdered frits and quartz were mixed together with water and possibly a binder and then shaped following the techniques used in ordinary faience. No examples of Variant D with obvious mould marks were found, but most of the pieces examined were large which tends to rule out moulding as a possibility since, in the 18th dynasty, the use of moulding seems to have been restricted to items less than 2-3cm across. Therefore most of the pieces of Variant D faience were probably individually hand modelled. Fine grained layers are quite common and were perhaps applied to some of the pieces because the finer grain size of the Variant A type layer would have been easier to subsequently work, both before and after firing, to give intricate detailing on the surface. In Khepresh crowns, as well as in other faience items intended for use in composite statues and other objects, the addition of a surface pattern to the damp core is quite common. Figure 6-10 shows a fragment of a Khepresh crown with the typical pattern of sets of concentric circles on its surface. All the sets of circles on this piece are identical and must therefore have been made with a tool that was pressed into the surface when it was damp. Unfortunately no example of this tool has yet been recovered.

Conclusion

In conclusion, glass has the wrong composition, particularly in terms of lime and colorants, to be used to create the body of the Variant D faience. The frits show a much closer fit, the glass matrix of each material being in some cases close to identical. If frit with the average bulk composition is mixed with a further 35% by weight of ground quartz, a composition that is very close to that of the bulk composition of the Variant D cores is achieved. The use of copper coloured frits could account for the copper in the core of some of the samples, or the copper might be incorporated in efflorescence glazing, a possibility discussed below. It seems possible, therefore, that the cores of the Variant D faience are created by mixing the powdered frits with extra ground quartz, with the variation in the amount of glass in the cores being explained by variations in the proportion of frit added to the body.

Glazing technique for the Variant D

If the cores are created as described above, then there still remains the question of how the Variant D faience was glazed, whether by cementation, efflorescence or direct application. Evidence from both the macroscopic and microscopic examination of the Variant D structure, along with compositional evidence can all be used to shed light on this problem. The three glazing techniques which are described fully above (6.4) each have their own range of characteristic features. Cementation, where the core is buried in a box of glazing powder and fired to create a glaze, is only capable of producing monochrome faience. Since a substantial proportion of the Variant D has polychrome inlaying or is itself an inlay, it cannot have been produced by cementation. It is possible, but this seems unlikely, that cementation was used for the monochrome pieces and some other method for the polychrome.

Two glazing techniques?

It is possible that two different glazing techniques were used on different pieces of Variant D faience, perhaps because the shape of the object made one or the other more applicable. The fact that some examples of the Variant D have very little copper in the core, while others have a significant amount does initially support this idea, with the low copper examples having a cobalt coloured core and an applied cobalt and copper glaze, whereas the high copper examples are formed by the mixing of a copper glazing powder into the core followed by efflorescence. However, one factor argues strongly against two different techniques being used and that is the close similarity of the glass matrices in the Variant D cores of both types (Table 6-5). It is difficult to imagine how the right amount of frit could have been omitted from one type and the appropriate compensating amount of glazing powder added so that the composition exactly matched in all the elements the second type of Variant D with just frit in the core and an applied glaze. It is also difficult to imagine a reason why such a balancing act would be worth doing. Therefore it seems unlikely that more than one technique was used.

One of the key observations in distinguishing between the use of efflorescence and application glazing concerns the copper content of the Variant D cores. Eight of the Variant D glass matrices analysed have copper contents of less than 0.15% CuO and averaging 0.07 % CuO, while others are higher, up to 3.0% in the case of sample 176. This might suggest that together they form a continuous series from very low copper to high copper contents and that, as such, they represent various stages in the efflorescence process. Thus, those with high copper in the cores have effloresced only poorly, while in those with low copper in the core, the copper has effloresced almost entirely to the glaze.

However, there are three points that argue against this interpretation. Firstly and most significantly, Table 6-5 shows that the average composition of the glass matrix of the Variant D with cobalt and copper in the core and those with cobalt, but without copper are very similar, particularly in terms of the alkalis and lime contents. If those with copper in the core were poorly effloresced versions of those without, then it would be expected that some of the former would also be higher in other elements of the glazing material, especially the alkalis and lime. The fact that the two groups have essentially the same composition with respect to these elements argues against the efflorescence hypothesis. The second point concerns the amount of copper that would be expected to remain in the cores if efflorescence is used as a glazing method. As discussed above, Vandiver (1983a, 1998) who has done a great deal of experimental work on efflorescence glazing, has shown that the most efflorescence occurs when there is a high surface to volume ratio. This is the case for the faience rings, which are thought to have been glazed by efflorescence (section 6.4.1), but Table 6-5 shows that in all examples of these rings, they still have very significant copper in their glass matrix after efflorescence and firing. However, some of the Variant D samples are very low in copper, 469 has no measurable copper, and 595 and 3897 are both below 0.05%. This seems to be far too low for efflorescence glazing considering that abundant glass with significant alkalis are typical of these samples. The final point concerns the lime content of the glass matrix. Lime is typically mixed with a glazing powder to give a resulting glaze resistance to attack by water (Chapter 1) and some of it remains after efflorescence in the core. This can be seen in the average analysis for the glass matrix of the faience rings in Table 6-5, where the average lime content is 2.2%, and the highest lime 3.4% in sample E5053/4 (Table 3-10). The highest lime content in the Variant D glass matrices is 1.8% for 1433 and the average 1.0%, which is much less and again argues against the addition of glazing powder to the core.

Glazing by application describes the process where a slurry containing raw or pre-fritted glaze components is painted or otherwise applied onto a dry, or nearly dry core. If a cobalt and copper rich glaze was applied to the Variant D body, prepared as proposed above, the penetration of the copper into the body would be limited and this would give the pattern of cobalt only in the core and cobalt and copper in the glaze that is seen in the majority of the Variant D. Alone, however, it does not explain why some of the Variant D have both copper and cobalt in the core and therefore offers only a partial explanation. However, if, as also proposed above, some of the cores were made using a mixture of copper and cobalt frits and others just cobalt frits, then application glazing of these bodies explains the full range of compositional variations seen in the Variant D faience.

Macroscopic structural evidence

Vandiver has stated that macroscopic evidence in the form of "the characteristic structures of various processes" and the thickness and continuity of the glaze is "the key to establishing the sequence of manufacture and the method of glazing" (Vandiver, 1998). As discussed above, in a faience piece that has been adequately fired, but not overfired, the presence of drips and runs, amongst other factors, is indicative of an applied glaze, whereas common firing marks and rough or raised areas are indicative of efflorescence glazing. Figure 6-10 shows a fragment of a Khepresh crown, one of the most common objects to be made from Variant D faience. The glaze has run down the side of the piece from the top, suggesting that this is an applied glaze. Examples of runs and drips are evident from a number of the pieces of Variant D which suggests that application glazing was practised on at least some of the pieces, if not the majority or all of them.

The microstructure of the Variant D is perhaps less useful in determining the method of glazing. In well preserved ordinary faience, such as the finger rings discussed above, examination of the thickness of the glazes, interaction zones and the amount of glass in the cores can all suggest one or the other of the techniques is more likely. However, in the Variant D samples, glass has been added to the core in the form of frit, so the last two of these criteria cannot be used. Also, the glazes are usually poorly preserved and in some cases even the interaction zone is affected by weathering, making thickness of glaze an unreliable criterion.

Conclusion

The use of two different glazing techniques for the two types of Variant D can be ruled out because of the great similarities between their chemistry. Many of the pieces are inlaid polychrome, and this rules out the possibility of using cementation for all the Variant D. Most of the cores have abundant glass containing significant amounts of alkalis, but some of these have very low copper contents. This argues against efflorescence, which would leave at least some of the copper in the abundant glass of the core. This lack of copper in the core of some of the pieces suggests glazing by application and this interpretation is supported by drips and runs in the glaze on some of the samples. Therefore, overall the most likely method of glazing for all the samples is application.

Cold working

Vandiver (1983a) has stated that some of the Variant D faience was worked after firing, specifically by grinding and drilling. Many of the examples of Variant D faience were used as components for composite statues or as inlays for architectural or other work. That some of the pieces might be ground when cold to make sure that they accurately fitted the hole or socket on a sculpture into which they were placed is perhaps not surprising. Figure 6-10 also shows an example of cold working where a Khepresh crown, perhaps intended as an architectural inlay, has been ground down on one side, leaving closely spaced vertical lines on the Variant A type layer. The interpretation that drilling was used in the crowns to form the central holes of the circles is based on the observation that "there is no glaze in the central hole" (Vandiver, 1983a). However, the same Khepresh crown shown in Figure 6-10 is cut to reveal a cross section of one of these central holes and shows clearly that in this case, the central depression still has some glaze in it. It is therefore probable that the drilling was not always totally efficient at removing the glaze.

Summary of Variant D production method

Comparison of the composition of the glass matrix of the cores of the Variant D and the glass matrix of the frits suggests that the frits were being used to provide glass in the cores of this sort of faience rather than glass itself as has been previously suggested. The majority of the frits are cobalt coloured, which perhaps accounts for the fact that the majority of the cores of the Variant D contain cobalt alone. However, Chapter 3 gives analyses of two examples of frits that are copper coloured. If occasionally these were mixed into the cores then they would result in a copper and cobalt colorant mix in the cores of some of the Variant D, which is the pattern observed in the ancient examples. The bulk composition of the faience contains more silica than the bulk frit, which suggests that powdered quartz was added to the ground frit before the body of the Variant D was formed. Similarly, the variation in the amount of glass in the cores can be explained by variations in the proportions of frit and quartz used.

Alternatively, this copper could have been incorporated into the core if a glazing powder was mixed into it prior to efflorescence glazing, but the absence of copper from the highly vitreous cores of the majority of the faience argues that efflorescence could not be responsible for both types. The use of efflorescence for one type and some other glazing method for the other can be ruled out because of the great similarities between the compositions of the two types other than the colorants. The most likely glazing method is therefore application, and this theory is supported by the presence of drips and runs in a number of the pieces. The experimental work discussed below tentatively suggests that the cobalt found in the ancient glazes has diffused from the glass matrix of the body, although it is also possible that it was included with a copper colorant in the applied glaze. Figure 6-11 is a flow chart summarising the proposed method for the production of Variant D.

Experimental replication of Variant D

In order to verify the conclusions concerning the method of production of the Variant D that were reached from the study of its composition and structure, as discussed above, attempts were made to replicate it experimentally in the laboratory. Some of the samples were glazed by efflorescence and some by application. For the samples to be glazed by efflorescence 75% by weight of the application core mixture (made up of equal amounts of frit F4 and powdered quartz) was mixed with 25% of a glaze powder of the composition 40% silica, 40% soda, 15% lime and 5% CuO. The glaze powder was made from powdered quartz, and the carbonates of sodium, calcium and copper, ground and thoroughly mixed, but not fritted. The components were mixed together with water, formed into shape and left to dry and effloresce. Several runs were conducted under different drying conditions, including hot draft, cold draft and ambient temperatures of both 20°C and 40°C.

Of the five cores to be glazed by application, four were made up from experimental frit F4 with either 60%, 50%, 40% and 30% by weight frit and one from experimental frit F6 (for the compositions of F4 and F6 see Table 6-1) with 60% by weight frit, the remainder of the core being made up of equal amounts of granular quartz and powdered quartz which was thoroughly mixed and sieved. The entire mixture was then thoroughly sieved and mixed, and the resulting powder was bound together with water and formed into balls about 1.5cm across. The glaze was made up to give a composition of 63% silica, 1.6% alumina, 2% iron oxide, 3.4% magnesia, 6% lime, 4% potash, 16% soda and 4% CuO and was then fritted at 750°C for 1 hour and powdered. This glaze powder was mixed with water and a little gum arabic and applied to the surface of the core with a glaze brush and left to dry. Both application and efflorescence samples were fired at 900°C for 2 hours.

The resulting faience was examined under the optical microscope and the SEM. Those glazed by efflorescence in general failed to effloresce well and had poor glazes. Rarely, those dried in a warm draft effloresced better and gave a thin glaze. When analysed, it was seen that the glass matrix of the finished beads contained significant copper, which in all of the efflorescence samples was greater than 1% CuO. The experimental replication therefore confirms that the observation from the ordinary faience that copper is left in the glass matrix of the core after efflorescence is applicable also to the Variant D samples. Since more than half of the Variant D samples examined had only trace amounts of copper in their cores, the replication evidence supports the theory that glazing by efflorescence cannot have been used for all the Variant D examined.

The experimental samples that were glazed by application showed a variety of microstructures when examined under the SEM. The top image of Figure 6-12 shows the experimental sample made up from 30% frit F4 and 70% quartz and illustrates the typical microstructure of those experimental samples made up from a relatively low proportion of frit in the core. Such samples tend to have a more open structure with moderate amounts of glass in the cores, and exhibit a microstructure which in many ways is similar to that seen in faience glazed by efflorescence. The bottom image of Figure 6-12 is of the experimental sample made up from 60% frit F6 and 40% quartz and shows that this results in a much more vitreous microstructure, with an interaction layer that grades into a core with extensive interstitial glass. Experimental faience with proportions of frit between these two extremes show microstructures that are in turn intermediate between microstructures in Figure 6-12.

To investigate the origin of the cobalt in the glaze, the experimental samples were produced with cobalt colorants in the glass matrix of their cores, but only copper colorants in their applied glaze slurry. A step scan across the glaze/body interface was then conducted and the results compared to a similar scan across the equivalent area in an ancient example. The results are shown in Figure 6-13 and Figure 6-14. The experimental sample shows a high content cobalt in the glass of the body which falls as the scan enters the glaze and drops to no cobalt at a depth of about 0.2mm into the glaze. This pattern is caused by the diffusion of the cobalt out of the body into the glaze. In the ancient example the cobalt content again falls as the scan enters the glaze, but perhaps due to the very fact that the glaze is very thin, never reaches zero. This could be either be caused by weathering and loss of cobalt from the surface of a glaze that contained cobalt when it was applied, or be due to diffusion of cobalt into a cobalt free glaze as in the experimental samples. However, the fact that the highest concentration of copper is at the outermost surface of the glaze in the ancient sample suggests that there has been minimal loss of copper by weathering. Since copper and cobalt weather from the glaze in similar ways this also implies that there has also been minimal loss of cobalt from the surface of the glaze. This tentatively argues that that the lower cobalt content of the glaze is not caused by weathering, but instead is due to the diffusion of the cobalt from the glass matrix of the body. This would suggest that the glaze slurry applied probably contained copper, but no cobalt.

In summary, the range of microstructures seen in the replicate Variant D faience glazed by application closely mimics that seen in the archaeological samples both in terms of the range of microstructures seen, from moderately vitreous cores with some voids to highly vitreous core with an almost continual glass matrix. This supports the suggestion that the ancient examples were probably glazed by application as was suggested by their chemistry and macrostructure discussed above.

Chapter 7:
The archaeology of production

INTRODUCTION

The technical information derived from the analysis of glass, faience and other vitreous materials provides a valuable insight into their production, but should not be considered in isolation. It is important to set this evidence in its archaeological context, both spatially and temporally, since only in this way can the various competing theories derived from the analysis be easily evaluated and the evidence assessed for relevance and importance. In addition, the material relating to the non-technical aspects of production, like organisation, living conditions and distribution can be studied. As laid out in Chapter 1, Amarna is an especially important site in this context, since it represents one of the very few examples of a preserved ancient city of this period. It is therefore potentially possible to set the production of vitreous materials in the context of the city, rather than having to consider it artificially removed from the infrastructure and other industries that would have controlled and influenced it.

With this in mind, various aspects of the archaeology of the vitreous materials were studied and are discussed in detail in this chapter. First, the layout of the city was briefly examined and the various areas devoted to religion, display, administration, production and habitation identified. Into this setting, an effort was made to identify and describe the sites of the 'glass factories', to use Petrie's term (Petrie 1894), by analysing the distribution of elements involved in the production, especially the clay moulds used in the production of amulets and rings. The organisation of the workforce is then discussed, drawing on a wide range of information from a number of different periods. This is followed by a survey of the distribution and type of faience objects produced, with an attempt to link the presence of faience to the status of their owners.

LAYOUT OF AMARNA

The city of Amarna was laid out between 14 boundary stelae, 11 on the east bank and 3 on the west, still known by the letters given them by Petrie in the first survey of the city (Kemp 1989, Petrie 1894). The site marked out by the boundary stelae is roughly 16km by 13km and included farmland as well as the city itself. From the fact that rare tomb scenes show the city to be on the waterfront, it can be deduced the Nile follows a course today not far removed from the course that it took in the 18th Dynasty (Kemp 1989). Figure 7-1 shows the surroundings of Amarna, indicating the positions of the boundary stelae (letters) and tombs, along with the high desert, the area of cultivation, the modern courses of the Nile, the Bahr Yusef (a small tributary to the Nile) and the various modern Arab villages in the area.

If the Bahr Yusef is taken as the western limit of the cultivation (a very conservative estimate) then the total area of cultivation contained within the boundaries has been calculated by Kemp (1989) to be 162km^2, or 59,200 aroura. Kemp gives a very approximate estimate of the carrying capacity of this area using the accepted ratios of 0.5 persons rural population, plus 0.25 persons urban population, per aroura. From this it can be calculated that the farmland could have supported a population of around 45,000, matching with estimates of the city population which have varied between 20,000 and 50,000 (Kemp 1989) and suggesting to Kemp that there is rough balance of city population and carrying capacity. Whether the city was supplied from outside the area of the boundaries is not certain, but the desire for a surplus might lead to estates outside the area supplying goods, and it is known from records that temples, for example Karnak, were receiving a wide variety of goods from some distance away (Kemp 1989).

The City

The appearance of Amarna during the 18th dynasty is drawn from a series of sources. Today it is very arid, windswept and inhospitable, with very high temperatures during the summer months making life difficult, but this may not always have been so since some pictorial scenes in the rock tombs in the cliffs to the east of the city have rare pictures of the city in amongst their habitual scenes of the life of the Royal family (Davies, 1908). While the topographical accuracy of such scenes is doubtful, in that they were never intended to be spatially accurate representations, they were intended to give a flavour of what life was like in Amarna and as such, they show frequent trees and gardens, filling many of the spaces between the buildings and perhaps indicating a greenness that is completely lost today (Kemp 1989).

The backbone around which the city was built is the Royal Road which stretches for some 3km almost due North from the Central City to the North City and this was the main artery running north-south between these areas. Kemp (1989) sees the two end points of the road as influenced by the topography of the area, the north end lying up against the steep cliffs rising to the desert to the north, and the south end on a slight rise, where the Central City was constructed. Figure 7-2 shows the excavated remains of the various central areas of special interest to this study, along with the encroachment of cultivation and modern Arab villages.

Central City
The Central City was built around the southern terminus of the Royal Road and its emphasis was religious and administrative, with the great majority of the City's administrative buildings so far excavated clustered together here. At the end of the Royal Road to the east is the Smaller Aten Temple (P42.4), followed by the King's House (P42.1), the main temple magazines and the Great Aten Temple (Q39.1), whilst to the west of the road is the Great Palace (O41.1 and O42.1-3) connected by a bridge over the Royal Road to the King's House on the opposite side.

The Great Palace is a series of courts and halls, often painted and decorated with colossal statues of the King and unlike many of the buildings in Amarna, significant amounts of it were built in stone. Originally the Palace was thought to have stretched down to the River, but now the cultivated area has extended so as to destroy most of the western half of the building, leaving only the eastern part hard up against the Royal Road. It was extensively excavated by Petrie (1894), who uncovered there the famous painted pavement, which is now in the Cairo Museum and it is thought it functioned as the main site for the receiving and entertaining of royal envoys and guests (Kemp 1989, Aldred 1988).

The two temples are the new open air type of temple first seen at Karnak and now built here for the worship of the Aten. Very little of the temples, other than their ground plan survives, most of the stone having been taken off for use in other later projects. Of particular interest are the magazines of the temples and the main bakeries of the city (P43.1) which stand beside them.

The King's House was excavated by the Egypt Exploration Society (Pendlebury 1951) and consists of a small palace, various storerooms, and a courtyard, which is where reward ceremonies associated with Egyptian Kingship took place. Behind the King's House are a series of small offices where the administration of the City was conducted. These, which were also excavated by Pendlebury and the Egypt Exploration Society, give some idea of how the administration worked. These offices are housed in small buildings, not all of which have been identified. The two that have are Q42.21, the "Bureau of the Correspondence of Pharaoh", where the diplomatic record known as the "Amarna Letters" was uncovered in the late nineteenth century, and Q42.19-20, the "House of Life", which contains information relating to medicine, religion and astronomy. Unfortunately, the other administrative offices of state remain elusive.

North of the Central City
At the other end of the Royal Road lies both the North Palaces and the North City. The North Riverside Palace is thought to be the main residence of the King, where he lived in relative seclusion away from the main areas of habitation and administration. It is partly lost under cultivation, but enough remains to show that it had thick, massive fortified walls (Jones 1983). It sits on a rise in the ground under the cliffs to the north that rise up to the desert beyond. It is surrounded by some of the largest houses found in all of Amarna, presumably belonging to the King's favourites. Along the Royal Road between the Central City and the North City, also lies the North Palace, thought to be the residence of the eldest princess, Meritaten, and one of the two main areas of habitation, the North Suburb.

The North Suburb lies close to the Central City and to the east of the Royal Road, the western half of the area and the central part of the Suburb being lost under modern cultivation and building, leaving now two isolated areas (Figure 7-2). It is likely that they were always separated, since the wadi that runs down through the area is ancient. What is left is built around two main roads, running roughly north-south and named by excavators East Road North and West Road North, themselves cut by other roads running east-west.

The Main City
To the south of the Central City lies the Main City, in which the majority of the population resided and in which, due to the evidence here of industrial activity, the main focus for this study lies. The planning that went into the construction of the Royal Road and the areas immediately associated with it, seems to have been less rigorous when it came to the Main City. The Royal Road ends, and in its place are two more-or-less straight broad roads running approximately north-south and named by excavators West Road South and East Road South (also known as High Priest Street). It should be noted that despite the names, they do not connect with West Road North and East Road North in the North Suburb. A further unnamed road lies to the west of these and runs past the modern water tower (see Figure 7-2), although little excavation has been done in this area. The whole Main City is cut by an ancient wadi that runs roughly east-west and that divided the Main City into two zones, the Main City North and the Main City South.

South of the Main City
To the south is the South Suburb which is separated from the Main City and which includes a set of gardens devoted to pleasure and prayer known as Maru-Aten and a further temple complex only recently uncovered known as Kom el-Nana.

Outlying areas - The Tombs and Workmen's Village
The tombs are split into three groups and their locations relative to the city are shown in Figure 7-1. The first two, the North Tombs and the South Tombs, were for courtiers and officials and are almost all unfinished, but they show representations of the royal family as their primary theme (Davies 1908). The royal tombs are in a particularly inaccessible spot some 11km to the east of the city itself and only one, Akhenaten's own, is anywhere near completion. This was of a similar design to royal tombs at Thebes, the one major difference being the provision within the tomb for the burial of one of the royal princesses. This tomb therefore seems to have been intended as a family vault as opposed to a tomb for the king alone (Aldred 1988). The village in which the workers who were building at least some of these tombs lived has also been excavated and bears some resemblance in terms of highly regulated house size and layout to the similar site at Deir el Medina near Thebes (Pendlebury 1933). There is, however, no evidence for the production of ceramics or anything else on the site and as such this village is one of a series of outlying houses and buildings that are only of peripheral interest to the investigation of the industries of Amarna.

Arrangement of houses and streets

The houses, or groups of houses in the case of small properties built close together, are given series of reference letters and numbers to uniquely identify them. The system follows that laid out by Paul Timme working with Borchardt and the early German expeditions (Timme 1917, Borchardt 1911), who divided the whole site of Amarna into 200m squares forming a grid, each square being given a letter denoting its place east-west and a number indicating its place north-south. The lettering starts with "A" in the far west and runs to "T" in the east, and the numbers run from 1 in the north to 51 in the south. Within each square of the grid, the houses are numbered sequentially in approximate order of excavation, the house of the chariotry officer Ranefer is thus N49.18, that is house 18 in square N49. Timme's survey was reviewed and anomalies in the original layout corrected by the survey of Kemp and Garfi (1993) and it is the designations detailed in this work that are used in the present analysis.

Design

The majority of the houses at Amarna have very similar ground plans independent of their size, although identical houses are rare and a number of key features can be seen in the average house. They are built around a square central living room, of varying size dependent on the overall size of the house, often with a low brick dais at one end where the owner and his wife could receive guests and a plastered washing place on one of the other walls. Around this central room are an outer reception room, connecting it to the outside, storage rooms and personal domestic quarters, usually consisting of a bedroom for the master of the house with a raised alcove for a single bed, and an associated bathroom and lavatory and often with further bedrooms and domestic rooms. The larger houses are set in their own grounds, again consisting of certain key features of which most houses show a selection, often with a high external surrounding wall, thought to have exceeded 3m in many cases. Within the compound are various groups of buildings and structures, including domed granaries for the storage of grain, typically 2.5m across and in groups of 4 to 5, animal byres, gardens, a well and a decorated garden shrine, set in formal grounds. Also in the compounds are rooms for kitchens and craft activities and these are frequently set at the southern end of the compound, where the prevailing northerly winds would have blown the smoke from any fires away from the house. There is often a difficulty in telling the kitchens and workshops apart, and several structures originally designated as ovens in the excavation reports have subsequently been re-excavated and are now thought to be kilns (Kemp 1989). An example of this is house P47.22, where a structure in the south-west corner of the house was designated "ein weiterer Backofen" in the original German report (Borchardt and Ricke 1980), which on re-excavation by the Egypt Exploration Society is now thought to be a "simple pottery kiln" (Kemp 1989). Thus, the attribution of ovens and kilns, and therefore kitchens and workshops, is often unclear.

Layout

The layout of the houses in relation to each other and to the streets and administrative buildings is evident from the excavation plans and perhaps not what would be expected. The larger houses were often built first along the main thoroughfares, often leaving a fair amount of space between their compound walls. This can still be seen in areas where space was apparently not at a premium, for example in the northern and eastern areas of the North Suburb (Figure 7-3) and the most southerly areas of the Main City east of the High Priest Street. In other areas, the gaps between these buildings have been filled in with smaller houses called "slums" by some of the excavators. Although adopting the same basic internal plan of the majority of the houses, houses in these areas tend to be flimsily built and sharing common walls. They often lean right up against the compound walls of the larger houses and are connected by small narrow alleys to the main thoroughfares and, due to their thin walls, they are often very poorly preserved (Figure 7-4). Thus there is a complex intermingling of the poorest and richest of the houses in some areas of the town, most notably the northern part of the Main City.

As Kemp (1989) points out, there was apparently no great attraction to live close to the Central City, since we know of senior officials who had their residences some distance from this area. For example, the Vizier Nacht had his residence some 2km to the south of the Central City (K50.1), and the Chief of the Seers of the Aten Pawah was a similar distance away also to the south (O49.1). It is evident therefore that they must have commuted to the centre, presumably by chariot, perhaps in a similar manner to the Royal progression from the North Riverside Palace to the Central City along the Royal Road.

Decoration of Royal and private buildings

The Great Palace was stretched out along the side of the avenue and comprised of outbuildings, the harem, the Palace proper, storerooms and the Coronation Hall. The decoration of the palace, temples and tombs was lavish and of high quality with much of the interior and exterior of the major public and royal buildings being painted and in the more prestigious areas of the court, glass and faience inlays to the painted limestone and mudbrick walls often being used. Although extensively robbed in antiquity with only chips and splinters remaining, enough has been recovered from the Great Palace to give some hint of its opulence. The capitals of the columns were inlayed with blue and red glazed chevrons set between ribs of gilded stone, with hieroglyphs and images cut from red quartzite, limestone, black granite and obsidian, white calcite and coloured glass and inlaid into stone of a contrasting colour (Aldred 1988). In addition to these inlays, glazed tiles were used, often painted with plants and flowers, fish swimming in tanks of water and the like, all the designs being similar to contemporary painted wall scenes and pavements.

The houses and private residences show some variation in the amount of decoration. They were often hastily built in mud brick and most have no stone in their architecture at all, this being particularly true of the smaller and presumably poorer houses. The walls are plastered and whitewashed, with little other decoration. In larger houses, there is more decoration, often with an attempt to imitate on a smaller scale the opulence of the Royal residences, but this has been systematically removed presumably by the owners to take to their new residences at the end of the active life of Amarna, this process being ably followed up by thieves and treasure hunters picking over the site over the succeeding millennia. Despite this, painted and inlaid capitals are seen, along with painted faience floor tiles and inlays for stone pediments and other areas of stone. Stone door frames sometimes remain, perhaps due to their weight, despite the prestige and re-use value of such objects meaning that many went with their owners when they departed. These door frames are of particular interest, since some are inscribed with the name and titles of the owner of the house, for example, P47.18, known from the frames to be the house of the general Ramose (Figure 7-5) and, listing his titles.

From this and other sources, the name and occupation of the owners of roughly fifteen of the houses are known to us, the most famous of these being the House of the chief sculptor Thutmose, P47.2, which was excavated by the German expedition in 1912-13.

THE LOCATION OF THE FAIENCE WORKSHOPS

As has been discussed earlier, Petrie (1894) stated that he had found "the sites of three or four glass factories, and two large glazing works... though the actual workrooms had almost vanished". Regrettably he does not state where these workshops were, the only hint being the suggestive word "moulds" that was written on his map (Petrie 1894, plate XXXV). This site would now be in grid square O45, in the Main City some distance from the areas that have been systematically excavated and right up against the area of modern cultivation. The Egypt Exploration Society decided to excavate the site and as a preliminary it was subjected to a proton magnetometer survey by Ian Mathieson which revealed significant anomalies that were subsequently excavated by Paul Nicholson (Nicholson 1995a, 1995b, 1998). Two kilns 2m in diameter were uncovered, described as thick walled and highly vitrified, with a sacrificial lining and associated with a large amount of *khorfush*, the local word for black slag. A third smaller kiln was found apparently associated with the other two and of a type recognised by Nicholson to be a pottery kiln. Associated with the site were frit, melted glass, glass rods and fragments of cylindrical vessels, similar to those described in Chapter 4. All of this suggests strongly that this site designated O45.1 was used for the manufacture of vitreous materials.

Petrie gives no clue as to the location of the other glass factories, but since large areas of Amarna have now been excavated, the search for them goes on. Ideally they could be recognised by high concentrations of glass and faience, glass making tools and the presence of kilns. Unfortunately, since the site was systematically cleared of valuables when it was abandoned, and has subsequently also been disturbed by treasure hunters and the like, anything brightly coloured stands a good chance of being picked up and large quantities of faience and glass therefore possibly would not survive. The kilns have not always been recognised as such by early excavators (as discussed above), and subsequent excavation of what were thought to be ovens have revealed that they were pottery kilns (Nicholson 1995a, Kemp 1989). What is needed to provide a clue to the presence of faience making is something connected with the manufacturing process that is small, common and of little value to any treasure hunter. The small baked clay moulds (Figure 7-6) used in faience making fit this criterion well and many of them have been recovered from the site during all excavations from Petrie's in 1893 through to the present. It is also the presence of the moulds that led to the uncovering of the O45 kilns and site, so these objects in large numbers do seem to indicate the presence of kilns and manufacturing. They are extremely common at Amarna, Petrie alone recovering many hundreds from the site and Peet and Woolley (1923) proposed that they were distributed through the population with each person or householder pressing an amulet or ring into some clay and then taking the resulting mould to the faience manufacturer for baking and filling with raw material, which would then be fired and returned to the owner. In this way Peet and Woolley attempted to explain the large number of moulds that they thought came from ordinary houses.

Given that the moulds are indeed used in faience making, which is not in doubt, a survey of the distribution of moulds throughout the site of Amarna might show clusters with larger numbers of moulds and suggest that the houses in that area were either workshops or the residences of the workers associated with faience workshops. This could give a suggestion of the position of the workshops in the town and possible further evidence of how faience manufacture fitted into the wider pattern of production and distribution of materials in Amarna. So with this in mind, a survey of the mould distribution was conducted.

Method

The survey maps of Amarna found in Kemp and Garfi (1993) were photocopied, scanned into the computer using a standard scanning program and compiled using CORELDRAW to enable their use as basemaps. The excavation reports were then consulted, Table 7-1 showing the various reports and the areas which they cover. Of these reports, Petrie (1894) was not of use since it fails to say precisely where each of the moulds were found and it was therefore disregarded. In the other reports, the details of each house were examined and the number of moulds excavated in them represented on the map by different sized circles, the presence of kilns by a star and other debris of interest as indicated on the key.

Table 7-1 : Excavation reports used and the areas that they covered.

Excavation report	Area covered
Petrie (1894)	Scatter of houses, some identites now lost
Peet and Woolley (1923)	Main City, Workmans' Village, South Suburb
Frankfort and Pendlebury (1933)	North Suburb
Pendlebury (1951)	Central City
Borchardt and Ricke (1980)	Main City, South Suburb

Table 7-2 : Results of the search through the various excavation reports. The table indicates the number of houses published and unpublished from the sources and the number of moulds recorded for each of the areas of the city.

Area	unexcavated	unpublished	*WIT*	*COA*	*JEA*	Total	Total published	no. of moulds
Central City	7	0	0	185	0	192	185	68
Main City North	6	232	125	24	0	387	149	221
Main City South	0	43	131	43	0	217	174	149
South Suburb	0	2	4	5	0	11	9	2
North Suburb	0	0	0	301	0	301	301	110
North City	1	0	0	0	29	30	29	0
Workmans' Village	15	0	0	48	0	63	48	0
					Total	1201	895	550

WIT - Borchardt and Ricke (1980), COA - Peet and Woolley (1923),Frankfort and Pendlebury (1933) and Pendlebury (1951), JEA - published in various copies of Journal of Egyptian Archaeology

Results

Table 7-2 shows the results of the survey in tabular form. Of the twelve hundred or so houses that have been given a grid reference number by Kemp and Garfi (1993), all except 15 have been excavated and three quarters of those excavated have been published, making a total of 895 houses to be checked. Those areas upon which valid interpretations can be drawn are Main City North, Main City South and the Northern Suburb, other areas having been discounted for reasons discussed below and the plotted results for these areas are shown in Figure 7-7, Figure 7-8 and Figure 7-16.

Many areas of Amarna have not be excavated and therefore cannot be included in this survey, whilst other areas have been excavated, but not published. This is especially true for the Egypt Exploration Society expeditions of 1923-5, which have never been published and the original records are believed lost (Kemp and Garfi 1993). This is a serious loss, since many houses were dug on this excavation in the northern part of the Main City North, near to the Great Aten Temple, and parts of the southern end of Main City South and the South Suburb. This means that these areas must be discounted, and they are not included in this survey.

In order for the distribution to be studied and any viable interpretation made, then all, or almost all, adjacent houses in an area must have been excavated. This rules out the South Suburb, where the reports of only a small number of scattered houses are available, but not enough to draw any conclusions; the Workmens' Village, which has no moulds at all; and the Central City, which has virtually no moulds, except for a number found together in one of the magazines.

Study of the plots for the other areas, Main City North, Main City South and the Northern Suburb, show that over most of the Main City area most houses have 0-2 moulds, the average being slightly higher in Main City North than Main City South where most have none. This is also true of the North Suburb, where by far the majority of the houses have no moulds at all, only 3 moulds being found in total from the houses to the East of West Road North, the newer part of this area. In all areas there are, however, definite clusters of houses with high numbers of moulds and very occasional isolated houses with high numbers of moulds that lie amongst houses with none or relatively few. Of all of these areas with high numbers of moulds, two are associated with a number of kilns (N50/M50 cluster and P46 cluster), two are associated with sculptors' workshops (P47.1 and O47) and one is associated with a small kiln (Q46.5), whilst in the North Suburb there is a large concentration, but no kilns preserved (T36).

N50/M50 cluster - High Priest Street and Street A (Figure 7-9)
Bordering onto the West side of High Priest Street (East Road South) are two houses (N50.27 and N50.25) with 10 and 6 moulds and three others (N50.23, N50.24 and N50.25) each with 3-4 moulds, along a street frontage of 100m (Figure 7-9). Excavated by the German expedition in 1912, this area is published in Borchardt and Ricke (1980) and N50.23, in the heart of this group, is perhaps the key to it. The plan of N50.23 is reproduced in Figure 7-10 which shows a fairly routine ground plan for an Amarna house with the domestic rooms 1-10 to the north (with a main living room (2), side rooms (1 and 7) and ante room (8)) onto which has been built a series of workrooms to the south (rooms 11-20), defined by the excavator as "a large workshop for the production of glass and faience". Rooms 11 to 17 were roofed and are believed to have been storage and preparation areas and "perhaps some finishing work was also done here", but their interconnections are not clear since the walls have decayed to a great extent. In the largest of these "preparation" rooms (16), there are "numerous trough like installations" perhaps used for the storage of materials. Room 11 has the remains of supports for a large table and two buried pots in the corners, one of which is also present in room 16. The exact interpretation of the use of all these structures is at the moment unclear. Rooms 18 to 20 were unroofed and separated from each other by walls with doorways, making them more yards than rooms, and in each of these yards is a kiln, two of which are square and about 1m across, the other too poorly preserved to be clearly drawn. Unfortunately no detail of these kilns is given in either Borchardt and Ricke (1980) or the excavation daybook, but they are interpreted by the excavators as

"for the production of glass and the firing of faience". The finds from N50.23 include, in addition to the moulds, two working pieces of glass, one red (11/557) and one blue (11/563), "a large number of broken pieces of glass rod 2mm in diameter", a total of 17 pieces of faience (which, despite the small size of the building, puts it in the top 5% in terms of number of faience finds in one building in Borchardt and Ricke (1980)) and various finished pieces of glass.

Surrounding N50.23 are the other houses with high mould numbers of more normal low class, thin walled house type, of which N50.27 is also in the top 5% in terms of numbers of faience finds. These buildings may by linked to another area of high mould density, some 100 m Northwest of the last group where, around two streets, 'A' and 'C', three houses (M50.13, M50.15 and N50.38) with 10, 5 and 9 moulds were found along with two others (M50.23 and M50.32) with 3-4, all of which were excavated by the Egypt Exploration Society in 1921 and 1922 and published by Peet and Woolley (1923). M50.14, described as the workrooms of M50.15 and M50.16, has a kiln, described as a "glaze kiln" by the excavators, which is a "pit cut in sand 1m in diameter by 0.5m deep full of burnt brick, glass and glaze slag and fragments of the pots used in the kiln for standing the glazed vessels on: the bottoms and sides of these are covered with tricklings of glaze", this last being a clear reference to vessels probably similar to the cylindrical vessels discussed in Chapter 4. It also contains a single clay mould and a limestone mould (22/586) of which no further details given. The plan of the site is very difficult to assess since it is "much ruined", making the house plan unrecognisable. M50.15, the dwelling quarters of M50.14, is also much ruined, "even entrance not recognisable", but has the remains of pigments in yellow, light green and blue in addition to the clay moulds. Many of the houses in this area are not well preserved, and this is certainly true of M50.13, N49.39 and N49.38 which are the next three houses along the street, two of these three have high numbers of moulds as mentioned above. Interestingly, N49.38 has a courtyard next to it (that may belong to an adjacent unexcavated house),with "pots sunk in three corners", a pattern very similar to the yard of N50.23 (room 11), which has pots sunk into two of the corners and was interpreted as fulfilling some function in glazing or manufacturing of faience. Peet and Woolley, excavating the area around these houses, found in 'Street A' onto which these houses all face, "large numbers of pottery moulds, pieces of rod-glass, imperfect glass beads, etc.", indicating that "one building in the neighbourhood was involved in the manufacture of glass". This find appears to be the rubbish tip of a glass/faience factory similar to that found by Petrie (1894), and is almost certainly related to the kiln in M50.14 and the houses with so many moulds adjacent to them.

N49.20, one of the houses with 3 moulds and within 50m of this group, is distinguished by being the house where the famous Amarna glass fish (EA55193) was found, along with two other glass vessels and other metallic items.

P46 cluster (Figure 7-11)

This is close to the "Weihnachtshaus" (Q46.1) in the heart of the city and shown in close up in Figure 7-11. Within an area of 100m x 50m, there are five houses with high numbers of moulds; 14 in P46.2, 13 in P46.1, 9 in P46 .5, 8 in P47.23 and 7 in P47.22, and around these there are others with 3 or 4 moulds (Q47.9 and P46.15). All these houses were excavated by the German expeditions and reported in Borchardt and Ricke (1980) and all except P47.22 are small size flimsy dwellings, sandwiched between larger houses set in their own grounds to the South and the yards of the Weihnachtshaus to the North. To the west there are several houses (P46.7-23), equally small, but excavated by the Egypt Exploration Society and published in Peet and Woolley (1923). These show moderate numbers of moulds, but several houses have finds of pigments (P46.11, P46.17, P46.20, P46.22), one (also P46.22) fragments of blue glaze and one, P46.11, "a mass of beads of glaze, carnelian and glass", no numbers given, but making up several different strings and hoarded under a brick. The remaining houses of P46 are unpublished (P46.24-32) or not excavated (the majority of P46). P46.1 consists of some 55 rooms, divided into 6 apartments, whilst the complex P 46.2 consists of about 20 rooms, which are divided into three living quarters and one workshop, which share one small common yard, accessible from the north-west. The apartments in this group are "small and shabby" according to the excavators. They believe that the occupiers would have "worked on one of the larger complexes, like other people...in the neighbourhood". All the dwellings are poor quality and are linked to High Priest Street by narrow alleyways running around the back of the larger houses. Two houses in this area are listed as having kilns, P46.2a having a round kiln associated with "traces of burning on the NE wall" and the presence of the moulds broken pieces of faience inlay are, according to the excavators "evidence for a workshop". The 1913/14 daybook for the excavation makes it clear that the moulds and faience were found around the kiln. The second kiln is in P47.22 (Nicholson 1995, Kemp 1989), a larger house set in its own grounds with finds of a potsherd with colorants (#1054) and a piece of crucible with glass (#1057). Unfortunately again this area runs up against unexcavated and unpublished houses which once again makes the interpretation difficult, but the presence of so many moulds, two kilns and other manufacturing debris is again suggestive of faience working as the excavators indicate.

Q46.5 - a possible jewellery workshop (Figure 7-12)

In Q46.5, a house situated just to the East of East Road South, in what is interpreted as one of the newer parts of the Main City (Kemp and Garfi 1993), 15 moulds were found, although the surrounding houses showed no particular increase in mould number. It was excavated in 1913 by the German expedition (Borchardt and Ricke (1980)), and is a series of small low class dwellings built with shared walls and very insubstantial. Q46.5b, however, has a kiln in Room 16, which the excavator states was previously a bedroom judging by the presence of a partially dismantled raised platform typical of a bed structure. The kiln, shown in Figure 7-12, is small and consists of an upside down clay pot with an internal diameter of 0.5m and a stoke hole entering from below, with fire marks and charcoal and was sufficiently unusual for the excavators to draw it in the 1912/13 daybook and for it to be repeated in Borchardt and Ricke (1980). Also found in the house were large quantities of faience, a blue glass rod (12/13/958) and a limestone mould with the figure of a displayed vulture carved into it (12/13/1234 and Figure 7-13). This mould is highly unusual, and the excavators believed (1912/13 daybook) that it was used for gold working, a thin sheet of the metal being pressed into it and trimmed to give the vulture shape. The mould has raised areas on the wing feathers and chest that would create depressions in the gold for inlaying of faience, precious stones or, most likely, glass, to create a finished object similar to that shown in Figure 7-14, a vulture Nekhbet pendant from the Tomb of Tutankhamen. This item is the same size as that which would be created by the mould and has similar inlaying in lapis lazuli and glass. Given that 12 of the small clay moulds are for jewellery (no further details given and the items missing), along with the presence of a glass rod and the vulture mould, plus the small kiln which could easily have been used for the melting of precious metal or the production of small items of faience, this area could tentatively be identified as the site for jewellery working and the production of faience on a small scale. This is an interpretation also reached by the excavators as recorded in the 1912/13 daybook.

Associated with sculptors' workshops; P47.1-3 (Figure 7-15) and O47

P47.1-3 is perhaps the most famous of all the house complexes at Amarna, being the workshops of the sculptor Thutmose, and famous for the many plaster faces and the Berlin bust of Nefertiti (Nr.21300) which were found on this site. Excavated by the German expedition of 1912/13 and published in Borchardt and Ricke (1980), P47.2 forms the house of Tuthmose himself and some of the workrooms of the complex, whilst P47.3 represents further workrooms and the houses of second rank sculptors, apprentices and servants. P47.1 is a late extension of P47.2 and P47.3, perhaps when there was a great need for more space (Figure 7-15). Once more, the accommodation is small and flimsy, just as in the houses of the N50/M50 and P46 groups discussed above. Borchardt and Ricke (1980) identify 13 small living and working areas for low ranking craftsmen built around a courtyard, plus a possible supervisor's house (P48.2). P47.1 has 30 moulds, the highest number of moulds in any one house or house group of this survey, but P47.2 and P47.3 have unexceptional numbers (2 and none respectively), showing that there was a good deal of partitioning of work on this site, with that involving faience apparently restricted to certain areas. These small houses in P47.1 also contained many unfinished pieces of sculpture in stone including pieces of arms and hands in quartzite and the head of the statue of the king, mostly intended for composite statues, along with 49 fragments of finished items of faience, the highest number found in any of the houses list by Borchardt and Ricke (1980). Given the poor nature of the housing and therefore presumably the low rank of the inhabitants, this suggests that the faience was not owned by the people here (the better housing of Tuthmose and the second rank sculptors have fewer finds), but may instead have been either produced here or incorporated into other objects like composite statues on the site. There no evidence for a kiln given in Borchardt and Ricke (1980), but re-examination of the site quoted in Kemp (1989) has shown that there maybe two kilns perhaps for pottery, immediately to the south of the workmans' area of P47.2, which could possibly have been used for faience as well.

A further cluster of houses with high numbers of moulds lies in square O47, 300m west of the previous P47.1 group, again in the heart of the City and next to the West Street South Sculptor's Workshop. Here, two houses, O47.8 and O47.6 have 10 and 5 moulds respectively, while four others have either 3 or 4, all of which fall into an area of 50m x 70m and were excavated by the German expedition of 1912/13 and published in Borchardt and Ricke (1980). It is smaller in terms of number of moulds than the P47.1 and also does not show evidence of a kiln, but it is right up against two unpublished sculptor's workshops in P47.16a and P47.20, both excavated by the Egypt Exploration Society and the plans lost. It is adjacent to further unpublished and unexcavated areas, so other features may be just off the edge of the known area or have been excavated and the information is no longer accessible, but this again appears to be large numbers of moulds associated with a sculptors workshop.

North Suburb, T36 cluster (Figure 7-16)

A further clear cluster falls in T36, the old part of the North Suburb to the West of West Road North. In a tight area of less than 100m square there are four houses - T36.44, 83, 62 and 71 - with 9, 8, 8 and 7 moulds, with a scatter of seven more houses with 3-4 moulds and of the 110 recorded moulds of the North Suburb, 90 come from this area (Figure 7-16). Two further houses, T36.41 and T36.72, each have two glass rods. It was excavated between 1928 and 1930, published by Frankfort and Pendlebury (1933) and described by them as slums, a series of small alleys and squares with small houses often poorly built and therefore also poorly preserved. The walls of the houses have often largely disappeared and the relationship of the houses internally and to each other is often difficult to define, which makes interpretation difficult. Boyce (1995), who used a similar mould survey to this study, but restricted only to the North Suburb, found the same concentration of moulds which he shows also corresponds to a concentration of amulets. He interprets this as a manufacturing or market area, but there is little evidence to say which, if either, of these interpretations is valid, since no kilns were recorded by the excavators and the preservation is so poor that it unlikely the they would have been recognised even if they were present.

Discussion and comparison

Both the Main City and the North Suburb show clear clustering of moulds, with six possible areas of interest, one in the North Suburb and five in the Main City. The North Suburb, while having a clear pattern of mould concentrations, has very little preserved on the ground, which makes interpretation very difficult and it will therefore not be considered, beyond the statement that the presence of the moulds suggest that there was probably some production of faience going on in this area, an interpretation also reached by Boyce (1995). The presence of four kilns in the N50/M50 area, along with the high numbers of moulds, glass pieces in the process of being worked, glass rods and high numbers of faience finds, all suggest that N50/M50 was involved in the making of both faience and glass. The scatter of moulds in houses adjacent to the house with the kiln could result from disturbance of the site, in other words all the moulds were initially around the kiln. This is possible, but the presence of large numbers of moulds in P47.1 and Q46.5 without surrounding scatter rather suggests that this is not the case and that the scatter is primary. If this is so, the scatter of moulds suggests perhaps that the houses in the vicinity of the house with the kilns are also involved in the production of faience. It is further possible that these houses represent the living accommodation for faience workers, perhaps with small workshops for the preliminary manipulation of the raw materials and moulds and the final finishing of the objects. This conclusion would make perfect sense, since houses in which the kilns were found are small and probably do not give enough space for preparation of faience prior to firing, which would take the vast majority of the time and labour in the production of the faience, the firing taking relatively little time at the end. It is possible that these former operations were carried out partly by workers living in neighbouring houses, who then merely transported the prepared faience to the adjacent kiln. If the houses with the moulds are taken as the residences and work areas for the faience workers, then in no cases, in all five of the areas discussed, would they needed to have walked more than 75 metres to the nearest kiln. The N50.23 set of three kilns in particular provides greatly excess kiln capacity if the possible number of residents of the N50.23 property alone are considered for the workforce, but including those houses spread around it with high mould numbers as residential and workshop preparation areas increases the possible workforce greatly and would more fully utilise the regular firing of three kilns. This proposed grouping of workers might also explain another anomalous find, the famous Amarna glass fish. Its find with other glass and metal objects under the floor of a small insignificant house (N49.20) led to speculation that the "cache may have been hidden away, perhaps by thieves, during the gradual abandonment of the City"(Aldred 1988). N49.20 is within the area of houses associated with the N50/M50 group, being only 60m from N50.23 and having 3 clay moulds and other glass fragments. This association therefore suggests that a more likely explanation than a thieves' cache is that the fish and the other glass vessels were part of the forgotten "stock" of finished items of the factory, hidden for safe keeping before handing on to a final user.

The P46 group appears to be a less well preserved example of the same sort of manufacturing areas as N50/M50. Although having only one kiln, its proximity to the edge of the excavated/published area means that further parts of the complex may be missing. The presence in the group of the higher status house P47.22 is also of interest as it is the only example of a large house associated with moulds, glass working debris and kilns. Maybe this property is the residence of a higher ranking faience worker, and this possibility is further discussed below.

The Q46.5 site is different to the N50/M50 and P46 groups in that it is much smaller and there is no evidence of the involvement of neighbouring houses outside those of the Q46.5 house group. This site could possibly be associated with a one family operation, perhaps dealing, as the excavators believe, primarily with jewellery and connected with the inlaying of glass and faience into metals, as the vulture mould suggests. Both Q46.5 and P46, tentatively proposed as production sites for faience, as well as the North Suburb (T36) concentration of moulds, have evidence of glass working too - P46 has a crucible with glass (#1057), Q46.5 has a glass rod (#958) and the North Suburb group has several glass rods and other glass fragments. It is probable therefore that the production of faience and glass at many of the proposed production sites at Amarna is intimately linked, using the same kilns and probably the same workers.

Of the two sculptor's workshops groups, P47.1 may have kilns associated with it and O47 is not published well enough to determine the possibility of the presence of kilns. P47.1 is thought to be the residence and workshops of the more junior sculptors in the Tuthmose complex, and from the numbers of moulds it is possible these were also involved in faience work, perhaps partly for inlaying into composite statues. Sculptor's workshops could be thought of as 'consumers' of faience, where it would be incorporated into inlayed objects of stone and wood, including tiling, column heads and composite statues, but the suggestion here at P47.1 is that they are also producers, and from the moulds are producing rings and amulets in addition perhaps to faience related to sculpture.

Confirmation from Amarna field walking

In order to test the identification of the areas suggested as sites for the manufacture of vitreous materials, as many as was possible were visited and the ground within 50m of them walked over briefly as a preliminary check for surface material. Q46.5 is now very poorly preserved and revealed nothing and P47.1-3, the House of Tuthmose, is heavily trampled over by tourists and is also unhelpful, but the two major production sites of N50/M50 and P46 revealed much more. A five minute search at N50/M50 uncovered two fragments of cylindrical vessels, five pieces of slag with glass adhering and one glass rod, while the same period at P46 showed the presence of three pieces of slag with glass adhering and much additional slag. These observations support the reality of the groups determined above and indicate that materials involved in the manufacture of faience and glass, probably lying on eroded waste heaps, are still present in the field today.

ORGANISATION OF THE WORKSHOPS AND WORKERS

Whilst there is some evidence for the persons in control of some of the crafts, sculpture for example, in the form of names and titles of officials, the precise organisation of the faience manufacture at Amarna including the number of workers, their remuneration and conditions and the officials that controlled them has not survived. Speculations on the organisation of this industry therefore must be derived from comparison with other industries and the little that is known of faience makers from other periods and sites. It is clear from texts describing the remuneration of workers that most crafts in Egypt, including pottery, sculpture, metalworking and others, were controlled by an official who was an administrator and part of the ruling élite (Kemp 1989, Bianchi 1998, Lesko 1994, and others). In some cases, officials of this administrator rank held the title of overseer, *imy-r*, (see Table 7-3) and there is one example of a person with such a rank at Amarna, Hatiay, "overseer of works" whose large house was T34.1 in the North Suburb (Frankfort and Pendlebury, 1933). Below this rank were the *ḥryw* for which the translation is usually "chiefs". There are three examples of chiefs known at Amarna; a chief builder, a chief sculptor, and Tuthmose whose titles read "the chief of works, the sculptor Thutmose". That holders of both of these ranks were powerful people cannot be denied from the size of their houses and comparison to other such officials whose tombs are known from outside Amarna.

Below the rank of *ḥryw* were the ordinary workers, who are attested to in many texts. Of particular interest due to its completeness is the record of the workers who were building the Royal tombs in the Valley of the Kings during the late New Kingdom and lived at Deir el Medina, studied by Cerny (1973) and Bogoslovsky (1980) and summarised in Kemp (1989). Here, the wages and conditions for the ordinary draughtsmen (*sšw-ḳdwt*) for the tombs, the chief draughtsmen (*ḥry-ḳdwt*) and both deputy chief draughtsmen and apprentices are known and can be compared, along with some of their living conditions and lives. Since they were working on a royal project, the chief draughtsmen reported directly to the Office of the Vizier (Lesko, 1994), as opposed to having an overseer

Table 7-3 : Titles of craftsmen at Amarna.

	title in translation	house/tomb	references
"Overseer"			
Hatiay [2]	Overseer of works	T34.1	Frankfort and Pendlebury (1933)
"Chief craftsman"			
Thutmose [1]	Chief of works, the sculptor	P47.1-3	Borchardt and Ricke (1980)
Iuty [3]	Chief sculptor	Tomb of Huya	Davies (1906)
Maanakhtuef [1]	Chief of builders	M47.3	Borchardt and Ricke (1980)

[1] Borchardt and Ricke (1980), [2] Frankfort and Pendlebury (1933), [3] Davies (1906)

Table 7-4 : Titles of faience makers.

	title	period	type of object
"Overseer of faience makers"			
Debeheni [1]	*imy-r ṯḥntyw*	13th Dynasty, King Hor	coffin inscripton
Qennou [2]	*imy-r irw ḥsbḏ*	19th Dynasty	papyrus
"Chief faience maker"			
Hatiay [3]	*ḥry irw ḥsbḏ n nb t3wy*	New Kingdom	funerary stelae
Ptahmose [4]	*ḥry jrw ḥsbḏ n nb t3wy*	New Kingdom	papyrus
"Faience maker"			
Rekhamun [5]	*jrw ḥsbḏ n Jmn*	19th Dynasty, Ramesses II	faience funerary stelae

[1] Bourraiu (1996), [2] Papyrus Vatican 64, Bellion (1987), [3] Cairo JE25641, Gaballa (1979), [4] Papyrus Krakow MNK IX-752/1-4, Luft (1977), [5] National Museum of Scotland A.1956.153, Friedman (1998)

belonging to the élite which might have been expected otherwise. Chief draughtsmen were also hands-on workers (Bogoslovsky 1980) and not merely administrators since their names appear on the records as actually carrying out some work in addition to supervising the technical work and the workers themselves, with the assistance of a scribe for record keeping (Lesko 1994).

None of these officials in other crafts at Amarna appears to have any connection with faience makers and textual references to such workers have been classified by Gaballa (1979) as "very rare", although Nicholson (1998) believes they are "not common, though... less rare than sometimes thought". Table 7-4 lists some titles of faience makers that are known, one dating from the 13th Dynasty, the rest to the New Kingdom. The words used for faience in this context include both *ḥsbḏ*, which was also used for lapis lazuli, and *ṯḥnt*, which is related to the words for "gleaming" or "shining" (Nicholson 1998). The titles using *ḥsbḏ* could refer to workers in lapis lazuli, but most authors agree that the faience translation is the more appropriate one in this context. The titles follow a similar line to those attested for other crafts, *imy-r irw ḥsbḏ*, for "Overseer of faience workers" (also known as *imy-r ṯḥntyw*, using the other word for "faience"), *ḥry irw ḥsbḏ* for "Chief of faience workers" and *irw ḥsbḏ* for "faience worker", suggesting that a three layer organisation was perhaps at work here as well. Figure 7-17 shows a hypothetical organisation chart based on these three ranks and drawing on the comparisons derived from the draughtsmen at Deir el Medina. The suggestion from Deir el Medina of Deputy chief craftsmen and apprentices have been carried onto this chart as a possibility, as it is clear that apprentices at least would have been needed and deputies would have been useful, but it should be noted that there is no record of either existing for faience workers. The Deir el Medina records list 97 draughtsmen, four of whom are chief draughtsmen, giving a supervisor to men ratio of just under 25:1, which is very close to that considered optimal by modern armed forces and certainly a workable situation. At Amarna several workshop areas have been speculated upon and they are scattered quite widely over the city from M50/N50 in the south to the North Suburb in the north. Assuming that a Chief faience maker would have similar hands-on technical and personnel supervisory roles as that seen in the draughtsmen, this is too wide an area for a single person to cover effectively. In addition to this, the number of workers involved, as speculated upon above, would possibly be too great for one man to control, so for these reasons, a more likely scenario would be one (or maybe more in the case of large facilities) Chief faience makers per factory area, with, since the Egyptian administration is far from top heavy, perhaps one administrative Overseer of faience workers controlling the whole operation at Amarna.

The link between faience making and sculpture is seen at Amarna with the finding of large numbers of faience moulds in P47.1, the residence and working areas of the lower ranking sculptors for Thutmose. The manufacture of composite statues would have required faience components, and the presence of the moulds suggests that they were perhaps made on site by resident faience workers working alongside the sculptors in stone that also lived there. As Bianchi (1998) points out, it would be routine in Egyptian crafts for several workers to work on the same object before it was completed, and certainly with composite statues several areas of expertise would have been necessary, so it is possible that there is a close link between the two crafts at Amarna. Friedman (1998) provides a further possibility of a close link between the two industries in her assessment of the stelae of Rekhamun, listed in Table 7-4. In addition to this stela listing Rekhamun as a faience maker, there is a second stela (Budischovsky and Genaille 1988) from the end of the 18th Dynasty which is stylistically similar, but in limestone, in the name of "Amunrekh" listed as a sculptor. Names in Egyptian were occasionally reversed and it is possible, as Friedman speculates, that "Amunrekh may be an earlier relative or possibly even the same man as the Rekhamun faience craftsman" (Friedman 1998). The Rekhamun faience craftsman, if he lived to old age and died in the first years of the reign of Ramesses II. as is suggested by the stelae , would have been the right age to be a young apprentice at the time of Akhenaten and Amarna, so fitting closely the period that is being examined and tentatively again suggesting a link between faience workers and, if he is in some way related to Amunrekh as Friedman speculates, sculptors.

DISTRIBUTION OF FINISHED PRODUCTS

The Egyptian economic system has long been known to be essentially an administered redistributive system, with the state controlling labour and the collection and distribution of resources (Kemp 1989). Kemp believes that this is largely the case at Amarna and illustrates this point with the bakery in the Central City, which shows the Egyptian approach to the problem of having to produce a product on an industrial scale. The modern western approach to the problem of bread production for a large city would be to increase the size of the ovens and so maximise the number of loaves that could be produced by the minimum number of man hours worked, but the Amarna approach was radically different. Instead of increasing the size of the ovens, excavation of the bakeries has revealed that they consist of the basic kitchen of a domestic house repeated many times until the required capacity was reached. This is evident from an artist's reproduction of the result in a relief is shown in Figure 7-18. Kemp points out that when labour is very cheap, then solutions contrary to the modern ideas of "efficiency" and "productivity" could be effective and continues that this can be thought of as "a paradigm for the whole ancient system" (Kemp 1989).

It is possible that a similar tight control was exerted over the production of other goods, and that at least some of the faience factories that have been discussed above were working (at least officially) primarily, or even exclusively, for the Court. Goods produced by the bakeries and other industries under central control would then be distributed by the state possibly in part in a ritualised way by the King personally. There are many representations in the tombs at Amarna (Davies 1908, Figure 7-19) showing the King dispensing gold and precious things from the "Window of Appearances" and he is known to have also given promotions and offices in the same way. Also pictured on the relief from the Tomb of Parennefer, in the rock tombs at Amarna, various items are shown as being given out, including, at the bottom, amphorae and baskets of food. This suggests to Kemp (1989) that the King is dispensing more basic commodities than just the high value prestige goods like gold and jewellery. The "Window of Appearances" is known to have been in the Kings House in the Central City and it is very likely that at least some of the items produced by the faience and glass factories were made for the King to distribute amongst his favourites and amongst the populace as a whole in this way.

In addition to this central controlling power, however, there are other factors possibly involved in the distribution of faience in the New Kingdom. Various authors (Lesko 1994, Kemp 1989, Janssen 1975) working on the records of craftsmen at Deir el Medina have calculated the pay and wealth of the draughtsmen working on the royal tombs and Lesko has pointed out that the early ideas that they were totally dependent on the state for their living, as propounded by Edgerton (1947) and others, is almost certainly incorrect. Ostraca from the site show that "many of the workers, apparently with no objection of the authorities, were producing saleable items and finding a market for them in Thebes" (Lesko 1994). Indeed, it seems that many of the crafts engaged in the tomb building were working in the private sector as well, for example, the carpenter who built the scaffolding also built coffins, chairs, boxes and tables to order and sculptors and painters hired themselves out for work on private tombs. It is clear therefore, that at least during the period of Deir el Medina and probably before, the craftsmen were free to spend some of their time working for private gain, and this should be taken into account when the distribution of faience is considered. This private work could well account for the finds of jewellery moulds in P47.1, the Thutmose workshop. The craftsmen here working on sculpture and composite statues would not have required jewellery in their work, but their easy access to kilns for use in other statuary linked faience objects, plus the access to raw materials and the skill needed, could mean that they were producing jewellery as well, perhaps in part for private gain.

NATURE OF THE FAIENCE PRODUCED

In order to obtain information on exactly what was being produced at the Amarna faience workshops, a survey of the finds of faience and glass throughout Amarna was conducted. The idea of the survey was to determine the types of objects that were produced and the range and proportions of colours used for these objects. The difficulty in conducting such a survey and interpreting the results are numerous, but if the limitations are taken into account, the information derived from a survey at Amarna can be very valuable, since most objects from Egypt come from burial sites, either tombs or graves, and Amarna is almost unique in being a town site. It could be argued that the number of objects in a town is likely to be much larger than the number made for tombs, so the distribution of object types and colours at Amarna is therefore much more likely to match the range of objects produced in the factories and in use in everyday life than the objects found in tombs, which are specially selected for their role in accompanying the dead and are not necessarily representative of the industry as a whole. Amarna, however, has its own special problems which must also be taken into account. It was abandoned by its owners, who were probably in no hurry to leave and subsequently had plenty of time to remove any goods that were of interest to them, leaving only the remnants for the excavator. This is illustrated in Figure 7-20, a flow chart demonstrating the loss of potential archaeological material from a typical house at Amarna through time. When it was in use the house could be considered as a number of discrete "zones", as shown, with the possibility of rubbish tips, hoards and lost material. When the owners move out they take with them the majority of their possessions and this along with subsequent occupation, weathering, theft and so forth severely depletes the remains. It should be especially noted that the chances of recovering first class (ie. the very best quality in terms of workmanship and material, available only in the largest of houses) or prestige (above average in materials or workmanship, but not necessarily both) goods intact are vanishingly small, unless they form part of a forgotten hoard or are broken or mislaid. The survival of even utilitarian goods is rare in the houses themselves. The evidence is therefore heavily skewed towards broken or unwanted items, a process that has been termed "negative selection" (Patch 1998), with an occasional utilitarian vessel and very rare, not necessarily typical, but frequently cited first class examples from hoards or mislaid small items. Much of the rest is missing and this should be borne in mind when interpreting subsequent distribution graphs.

Method

The find lists of Borchardt and Ricke (1980) were compiled into an ACCESS database, with each faience or glass object found listed under a type (jewellery, figurine, tile, etc.), subtype (for jewellery this included ring, amulet, bead and so forth), colour, and house where it was found. The types and subtypes are fairly arbitrary, for example, scarabs are not counted as jewellery, but amulets are, due to the restriction of the completeness of the data, which often listed such an item merely as "scarab" or "amulet". Where two or more beads were found, apparently from the same necklace, these were counted as one and, similarly, two fragments of the same vessel were counted as one, the aim being to give some idea of the number of finished objects (even if these had more then one component part). ACCESS then allowed the compilation and analysis of the data using various criteria.

Results

Borchardt and Ricke (1980) list 1567 pieces of faience in their find lists, all accurately pinned down to a particular house or group of houses within the city. Table 7-5 shows the summary for the faience divided into types and colours. Figure 7-21 shows a pie chart of the different types of faience (top of figure) and glass (bottom of figure) that this represents. As can be seen, over half the objects were jewellery (874 pieces, 56%), with roughly equal numbers of scarabs and inlays (both about 10%), and lesser numbers of tiles and items modelled in the round, like animals plants and figures. Rarer examples of other items, such as boomerangs, playing pieces, writing instruments and so forth were also found. The jewellery was divided into subtypes, including beads, rings, amulets and ear-rings, as displayed in Figure 7-22, revealing a high proportion of rings (bezels and shanks accounting for a total of 44% of the jewellery, or one in four of all the pieces of faience found by the expedition).

Figure 7-23 shows a pie chart of the colours of the faience, revealing predictably that blue is the most popular colour at 67% of the items. Unfortunately, no division is made in Borchardt and Ricke (1980) between cobalt and copper blue, so this area includes both. They do, however, distinguish a "blue-green" which accounts for 3% of the pieces and is associated specifically with copper

Table 7-5 : Summary table of the faience survey from Borchardt and Ricke (1980), giving the types and colours into which the faience was classified and their totals.

type	blue	blue-green	green	grey-blue	yellow	white	red	violet	polychrome	colourless	not stated	total
scarabs	118	14	27	3	1	1	0	0	0	0	1	*165*
jewellery	599	31	87	11	19	13	8	9	88	0	9	*874*
inlays	69	1	4	2	2	3	5	1	70	0	6	*163*
tiles	29	1	6	0	0	0	0	1	31	0	1	*69*
vessels	20	1	5	0	2	1	0	0	14	0	1	*44*
plant models	58	0	2	1	1	1	1	0	10	0	1	*75*
animal models	68	3	1	1	0	0	2	0	3	0	0	*78*
figurines	25	0	1	0	0	0	0	0	4	0	0	*30*
not stated	7	0	2	1	1	0	1	0	4	0	0	*16*
others	42	1	1	1	2	1	1	1	3	0	0	*53*
total	*1035*	*52*	*136*	*20*	*28*	*20*	*18*	*12*	*227*	*0*	*19*	*1567*

Table 7-6 : Summary table of the glass survey from Borchardt and Ricke (1980), giving the various types and colours into which the glass was classified and their totals.

type	blue	blue-green	green	grey-blue	yellow	white	red	violet	polychrome	colourless	not stated	total
scarabs	0	0	0	0	0	0	0	0	0	0	0	*0*
jewellery	20	0	3	1	2	9	4	0	14	2	16	*71*
inlays	1	0	0	0	0	1	0	0	0	0	1	*3*
tiles	5	0	0	0	0	0	1	0	1	0	0	*7*
vessels	5	0	0	0	0	0	0	0	11	0	3	*19*
plant models	0	0	0	0	0	0	0	0	0	0	0	*0*
animal models	0	0	0	0	0	0	0	1	1	0	0	*2*
figurines	0	0	0	0	0	0	0	0	0	0	0	*0*
others	1	0	0	0	1	1	0	0	2	0	0	*5*
manufacturing debris	8	0	1	1	4	0	2	1	1	0	0	*18*
not stated	1	0	1	0	0	0	0	0	3	1	0	*6*
total	*41*	*0*	*5*	*2*	*7*	*11*	*7*	*2*	*33*	*3*	*20*	*131*

colouring. The green coloured pieces could be coloured by lead antimonate plus copper, or conceivably by just copper alone, but these colorants give very different results, the latter giving only a turquoise-green, which can, with practice, be distinguished from the deeper greens of the lead antimonate plus copper. In order to test which was being cited here, 20 samples of green faience, selected at random from amongst those listed in the text, were examined in hand specimen and 9 of them were the deep green that placed them firmly into the lead antimonate plus copper group, 7 were copper coloured and 4 may have been either, showing that the use of the lead antimonate plus copper recipe does not account for all the "green" examples. The unusual colours of grey, white, yellow, red and violet each accounts for between 1-2% of the pieces, but the most surprising pattern is the number of polychrome pieces. The polychrome samples accounted for 14.5% of the total faience, again perhaps higher than might have been expected and about equal to all the non-blue/blue-green/green faience in total.

Table 7-6 lists the finds of glass in Borchardt and Ricke (1980) by type and colour and these results are also plotted in Figure 7-21, Figure 7-22 and Figure 7-23 alongside the faience equivalent for comparison. The overall finds of glass number only 131, meaning that faience finds outnumber glass by twelve to one. The percentage jewellery in the case of glass and faience is very similar at around 56%, but as can be seen from Figure 7-22, this percentage is made up of different types of jewellery, the glass having very few rings and fewer amulets, but a higher percentage of ear-rings and beads. The glass is not used for the manufacture of tiles and Borchardt and Ricke (1980) apparently do not list any examples of glass being used in the manufacture of plant models or figurines with only a solitary animal model (a violet duck) being recorded. Vessels, which are the principal use of glass other than jewellery, make up a higher proportion of the glass finds than they do in the faience finds (15% as compared to only 3%). No manufacturing debris is recorded for faience, but 18 examples classified as "rods" or "Werkstück" were found among the glass, not counting the find of "a large number of glass rods" from N50.23, the number of which was not given and which have not therefore been recorded in the tables above. Blue glass, is the most common, and polychrome the second most common, just as they are in faience, but polychrome glass is relatively much more abundant, at 30% being nearly as common as the blue glass (36%). When it is taken into account that the blue glass contains four blue glass bases and necks for glass vessels, which would have almost certainly have been polychrome (although the surviving fragments contains only blue) and that eight of the blue samples are manufacturing debris and not finished, then in terms of finished products, polychrome glass objects are perhaps more common than blue glass objects. There are also subtle differences in the proportions of the other colours, with green, yellow, white and red glass being about as common as each other at 5-10% and no evidence of the preponderance of green that is observed in faience.

Interpretation

The reasons for the differences in the use of glass and faience at Amarna could be either artistic (the appearance of the object), technological (varying difficulties in the production of different colours or materials) or social and economic (status and/or cost). To the modern eye, a complete faience object is often very hard to distinguish from a glass object, which would suggest at least that artistic reasons are not the driving factor in the selection of materials. It is very difficult to appreciate whether there was any difference to the ancient eye, but again, visually at least there is no reason to suspect that was much distinction made. All the colours are available in both media, and there is no reason to suspect that is substantially more difficult to employ a particular colour relative to the others in either glass or faience. Artistic preferences therefore appear to govern the choice of colour to a certain extent, since one colour is often used for the majority of the examples of some objects, crown and wig inlays, for example, are almost always blue. Choice of the colour is therefore artistic rather than technological. As discussed in Chapter 6, glass and faience are worked in very different ways, glass of the period not being routinely moulded, but rather worked plastically into shape when hot and perhaps engraved when cold. To produce an inscription or pattern on the glass, as is common in moulded faience rings, for example, is very difficult requiring a skilled glass worker who works the material when plastically hot and engraves it when cold, a much more lengthy process compared to the mass production of faience by open moulding and efflorescence, processes which can be carried out at a much lower skill level, and need only one firing. Therefore inscribed faience scarabs, rings and amulets can be relatively easily produced in large numbers, whereas an equivalent glass object has to be individually worked by a skilled glass-worker. This is also true of animal, plant and figurine models too, which again could be much more easily made in faience by open or hand moulding than in glass with again the difficulties of hot plastic working followed by cold carving. Ear-rings, on the other hand, being generally simple rings, and not inscribed, could be made without too much effort by a glass-worker and this is perhaps the reason why they exist in larger numbers.

The differences in use of materials for jewellery can therefore perhaps be explained in terms of the differing technologies required to manipulate them, making one more appropriate than the other to mass production on the scale apparently required at Amarna.

Another trend seen is the use of polychrome glass, which, as argued above, could actually be more common than blue in total glass production. As will be argued below, vessels of glass are probably some of the most prestigious of all the vessel types, perhaps ranking alongside vessels in precious metals, and were almost certainly used for valuable perfumes and cosmetics by the very top of the Egyptian social hierarchy. They were individually made in relatively small numbers and each almost certainly took very many hours to make, the time and effort involved in their production being less of a factor than for items of faience that were needed in greater numbers. As discussed at greater length in Chapter 8, the Egyptians used colour very widely and brightly, as is typified by the painted pavements of the Great Palace and the fact that many of the statues and buildings were extensively painted (Baines 1985). The desire amongst the élite was therefore for bright colours, and glass was a medium that gave them exactly that. In making a glass vessel, it is a lot easier to trail and marver bright colours into the glass than to decorate them in any other way and it is an easy step to imagine that once the technology for making a core-formed vessel, involving the coiling of hot plastic glass has been developed, it does not take a great leap of imagination to add different colours to the coils and produce polychrome vessels. Indeed, glass vessels that are not polychrome are relatively rare in Egypt in this period.

The relative ease of adding colours to a monochrome glass probably accounts for the abundance of polychrome glass and the bright colours for its popularity. The range of patterns associated with this technique are, however, very small, with swirls, dots, 'U' and 'V' shapes and geometric patterns being about the extent of the possibilities. There are rare examples of attempts to produce cartouches on glass vessels by marvering, for example, in the Munich vessel (Munich ÄS630) inscribed for Tuthmosis III, but these tend to be crude and of limited scope, and certainly cannot match the quality and precision that is achieved in faience. The very small numbers of polychrome glass tiles compared to the large numbers of polychrome faience tiles is probably linked to this fact. The great majority of polychrome faience tiles show scenes from nature, most commonly including depictions of plants, animals and water that would be immensely difficult to reproduce in glass and would probably result in an effect that was nowhere near as good as for the faience examples.

The link to status

The question of whether there is any link between amounts glass or faience at Amarna and the status of the house has long been speculated upon by various authors. In the case of glass, Petrie (1894) started this speculation by uncovering 38 pieces of glass vessel from the Great Palace, along with 750 pieces from the Palace Dumps (interpreted as the depository for rubbish from the central palace areas), "and none from elsewhere", the clear implication being that the glass was used almost exclusively in the Palace. As discussed above, subsequent excavation has shown glass vessels scattered across the town, but only in small numbers, only 19 fragments of glass vessels being recorded in Borchardt and Ricke (1980) for the entire German Expeditions. This suggests that while glass vessels are not entirely restricted to the palace, this is the area where the great majority of the material went and hence the Palace consumption was the driving force behind the manufacture of glass vessels. Archaeological evidence therefore fairly securely places glass vessels as high prestige objects, which perhaps is not entirely surprising given their colour and style.

At the other end of the scale, Patch (1998) has argued from a study of 26 houses of varying sizes reported in Borchardt and Ricke (1980) that "the smaller house structures and compounds yielded more pieces of jewellery than the larger ones", although she does qualify this by adding that "they may have been found at their site of manufacture". In order to try to determine if this observation is valid, and whether there are any other patterns that relate the finds of faience to status, as there seems to be for glass vessels, each of the houses in Borchardt and Ricke (1980) was rated according to size and method of construction following the guidelines laid out in Table 7-7 and this status was related to the ACCESS database for faience discussed earlier. Kemp (1989) has convincingly demonstrated that the size of the house at Amarna is directly related to the rank of the occupant, so relating the finds of faience to the size of the house indirectly links them to the status of the owner.

In order to try to ensure as far as possible that it was the faience in use that was studied rather than that from the manufacturing areas, as proposed above, the finds from all those houses with high numbers of moulds were ignored, along with those that have been speculated upon as involved in manufacturing in the study above. This leaves a total of 228 houses to be examined. As noted above, some of the "houses" listed in Borchardt and Ricke (1980) are groups of houses rather than single dwellings, P46.4 for example is actually four small houses in a group, numbered P46.4a to P46.4d by Borchardt and Ricke (1980). Where these house groups occur, the total number of all houses rather than the number of groups is used in the tables and statistics below. Large numbers of objects are needed to make the statistical validity of a pattern evident, so this really restricts the search for slight changes in distribution to

Table 7-7 : Description of ascribed house status's.

house status	description
1	top quality large house, strongly built with large enclosure and outbuildings
2	large house with small enclosure and minor outbuildings
3	house with less well built walls, no enclosure or outbuildings
4	flimsy house, but retaining some of the standard layout of an Amarna house
5	flimsily built, with little or no ground plan, often a "lean-to" to other houses

Table 7-8 : Objects and faience shown according to the status of the house in which it was found.

property status [1]	houses [2]	objects [3]	faience [4]	objects per house [5]	%faience [6]
1	12	600	157	50	0.26
2	16	485	121	30.3	0.25
3	61	884	283	14.5	0.32
4	111	748	257	6.7	0.34
5	186	1093	354	5.9	0.32

[1] after Table 7-7, [2] number of properties included, explained below, [3] total number of objects listed in the houses, [3] total items of faience listed in the houses, [4] column 3 divided by column 2, [5] column 4 divided by column 3,

Table 7-9 : Faience shown according to the status of the house in which it was found. The columns to the left list the types as a percentage of the total faience, those to the right as a percentage of the faience jewellery.

		% of total faience			% of total jewellery		
status [1]	no. faience	"blue" [2]	non-blue [3]	jewellery	rings	amulets	other
1	157	0.83	0.11	0.62	0.52	0.28	0.20
2	121	0.85	0.11	0.56	0.54	0.20	0.26
3	283	0.79	0.18	0.62	0.37	0.38	0.25
4	257	0.72	0.23	0.51	0.46	0.38	0.16
5	354	0.77	0.19	0.53	0.47	0.35	0.18

[1] after Table 7-7, [2] including blue, blue-green and green, [3] excluding blue, blue-green, green and polychrome

the jewellery objects and the colours. Selected results from this survey are shown in Table 7-8 and Table 7-9.

From the results, it is clear that the higher the status of the property, the larger the number of objects found within it, from 50 objects in the average status 1 property, down to an average of 5.9 objects the flimsy status 5 houses. The percentage of faience in the objects is slightly lower in the high status properties than in the lower status, but this pattern is weak and should not be relied upon and overall the percentage of faience in the various house groups looks quite similar at between 25% and 35% of the objects found.

Most of the results showed that there was no relationship between type or colour of object and the status of the house. In fact in most cases, so far as can be judged from the small numbers of finds in some categories, the distribution is fairly uniform between the different house statuses. Despite Patch's suggestion that there is more faience jewellery in small houses, the pattern shown by the study of all the data tentatively suggests the opposite, with a more of the faience being jewellery in high status houses (62% in status 1 and 3, and 56% in status 2), than low status houses (51% in status 4 and 53% in status 5), but the pattern is far from marked. The distribution within the jewellery shows a slight preference for rings in high status houses and for amulets in low, but again there is no clear distinction. In terms of colour, the copper and cobalt coloured blues (classified "blue", "blue-green" and "green" in Borchardt and Ricke (1980)), are slightly more common in the higher status houses, and the more unusual colours (red, white, violet and yellow) more common in lower status properties. This is perhaps the opposite of what would be expected if the colours were difficult to produce and therefore rare and in high demand, and perhaps reflects that these colours are used in imitation of other more valuable materials (yellow for gold, violet for amethyst, for example), which was more available to those of high status, so they used less of the unusually coloured faience and therefore a higher proportion of the faience found there is blue.

The outcome of this study of status and faience is that there appears to be very little variation in its use across the social spectrum, at least in terms of what items are used and in what proportion. It seems reasonable to expect therefore that the bulk of the faience that is preserved at Amarna, including all the most common types of jewellery, scarabs, vessels and inlays, were not particularly valuable or prestigious. It seems likely that as discussed before, the owners have comprehensively removed any thing of great value, perhaps even the broken examples, and what is left is almost entirely the more inferior and less desirable pieces of faience. The fact that so many of the items left are inscribed for Akhenaten or otherwise tainted by the "Amarna heresy" as it came to be regarded, soon after the death of Akhenaten himself, may not be entirely coincidental.

SUMMARY OF THE ARCHAEOLOGY OF PRODUCTION

The ancient city of Amarna was laid out on the East Bank of the Nile on a plot marked by boundary stelae that included both the City itself and the farmland surrounding it, estimates of the carrying capacity of which, and other methods, have put the city population at 20,000 to 50,000. The City can be divided into areas, most of the religious and administrative building lying in the Central City, while the major areas of habitation lie in the Main City to the south of this and in outlying suburbs to the north and south. The principal areas of interest to this study are the Main City and the North Suburb, where major areas of habitation coincide with areas of manufacturing.

The presence of moulds led to excavations that revealed the site of three kilns in O45.1, which were subsequently excavated and are thought to be for the manufacture of glass and glazed material. A survey of the distribution of clay moulds and other manufacturing debris has revealed the possible presence of six more sites where manufacturing may have taken place. At two of these (M50/N50 and P46), there are a number of kilns associated with moulds and glass and faience debris, while at a third (Q46.5) there is also tentative evidence for the working of precious metals. Two further sites have moulds and faience without the presence of large amounts of glass and are associated with sculptors' workshops. The sixth site in the North Suburb is poorly preserved but again suggests a link between glass and faience making.

The houses surrounding those containing kilns also have high numbers of faience moulds and it is hypothesised, that in these areas, the workers operating the kilns lived and perhaps did preliminary preparation and finishing of objects. These houses are almost entirely of low class and poorly constructed, accommodation being limited to a small number of rooms and suggesting that workers here were of low status. The one exception to this pattern is P47.22, which is larger, better built and set in its own grounds and might possibly be the residence of a higher ranking craftsman. The actual ranks of the faience workers can be traced from inscriptions on stelae and other objects from other New Kingdom sites, showing that there were at least three ranks involved; faience worker, chief faience worker and overseer of faience workers, the last of which would probably have been a member of the ruling administrative élite. There is no textual evidence for these ranks at Amarna, but their equivalent in other crafts is known, which suggests that they are applicable.

Two methods for the distribution of faience and glass from the factories are envisaged. The first is central commissioning and control followed by redistribution which is classically regarded as the dominant force in Egyptian economics and the second is production and sale in the private sector, which from the Deir el Medina records is perhaps more important than had first been thought. A combination of these two methods led to a wide distribution of glass and faience across the site, with jewellery making up over half the finds both of glass and faience, with smaller numbers of inlays, tiles, vessels and figurines along with occasional examples of gaming pieces, boomerangs and other rarities. Blue faience is the most common, followed by polychrome, with lesser amounts of other colours, whereas in glass, polychrome appears almost equally abundant as blue, the abundance of polychrome glass vessels accounting for the high proportion of polychrome glass. From the number that were found in the Palace Dumps, it has been proposed that such vessels are high status objects. The presence of faience seems to be virtually independent of the status of the house, the larger houses having more objects, but the relative proportions of different types and colours of faience not varying greatly between the houses of very different size and therefore status. This suggests that the great majority of faience objects were not of high status and were available to, and used by, all ranks of society, with high status individuals having more objects, but of the same type. It is clear that rare high status faience would have existed, but this may have been restricted to royal use and, along with any in private hands would have been removed on the abandonment of Amarna, making the remaining fragments so rare as to make their occasional finds on the site impossible to interpret in terms of the status of an individual house.

Chapter 8: Conclusions

INTRODUCTION

This first part of this chapter draws together the results and conclusions from each of the previous chapters as derived from using the techniques described in Chapter 2. It starts with a brief review of the vitreous materials themselves (Chapter 3) and then goes on to discuss the raw materials (Chapter 5) and production methodologies (Chapters 4 and 6) that are interpreted as having been used in their manufacture. The archaeology and organisation of the workshops (Chapter 7) is then summarised, along with an account of the distribution of the finished products of these workshops.

The second part of the chapter examines the wider implication of the results, especially as they relate to the history of vitreous materials. The significance of the different types and colours of glasses that have been examined for establishing the source of the earliest glass in Egypt is discussed first. This discussion draws on comparative material from tomb paintings and texts that was outlined in Chapter 1 to shed light on the relationship between Egyptian glass and Mesopotamian glasses in the New Kingdom. Secondly, the relationship between the various innovative materials (*i.e.* glass, vitreous faience and the new colorants) that appear in the New Kingdom is covered, with an emphasis on how the discovery of each of these materials impacts on the others, both in terms of order of discovery and their causal relationships.

THE VITREOUS MATERIALS

Samples of glass, frit, faience and ceramics derived from museum and excavation specimens were used as the basis for the analysis in this work. Of particular importance was the investigation of the significance of the frits, consisting of quartz grains cemented usually by a cobalt coloured glass matrix, that have never been found used alone in a finished object and are therefore thought to have been an intermediate stage in the production of some other vitreous material (pages 19-23). Also of importance was the investigation of the faience, particularly that classified as Variant D since it is an 18th Dynasty innovation. Variant D faience has a hard vitreous core that is coloured distinctly blue. Analysis of this Variant showed that it fell into two distinct groups, one of which had a core coloured only by cobalt and the other that had both cobalt and copper colorants in the core. Both groups were covered by a glaze that contained both cobalt and copper (page 25).

Raw materials

To produce the finished products, the vitreous materials workshops at Amarna were obtaining raw materials from a wide range of areas both local to Amarna and distant (Chapter 5). All the vitreous materials produced required silica, some form of alkali and a colorant. Sources of silica in the form of sand and quartz pebbles are freely available in and around Amarna and there is no evidence to suggest that these materials were imported. The most likely source of silica for the glasses and for some of the faience is ground quartz pebbles, but it is probable that sand was used for materials where impurities are not critical, for example the bodies of faience tiles (page 44). The source of alkali for the faience is not clear, but the glass seems to divide into two types, that made from natron based alkali, which includes most of the cobalt coloured examples, and that made from plant ash alkali, which includes all of the copper coloured examples and one that was cobalt coloured. Plant ash could have been obtained locally from the desert plants growing around Amarna, although Nicholson (pers comm. 1998) has failed to discover any suitable plants in recent searches that he has conducted. The plant ash could also have been transported in from some distance. Natron was readily available from the Wadi Natrun natron from this source being known to have been exploited for use in medicines and mummification during the New Kingdom. This would require transport over some distance to bring it to Amarna, the most probable route being up the Nile by boat.

By far the most interesting of the raw materials are the colorants which, while they are present only in minor concentrations, generally are not available locally at Amarna and their presence in the glass and faience can suggest a long and involved procurement and transport process. The most common colorant in all the vitreous materials is copper (pages 46-7). Analysis of the glasses suggests that this was introduced either as copper (or a copper compound), or as bronze. Those glasses with low concentrations of copper and tin could be accounted for by accidental contamination of the molten glass by bronze implements, perhaps during stirring of the melt. However, other glasses have levels of copper and tin in them that are too high to be accounted for by contamination and suggest its deliberate addition, perhaps in the form of scrap bronze from the metalworking areas. The glasses with the highest copper contents, but no cobalt, possibly have either copper or both copper and bronze added and stand apart from the other glasses in this respect. The second most common colorant in the glass and faience at Amarna is cobalt (page 47). Where cobalt occurs in New Kingdom Egyptian vitreous materials, its presence coincides with elevated levels of alumina, magnesium, manganese, nickel and zinc. This suite of elements is similar both in terms of their presence and the ratios of one to another to the cobalt bearing alums of the Kharga Oasis in the Western Desert of Egypt and suggests that this was the source of the colorant.

The yellow glass and faience is coloured by lead antimonate and the source of the lead in this colorant has been the subject of some debate (pages 50-2). The lead isotopic study of the Amarna yellow glass and faience show that in a plot of the lead isotopic ratios, the majority of the samples fall in a field with lead isotope ratios within or slightly higher than, but still close to, that for lead from Gebel Zeit on the Red Sea coast of Egypt. However, a single sample (a glass rod) has different lead isotopic characteristics, its lead isotopic ratios suggesting that its lead had a Mesopotamian source. Therefore, it seems that, at the time of Amarna there were perhaps two sources of lead used for the vitreous materials, the most common being probably Gebel Zeit and a rarer Mesopotamian source which is as yet unidentified. The higher ratios of some of the samples that plot close to the field for Gebel Zeit may be due to some mixing in of lead with Mesopotamian characteristics.

The fact that non-local colorants were extensively used in the manufacture of vitreous materials at Amarna argues that the officials who controlled the industry were able to organise a long distance transport for specialist colorants, especially for the cobalt. This argues for the idea that a significant proportion of the industry worked for, and was controlled by, the state rather than operating privately, since only the state would have been capable of controlling such long trade routes.

Production methodologies

Petrie's interpretation, that the frits found at Amarna were an intermediate stage in the production of glass, has been widely accepted and repeated by subsequent authors (Turner 1956a, Tait 1991). However, close examination of the microstructure of the frits and the composition of the glass phase show that the frits are too rich in silica and colorants and low in alkalis and lime to produce a glass if they were completely melted (pages 54-5). While the removal of all the remaining quartz grains does make the resulting composition closer to that of the glasses, it is still significantly different. It is therefore probable that the frits were not an intermediate stage in the production of glass as has previously been concluded. This new conclusion is supported by temperature estimates for the formation of the frits, derived from their composition and the microstructure of the vessels associated with them. These estimates suggest that the frits were formed at 1100-1200°C, far higher than the suggested maximum of 750°C for a glass making frit. A further possible explanation for the use of the frits is that they are the "colorant carrier" by which the cobalt colorant was introduced to a colourless glass cullet. The composition of the frits is once again wrong for this interpretation to be valid, the frits being too low in cobalt and too high once again in silica. This information, added to the lack of finds of colourless glasses, argues against the hypothesis that the frits are used only to add colour to the glasses. Experimental reproduction of glasses in both the laboratory and the field suggests that it is possible to produce a good glass straight from the raw materials in a one stage process, without the necessity of a fritting stage. It is also possible that if a frit was produced at a temperature of 750°C, as suggested by Turner (1954), the resulting frit may be friable and therefore would not be well preserved in the archaeological record. Therefore a fritting stage may have been used for the production of glass, but the frit has not survived. Whether there was a fritting stage or whether glass was produced in a one stage process, it is clear that the frits in this study are not an intermediate stage in the production of glass.

Since there are no finished objects consisting of these frits, they must be involved in the production of some other vitreous material. It has been suggested that glass was mixed with crushed quartz or quartz sand to produce the vitreous cores of Variant D faience (Vandiver 1983), but analysis of the Variant D faience (page 55) has shown that glass has the wrong composition, particularly in terms of lime and colorants, to be used to create the body of the Variant D faience. The glass matrix of the faience is, in fact, much closer in composition to that of the frits and in some cases the glass matrix of each material is close to being compositionally identical. If frit with the average bulk composition is mixed with a further 35% by weight of ground quartz, a composition that is very close to that of the bulk composition of the Variant D cores is achieved. Further, with the occasional use of copper coloured frits, the copper in the core of some of the samples is readily accounted for. It is likely that only one technique is used for glazing the cores, the use of two different glazing techniques (one for the cobalt coloured cores and one for the cobalt and copper coloured cores) being ruled out because of the great similarities in the chemistry of all of the Variant D samples. In considering which of the three glazing techniques (i.e. cementation, efflorescence and application) has been used, the fact that many of the pieces are inlaid polychrome rules out the possibility of using cementation for all the Variant D. In addition to this, most of the cores have abundant glass containing significant amounts of alkalis, but some of these have very low copper contents. This argues against efflorescence, which would leave at least some copper in the abundant glass of the core. This lack of copper in the core of some of the pieces suggests glazing by application and this interpretation is supported by drips and runs in the glaze on some of the samples. Therefore, overall the most likely method of glazing for all the samples is application using a prefritted glaze.

It is therefore likely that the frits were being used to increase the amount of glass in the cores of this sort of faience rather than ground glass itself, as has been previously suggested. Occasional mixing of copper coloured frits into the cores would then result in a copper and cobalt colorant mix in some of the Variant D, which is the pattern observed in the ancient examples. The bulk composition of the faience contains more silica than the bulk frit, which suggests that additional powdered quartz was added before the body of the Variant D was formed. The most likely glazing method is application as indicated by the presence of surface marks in a number of the pieces.

VITREOUS MATERIALS WORKSHOPS

Petrie's excavations at Amarna in 1893 found some of the earliest evidence for the manufacture of glass and faience. He uncovered what he described as "the sites of three or four glass factories, and two large glazing works... though the actual workrooms had almost vanished" (Petrie 1894). These sites were identified from their refuse tips, which contained abundant clay moulds for faience, glass rods and ceramics associated with manufacturing of these materials. Regrettably, Petrie neglected to record where these workrooms were on his map of the site, just applying the label "moulds" to one area. Recent ongoing excavations by Paul Nicholson and the Egypt Exploration Society in this area, now given the reference O45.1, have uncovered kilns and kiln debris that appear to be from both a pottery workshop and a faience and glass workshop which either existed together or followed one another closely (Nicholson 1996). A study of the distribution of the clay moulds for faience and other debris excavated by the post-Petrie excavations (pages 63-68) has identified sites of the other possible manufacturing areas, some of which may be the "factories" mentioned by Petrie. Most of these sites are scattered throughout the Main City, although one is in the North Suburb (page 69). Examination of the finds from these sites show that in some cases the faience manufacturing could have been intimately linked with the manufacture of other objects and materials. Two of the sites (N50/M50 and P46) reveal the manufacture of faience and glass alone, having moulds associated with high numbers of faience fragments, glass rods, working pieces of glass and cylindrical vessels with glass adhering to them. In these two sites glass and faience manufacture appear to have been going on side by side, possibly carried out by the same workers.

Q46.5 seems to show a different pattern. In this house, a small kiln is associated with clay moulds for the manufacture of faience jewellery and a further large limestone mould in the shape of the vulture goddess Nekhbet, for the shaping of gold sheet to create an inlaid gold pectoral pendant. The excavators interpreted this as the remains of a jewellery workshop, where faience is combined with precious metals for the manufacture of a range of prestige goods. A combination of materials is also seen in the sites of P47.1-3 and O47, where faience moulds , but no glass manufacturing debris, were found in sculptor's workshops, which were producing composite statues and other items in stone and plaster, many of which were inlaid with faience.

Organisation of the workshops

The evidence of the workshops therefore points towards very close liaison between industries for the manufacture of pottery, statuary, jewellery, faience and glass. It also suggests that the modern division of these industries into groups by the materials that they use (metalworkers, potters, glassmakers and so forth) may not have been how these industries were viewed in the New Kingdom. Instead, the observed distribution pattern suggests that the industries were not organised by materials, but by finished products. Therefore, instead of metalworkers, potters and glassmakers we have workshops for jewellers, vessel makers, statue makers and so on, each of which contained artisans "in house" who were capable of working in all the various materials necessary to produce their finished product. This pattern is quite different from the modern western organisation of industries where the subcontracting out for component parts is considered the most efficient.

The size of the different workshops can give some indication of the overall organisation of the industries and the arguments concerning the degree of state control in the manufacturing. Amarna has several examples of large bakery complexes, consisting of many copies of the conventional kitchen of a house built next to each other. It is by this process of the duplication (*i.e.* many times the small kitchen) that the problem of producing the required amount of bread is overcome. Such complexes were almost certainly state controlled and it can be argued that even if (as seems likely) some bread was produced privately in the kitchens of individual houses, a significant proportion of the bread must have been produced under state control in these large, purpose-built bakeries. The multiple duplication of the small kitchen layout once again contradicts the modern ideas of efficient production, which would solve the same problem of high volume production by the building of a few larger ovens with a greater capacity. This in turn would employ fewer people and hence, to the modern eye, be more efficient and preferable.

In the case of the production of vitreous materials, the workshops are scattered throughout the city and none of them appear to be particularly large. The production of vitreous materials therefore does not seem to obviously follow the state organised pattern that is seen in bread making, perhaps suggesting less rigid control of this industry by the state.

Further information concerning the degree of state control in New Kingdom manufacturing industries can be drawn from the records of Deir el-Medina concerning the organisation of draughtsmen working on the royal tombs in the Valley of the Kings. These records indicate that private contracts were undertaken alongside the official work and quite a large proportion of a worker's income could come from this private source. It is probable that a similar system was in operation in the vitreous materials industries of Amarna, where objects for royal consumption may have been produced alongside those for private exchange. The P47.1-3 sculptor's workshop has a number of moulds for rings which would not have been of any use in the production of statuary, but may indicate the use of facilities for private production, which would then be exchanged on the open market. The records of Deir el-Medina make it clear that this private production and sale went on with the knowledge, and hence presumably the agreement, of high ranking officials who controlled all aspects of the production of the royal tombs.

A few examples of the names, ranks and titles of officials concerned in the manufacturing industries of Amarna have survived, enough to show that the ranks at Amarna followed the standard New Kingdom pattern which was common to all trades and professions. The top rank was overseer, *imy-r*, an administrator who was part of the ruling elite. Below this rank were the *ḥryw* for which the translation is usually "chiefs" and below *ḥryw* were the ordinary workers. The titles of faience makers that are known, dating from other 18th Dynasty contexts, follow a similar line to those attested for other crafts, *imy-r irw ḥsbḏ*, for "Overseer of faience workers", *ḥry irw ḥsbḏ* for "Chief of faience workers" and *irw ḥsbḏ* for "faience worker", suggesting again a three layer organisation. A Chief faience maker would probably have had similar hands-on technical and personnel supervisory roles as that seen in other equivalent ranks (for example the draughtsmen at Deir el-Medina) and it is possible that one official of this rank (or maybe more in the case of large facilities) would control each proposed workshop site, with perhaps one administrative overseer of faience workers controlling the whole industry at Amarna.

The distribution of products from the workshops

Textual and archaeological information suggests that two methods for the distribution of faience and glass from the factories were present at Amarna (pages 72-5). The first is central commissioning and control followed by redistribution which is classically regarded as the dominant force in Egyptian economics and appears significant in the production of vital commodities, for example bread. The second is production and sale in the private sector, which from the Deir el Medina records is perhaps more important than had first been thought. A study of the typology of the vitreous materials suggests that a combination of these two methods resulted in the observed wide distribution of glass and faience across the city, with jewellery making up over half the finds both of glass and faience and with smaller number of inlays, tiles, vessels and figurines as well as occasional examples of gaming pieces, boomerangs and other rarities. Blue faience is the most common, followed by polychrome, with lesser amounts of other colours. In glass, polychrome appears almost equally abundant as blue, the fact that most glass vessels are polychrome increasing the proportion of polychrome glass.

Examining the finds in a large number of different houses has shown that the presence of faience seems to be virtually independent of the status of the house (pages 75-6), the larger houses having more objects, but the relative proportions of different types and colours of faience not varying greatly between the houses of very different size and therefore status. This suggests that the great majority of faience objects were not of high status and were available to, and used by, all ranks of society, with high status individuals having more objects, but of the same type. In most cases, glass is too rare to make any possible link to status, but an exception to this is the large number of glass vessels that were found in the Palace Dumps which suggest that such vessels were high status objects.

IMPLICATIONS FOR THE EARLIEST GLASS

Glass making at Amarna

From Petrie's first report on Amarna (Petrie 1894) right up to modern textbooks on glass (Newton and Davison 1989, Tait 1991 and others), Amarna has been cited as one of the earliest sites where there is evidence for glass making. This interpretation has largely been based on the fact that Petrie thought the frits that he found at Amarna were an intermediate stage in glass making. Unless the frits were imported, which seems unlikely, they would provide strong evidence for the actual manufacture of glass on site. However, the conclusions of this work indicate that it is very unlikely that the frits do have this intermediate place in glass making and instead are more likely to be involved in the manufacture of Variant D faience. Given this new interpretation, it is necessary to review what evidence is left to indicate that Amarna was a centre for glass making and whether instead only glass working was undertaken at Amarna.

It is clear from the evidence of glass rods, half finished working pieces of glass and glass that bears the marks of pincers and other tools, that glass working was being carried out at Amarna and from the distributions of these objects, it was almost certainly occurring in more than one area of the city. The cylindrical vessels with glass adhering to them could also be evidence for glass working, although they perhaps may alternatively, or additionally, have been used for glass making as well. Nicholson (1996) suggests that they are also used as moulds for the shaping of ingots (section 4.2.4), which also seems likely and may represent a third use for these vessels. With the frits involved in other materials, there appears to be little direct evidence for the manufacture of glass at Amarna, but in order to consider the likelihood of glass making, the question of how much evidence would be expected to survive from this process into the archaeological record must be addressed.

Firstly, if fritting was carried out at only moderate temperatures around 750°C (as suggested by Turner 1956c), then the resulting frit would be very friable and would not preserve into the archaeological record at all easily. The lack of its presence therefore does not indicate that there was no glass making taking place. Secondly, there is good evidence, in the form of fused clay and vitrified kiln wall linings, that some high temperature process requiring temperatures of 1100-1200°C was taking place. This could be associated with the production of Variant D faience, but the new interpretation of the methodology of glass making (section 6.3.4) suggests that there is not actually a need for a fritting stage, and that it is possible to obtain a good clear glass by a one stage process in a replicate kiln directly from the raw materials (Nicholson 1998). However, although there is no direct incontrovertible evidence for glass making at Amarna, evidence is at least such that glassmaking certainly could have been taking place in the city.

In summary, therefore, although there is little evidence preserved at Amarna that directly indicates that glass making was being conducted on the site, the current theories for how glass was manufactured in the Egyptian New Kingdom (section 6.3.4) would suggest that even if a fritting stage was used, little evidence is likely to be preserved. Therefore, the lack of evidence does not indicate that there was no glass making taking place.

Glass making in Egypt

The conventional theory for the origins of early glass is that glass making began in Mesopotamia (Moorey 1994) and spread to Egypt at around the time of Tuthmosis III, perhaps in connection with tributes paid by the conquered peoples of that area after Tuthmosis's successful campaigns against the Mitanni (Nicholson 1995a). It is assumed that the glass makers themselves were soon brought into Egypt, so that at least by the time of Amenophis III, the local production of glass was established at Amenophis III's palace at Malkata (in western Thebes). However, given that there is no direct evidence for glass making at Amarna and that the finds from Malkata have largely been lost and therefore not subjected to modern examination (Keller 1983), it becomes necessary to look once again at the evidence of glass making in Egypt.

The primary evidence to suggest that glass was made in Egypt is that Egyptian raw materials were used in its production. The cobalt coloured glasses from Amarna examined (pages15-19 and 47-9) have a cobalt colorant that can be sourced to the Kharga Oasis in the Western Desert and a natron based alkali the most likely source of which was the Wadi Natrun in Lower Egypt. The lead isotope study suggests that the lead in yellow glass and faience may have

Table 8-1 : Average compositions of glasses from Egypt and Nuzi

Type	SiO_2	Al_2O_3	CaO	MgO	Na_2O	K_2O	CoO	CuO	MnO
cobalt blue									
Amarna	64.1	2.2	7.9	3.6	19.2	1.2	0.16	0.17	0.31
Tuthmosis III [1]	62.2	3.4	7.8	4.2	17.8	1.0	0.19	0.35	0.30
copper blue									
Amarna	65.9	1.1	8.5	3.8	15.2	2.7	0.02	1.54	0.04
Tuthmosis III [1]	63.5	1.0	7.0	4.5	16.7	3.5	0.00	0.85	0.04
Nuzi [1]	68.4	0.7	7.7	4.5	15.5	2.8	na	na	na

[1] *Lilyquist and Brill (1995)*

come mostly from Gebel Zeit on the Red Sea Coast. It seems unlikely that these raw materials would have been exported from Egypt for glass making and the raw glass subsequently re-imported for working on sites like Amarna and Malkata. A more reasonable explanation would seem to be that these local raw materials were being used to produce glass within Egypt and that the procedure used has left no incontrovertible trace, which fits in well with the current theories for the method of glass production discussed above. An alternative solution, but less likely, is that the sites where glass was being made have all been destroyed or have not yet been excavated.

In contrast to the cobalt blue glass, the copper blue glass from Amarna has the higher potash content and the high magnesia contents associated with the use of a plant ash alkali (Table 8-1). This same pattern of potash and magnesia compositions can be seen in the early Pre-Amarna glasses reported in Lilyquist and Brill (1995) and also summarised in Table 8-1. In glasses confidently dated to the reign of Tuthmosis III, three glasses containing cobalt average a K_2O content of 1.0% and a MgO content of 4.2% and the presence of relatively high concentrations alumina, manganese, nickel and zinc suggests that cobalt from the Kharga Oasis was already being used and therefore that the magnesia may once again have a contribution from the alum. In contrast, the four Tuthmosis III glasses coloured only by copper average a K_2O content of 3.5% and a MgO content of 4.5%. This pattern therefore suggests that even during the reign of Tuthmosis III, the glasses containing a cobalt colorant also had a natron based alkali, whereas those coloured only by copper contained plant ash alkali. Why two different alkalis were used for different colours is a difficult question and may have a number of explanations.

First, there is the possibility that the procedure by which the cobalt colorant was obtained from the alum involves the use of natron. Kaczmarczyk speculates that dissolving and reprecipitating the raw cobalt bearing alum in water made alkaline by the presence of natron may be a way of concentrating the cobalt colorant. However, he has not done this experimentally and the results from the present work (section 5.3.3.2) seem to indicate that the alum does not need to be concentrated in any way to give the level of colorants seen in the glasses. Second and more likely is that the different colours and alkalis are a response to the availability of these materials to the glass makers. The two colours of glass could be from two areas with contrasting traditions, one where natron and cobalt colorants were available, and one where they were not, so plant ash alkali and copper were used in their place. The implications of this theory are that the former group of glasses was locally made, either at Amarna itself or at least in Egypt, and that the latter group was made at a site some considerable distance from Amarna, and perhaps even outside Egypt altogether. This idea is supported by a comparison of the composition of the copper coloured group with the compositions of glass from Nuzi (Table 8-1). This shows that, to all intents and purposes, all the copper blue glasses are identical in composition, suggesting that the copper coloured glass from Egypt may either have been made in Mesopotamia, or alternatively they were made in Egypt following Mesopotamian ideas.

Chapter 1 discussed the decoration of the tomb of Rekhmire (Thebes, TT100), which depicted two glass vessels being given to Tuthmosis III in tribute. The vessels depicted are quite similar to two real glass vessels found in the Wadi Qirud tomb of the Foreign Wives of Tuthmosis III and give support to the idea that there were foreign imports of glass coming into Egypt at this time. Lead isotopic analysis of one of these vessels (Lilyquist and Brill 1995) also suggests a non-Egyptian source. Evidence for glass imports later during the reigns of Amenophis III and Akhenaten comes from the diplomatic correspondence known as the "Amarna letters", also discussed in Chapter 1. These have yielded a number of mentions of *eḫlipakku* and *mekku*, materials that are thought, by reference to other contexts to be raw glass. In these letters, Amenophis III and Akhenaten are demanding raw glass from Palestine and Syria, suggesting an import of glass into Egypt even as late as the Amarna period and, as Oppenheim says, "this is in spite of the fact that all ingredients necessary to produce glass objects were abundantly available in Egypt".

It is therefore possible that in Egypt there were two contrasting traditions in the production of glass. One tradition used cobalt colorant from the Kharga Oasis, either alone or with copper, and natron from the Wadi Natrun and was being followed at least as early as Tuthmosis III, since it is incorporated into glass vessels dated stylistically and by provenance to this reign. The raw materials indicate that it is likely that this glass was being made in Egypt. A second tradition was manufacturing glass using a plant ash alkali and a copper colorant. This glass is compositionally identical to glass from Nuzi, and was also present in Egypt from the reign of Tuthmosis III. Lead isotopic studies, tomb paintings, the Amarna letters all indicate that at least some of this latter glass was being imported from Syria and Mesopotamia in the form of gifts and

Table 8-2 : Earliest examples of various innovative materials and techniques in Egypt.

innovation	when first used	reference
cobalt in vitreous materials	Tuthmosis III (1479-1425BC)	Lilyquist and Brill (1993)
cobalt on blue painted pottery	?Amenophis III (1391-1353BC)	Boston (1982)
lead antimonate	Tuthmosis III (1479-1425BC)	Lilyquist and Brill (1993)
calcium antimonate	Amenophis II (1427-1401BC)	Lilyquist and Brill (1993)
Inlaid tiles	?Amenophis III (1391-1353BC)	Boston (1982)
Variant D vitreous faience	?Amenophis III (1391-1353BC)	Kaczmarczyk and Hedges (1983)
glass	Ahmose (1550-1525BC)	Lilyquist and Brill (1993)
glass vessels	Tuthmosis III (1479-1425BC)	Lilyquist and Brill (1993)

tributes to the Pharaoh and these gifts continued at least through to the Amarna period. Alternatively, some glass of this latter type would have been made in Egypt by foreign workers, following the Mesopotamian tradition. These two traditions continue in parallel at least as long as the Amarna period and in some cases glass from both traditions were used within the same vessel. An example of this is Brooklyn 53.176.4 (Lilyquist and Brill 1995), which is a vessel of turquoise blue glass with a chemistry typical of copper colorants and plant ash alkali, decorated with yellow glass containing a lead of Mesopotamian type and a black glass that contains a cobalt colorant of Kharga Oasis composition (as well as a high copper content) and chemistry typical of natron. The likelihood for these vessels is that they were worked in Egypt from a mixture of imported and local glasses.

PRODUCTION METHODS - 18TH DYNASTY INNOVATIONS?

As discussed in Chapters 1 and 3, glass and Variant D faience first appear in the 18th dynasty, and faience tiles, while some were made sporadically in early periods, become much more abundant in the same period and show a standard of artwork and production that no previous examples even approach. Cobalt, lead antimonate and calcium antimonate colorants are also first used at the same time. It seems clear that at least some of these discoveries and materials must be in some way linked, but how they are linked and the order of their discovery are far from clear.

Difficulties in dating

Accurate dating of the first use of each of the innovations would help greatly in showing whether there might be any evolutionary relationships between them, but unfortunately the dating evidence is poor.

Table 8-2 shows the evidence that can be deduced from objects that can be dated stylistically or from their contexts. The dates for some of the materials (for example glass vessels) are accurate and reliable because they are based on the sudden appearance of quite a number of examples. However, some materials, particularly the Variant D faience has often not been recognised in the archaeological record, so the dates given in Table 8-2 are probably later than they should be, and are marked by a question mark as such. There is also a problem with the tiles and the blue painted pottery, due to many of the pieces being removed from their contexts and thus dating evidence being lost. These are therefore also marked in Table 8-2 by a question mark. Overall, the dating evidence is poor, but tentatively suggests that the glass and colorants are the earliest and that the tiles and use of cobalt on pottery later.

A possible interpretation

The earliest glass seems to have been produced in Mesopotamia and imported into Egypt, and included yellow glasses with Mesopotamian lead isotope characteristics which suggests that the first use of lead antimonate was also in Mesopotamia. If the discovery of calcium antimonate is linked to that of lead antimonate, then it is possible that calcium antimonate is also a Mesopotamian innovation. The details of the discovery of glass in Mesopotamia and the first production of glass in Egypt are obscure and lie beyond the scope of this work, but if the Mesopotamian copper coloured glass came first, then the cobalt coloured glass with Egyptian raw materials may be a local attempt to copy the Mesopotamian imports. Experimentation in glass production may then have led to the production of frits and their addition to ordinary faience to produce the Variant D with its hard, blue cores.

The rapid rise in the apparent popularity of glass and its spread during the 18th dynasty may well be linked to the discovery of the new colorants. The colour palette for glass before the discovery of the new colorants would have been restricted to those colours that could be made from the colorants applied to pre-New Kingdom faience, that is to say copper and manganese giving blue-green, black and possibly purple. With the discovery of the antimonates and cobalt blue, the possible colour palette for glass increases from just three up to seven (blue-green, dark blue, yellow, white, green, purple and black). Colour was very important to the ancient Egyptians and with bright colours being particularly prized and used extensively over statues, tombs and temples (Baines 1985). The additional possibilities that the new colorants gave to vessels in glass may well have been responsible for the rapid rise in the apparent popularity of glass and its spread during the 18th dynasty. The new colorants also allow the production of three colours (deep blue, white and a good green) which are extensively used on the faience tiles. Here, they are used for the depiction of plants, water, animals and other natural themes that are very common in wall paintings throughout Egyptian history, but not possible in faience prior to the New Kingdom because of the limited palette. It is possible therefore that the rise in the use of tiles in the New Kingdom is a response to the opening up of new artistic possibilities, echoing that proposed for glass vessels.

Appendix A: Bibliography

Aldred, C., 1988, *Akhenaten, King of Egypt*, London.

Andrews, R.W., 1962, *Cobalt*, Overseas Geological Survey, Mineral Resources Division, HMSO, London.

Arnold, Dorothea, and Bourriau, J., 1993, *An introduction to ancient Egyptian pottery*, von Zabern, Mainz.

Azer, N., 1966, Remarks on the origin of Precambrian mineral deposits in Egypt (U.A.R.), *Tschermaks mineralogische und petrographische Mitteilungen*, 3rd series, **11**, p41-64.

Baines, J., 1985, Color terminology and color classification: ancient Egyptian color terminology and polychromy, *American Anthropologist*, **87**, p282-297.

Barag, D., 1962, Mesopotamian glass vessels of the second millennium BC, *Journal of Glass Studies*, **4**.

Barag, D., 1970, Mesopotamian core-formed glass vessels (1500-500 BC), *in* Oppenheim, A.L., Brill, R.H., Barag, D. and von Saldern, A., *Glass and glass-making in ancient Mesopotamia*, New York, Corning Museum of Glass.

Beadnell, H.J.L., 1901, *Dakhla Oasis: its topography and geology*, Cairo.

Beadnell, H.J.L., 1909, *An Egyptian oasis, An account of the Kharga Oasis in the Libyan Desert, with special reference to its history, physical geography and water supply*, J. Murray, London.

Bellion, M., 1987, *Catalogue des manuscrits hiéroglyphiques et des dessins, sur papyrus, cuir, ou tissu, publiés ou signalés*, Pau.

Bianchi, R.S., 1998, Symbols and meanings, *in* Friedman, F.D. (ed), *Gifts of the Nile -Ancient Egyptian faience*, Thames and Hudson, London.

Bimson, M. and Freestone, I.C., 1988, Some Egyptian glasses dated by royal inscriptions, *Journal of Glass Studies*, **30**, p11-15.

Bimson, M., and Freestone, I.C. (eds.), 1987, *Early vitreous materials*, British Museum Occasional Paper, **56**.

Borchardt, L., 1907-14, Ausgrabungen in Tell el-Amarna, *Mitteilungen der Deutschen Orient-Gesellschaft*, **34, 46, 50, 52, 55**, Berlin.

Borchardt, L. and Ricke H., 1980, *Die Wohnhäuser in Tell el-Amarna*, Deutschen Orient-Gesellschaft, Berlin.

Boston Museum of Fine Arts, 1982, *Egypt's golden age: the art of living in the New Kingdom 1558-1085BC*, Boston.

Bourriau, J., 1996, The dolphin vase from Lisht, *in* Manuelian, P.D., *Studies in honour of William Kelly Simpson*, **1**, p101-116, Boston.

Boyce, A., 1995, Collar and necklace designs at Amarna: A preliminary study of Faience Pendants, *in* Kemp, B.J. (ed), *Amarna Reports*, ***6***, p336-71, London.

Brill, R.H., 1962, A note on the scientist's definition of glass, *Journal of Glass Studies*, **4**, p127-138.

Brill, R.H., 1970, The chemical interpretation of the texts, *in* Oppenheim, A.L., Brill, R.H., Barag, D., von Saldern, A., *Glass and glass-making in ancient Mesopotamia*, New York, Corning Museum of Glass.

Brill, R.H., Barnes, I.L. and Adams, B., 1974, Lead isotopes in some ancient Egyptian objects, *in* Bishay, A. (ed.) *Recent advances in the science and technology of materials*, **3**.

Budischovsky, M-C. and Genaille, N., 1988, Les stèles égyptiennes du Musée de Rennes (suite) Imen-Rekh, *Annales de Bretagne et des pays de l'Ouest*, **95**, no.3, p225-246.

Caley, E.R., 1962, *Analysis of ancient glass*, Corning

Campbell Thompson, R., 1925, *On the chemistry of the ancient Assyrians*, London.

Castel, G., Pouit, G. and Soukiassian, G., 1988, Les mines de galene pharaonique du Gebel Zeit (Egypte), dans le Miocène du rift de la Mer Rouge, *Chronique de la Recherche Minière*, **492**, p19-32.

Cerny, J., 1973, *A community of workmen at Thebes in the Ramesside period*, Bibliothèque d'Etude, **50**, Institut

Francais d'Archéologie Orientale, Cairo.

Chambers, A., and Rigg, J.F., 1925, Notes on antimony yellows, Part 1, *Transactions of the Ceramics Society,* **25**, p101-107.

Davies, N. de G.D., 1908, *The Rock Tombs of el Amarna 6,* Archaeological Survey of Egypt, **18**, London.

Davies, N. de G.D., 1935, *Paintings from the tomb of Rekhmire,* Publications of the Metropolitan Museum of Art Egyptian Expedition, **10**, New York.

El Goresy A., Tera, F., Schlick-Nolte, B. and Pernicka, E., 1997, Chemistry and lead isotopic compositions of glass from a Ramesside workshop at Lisht and Egyptian lead ores: A test for a genetic link and for the source of glass, *Proceedings of the seventh annual congress of Egyptologists 1997,* in press.

Farnsworth, M. and Ritchie, P.D., 1938, Spectrographic studies on ancient glass. Egyptian glass mainly of the 18th Dynasty with special reference to its cobalt content, *Technical studies in the field of fine arts,* **6** (3)

Frankfort, H. and Pendlebury, J.D.S., 1933, *The City of Akhenaten II*, Egypt Exploration Society, memoir **38**, London.

Friedman, F.D. (ed), 1998, *Gifts of the Nile - Ancient Egyptian faience*, Thames and Hudson, London.

Gaballa, G.A., 1979, False-door stelae of some Memphite personnel, *Studien zur altägyptischen Kultur* **7**, p41-52.

Giddy, L.L., 1987, *Egyptian Oases*, Aris and Phillips.

Gielmann, W., 1955, Beitnege zur Kenntnis alter Glaser III, *Glastechnische Berichte*, **28**, p146-156.

Harris, J.R. (ed.), , *The legacy of Egypt.* Oxford University Press.

Hatcher, H., Tite, M.S. and Walsh, J.N., 1995, A comparison of ICP-ES and ICP-AAS analysis on standard reference silicate materials and ceramics, *Archaeometry* **37 (1)**, p83-94.

Hume, W.F., 1937, *The Geology of Egypt,* **2**, Part 3, Egyptian Government Press, Cairo.

Janssen, J.J., 1975, *Commodity prices from the Ramesside Period: An economic study of the village of necropolis workmen at Thebes,* Leiden.

Kaczmarczyk, A., 1986, The source of cobalt in ancient Egyptian pigments, *in* Olin, J.S. and Blackman, M.J. (ed.), *Proceedings of the 24th International Archaeometry Symposium*, Smithsonian, Washington, p369-376.

Kaczmarczyk, A. and Hedges R.E.M., 1983, *Ancient Egyptian faience*, London.

Keller, C.A., 1983, Problems of dating glass industries of the Egyptian New Kingdom: Examples from Malkata and Lisht, *Journal of Glass Studies,* **25**, p19-28.

Kemp, B. J., 1989, *Ancient Egypt: Anatomy of a civilization,* Routledge.

Kemp, B. J. and Garfi, S., 1993, *A Survey of the ancient city of el Amarna,* The Egypt Exploration Society Occasional Publications, **9**, London,

Kozloff, A.P. and Bryan, B.M., 1992, *Egypt's Dazzling Sun: Amenhotep III and his World*, Cleveland Museum of Art, Cleveland.

Lakatos, T., Johansson, L-G., and Simmingskold, B., 1972, Viscosity temperature relations in the glass system SiO2-Al2O3-Na2O-K2O-CaO-MgO in the composition range of technical glasses, *Glass Technology,* **13** (3).

Lambert, J.B. and McLaughlin, C.D., 1978, Analysis of early Egyptian glass by atomic absorption and X-ray Photoelectron spectroscopy, in Carter, G.F. (ed.), *Archaeological Chemistry II*, American Chemical Society; Washington DC, Advances in Chemistry Series, **171**.

Lesko, L.H., 1994, *Pharaoh's workers, the villagers of Deir el Medina*, Cornell University Press, Ithaca and London.

Lilyquist, C., and Brill, R.H., 1995, *Studies in Ancient Egyptian Glass,* New York, Metropolitan Museum of Art.

Lucas, A. and Harris J.R., 1962, *Ancient Egyptian materials and industries*, 4th edition, London.

Luft, U., 1977, Das Totenbuch des Ptahmose, *Zeitschrift fur ägyptische Sprache und Altertumskunde,* **104**, p47-75.

Maniatis, Y. and Tite, M.S., 1979, Technological examination of Neolithic-Bronze Age pottery from central and south-east Europe and from the Near East, *Journal of Archaeological Science*, **8**, p59-76.

Moorey, P.R.S., 1994, *Ancient Mesopotamian materials and industries*, Oxford.

Newton, R.G. and Davison, S., 1989, *Conservation of glass,* Oxford.

Nicholson, P.T., 1993, *Ancient Egyptian Faience and Glass,* Shire Egyptology, London.

Nicholson, P.T., 1995a, Recent excavations at an ancient Egyptian glassworks: Tell el-Amarna 1993, *Glass Technology,* **36** (4), p125-128.

Nicholson, P.T., 1995b, Glass-making and glass-working at Amarna: some new work, *Glass Technology,* **37**, p11-19.

Nicholson, P.T., 1998, Materials and technology, *in* Friedman,

F.D. (ed), *Gifts of the Nile - Ancient Egyptian faience*, Thames and Hudson, London.

Nicholson, P.T. and Jackson C.M., 1998, "Kind of blue" - glass of the Amarna period replicated, *The prehistory and history of glassmaking technology*, The American Ceramics Society, USA.

Nicholson, P.T., in press, The place of glass technology in XVIIIth Dynasty Egypt.

Nickel, E.H. and Nichols, M.C., 1991, *Mineral reference manual*, Reihold, New York.

Noll, W., 1981, Mineralogy and technology of the painted ceramics of ancient Egypt, in *Scientific studies in ancient ceramics*, (ed. Hughes, M.), British Museum Occasional Paper, **19**, London.

Oppenheim, A.L., 1973, Towards a history of glass in the ancient Near East, *Journal of the American Oriental Society*, **93**, p259-266.

Oppenheim, A.L., Brill, R.H., Barag, D., von Saldern, A., 1970, *Glass and glass-making in ancient Mesopotamia*, New York, Corning Museum of Glass.

Patch, D.C., 1998, By necessity of design: faience use in ancient Egypt, *in* Friedman, F.D. (ed), *Gifts of the Nile - Ancient Egyptian faience*, Thames and Hudson, London.

Peet, T.E., and Woolley, C.I., 1923, *The city of Akhenaten I: excavations of 1921 and 1922 at el-Amarna*, Egypt Exploration Society, London.

Peltenberg, E.J., 1987, Early faience: recent studies, origins and relations with glass, *in* Bimson, M. and Freestone, I.C. (ed.), *Early vitreous materials*, British Museum Occasional Papers, **56**.

Pendlebury, J.D.S., 1951, *The city of Akhenaten III*, Egypt Exploration Society, memoir **44**, London.

Petrie, W.M.F., 1894, *Tell el-Amarna*. London.

Petrie, W.M.F., 1909, *The arts and crafts of ancient Egypt*. T N Foullis Ltd, London.

Redford, D., 1985, *Akhenaten - The Heretic King* Chicago.

Reed, S.J.B., 1995, Electron probe microanalysis, in *Microscope techniques in the Earth Sciences, (ed.* Potts P.J. et al), London.

Richmond, H.D. and Orff, H., 1892, Indications of a possible new element in an Egyptian mineral, *Journal of the Chemical Society*, **61**, p491-5

Riederer, J., 1974, Recently identified Egyptian pigments, *Archaeometry* **16 (1)**, p102-109

Sayre, E.V. and Smith, R.W., 1967, Some materials of glass manufacturing in antiquity, *in* Levey, M. (ed.), *Archaeological Chemistry*, Philadelphia.

Sayre, E.V. and Smith, R.W., 1974, Analytical studies in ancient Egyptian glass, *in* Bishay, A. (ed.), *Recent Advances in the Science and Technology of Materials: Vol.3*. New York.

Segnit, E.R., 1983, Evaporite minerals from the Dakhleh Oasis, *in* Edwards, W.I., Hope, C.A., and Segnit, E.R. (ed.), *Ceramics from the Dakhleh Oasis*, Victoria College, p97-102.

Stos-Gale, Z.A., and Gale, N.H., 1981, Sources of galena, lead and silver in Predynastic Egypt, Actes du XXème Symposium International d'Archéometrie, *Review d'Archéometrie*, **5**, p285-295.

Tait, H. (ed.), 1991, *Five thousand years of glass*, British Museum Press.

Tite, M.S., 1987, Characterisation of early vitreous materials, *Archaeometry*, **29** (1).

Tite, M.S., and Bimson, M., 1986, Faience: an investigation of microstructures associated with the different methods of glazing, *Archaeometry* **28 (1)**, p69-78.

Tite, M.S., Maniatis, Y., Meeks, N.D., Bimson, M. Hughes, M.J. and Leppard, S.C., 1982 Technological studies of ancient ceramics from the Near East, Aegean and Southeast Europe, *in* Wertime, T.A., and Wertime, S.F. (ed.), *Early pyrotechnology*, Smithsonian Institution Press, Washington DC, p61-71.

Tite, M.S., Freestone, I.C. and Bimson, M., 1983, Egyptian faience: an investigation of the methods of production, *Archaeometry* **25 (1)**, p17-27.

Turner, W.E.S., 1954, Studies in ancient glasses and glass making processes. Part I. Crucibles and melting temperatures employed in Ancient Egypt at about 1370 BC., *J. Soc. Glass Technology*, **38**

Turner, W.E.S., 1956a, Studies in ancient glasses and glass making processes. Part III. The chronology of glass making constituents, *J. Soc. Glass Technology*, **40,** p39-52.

Turner, W.E.S., 1956b, Studies in ancient glasses and glass making processes. Part IV. The chemical composition of ancient glasses, *J. Soc. Glass Technology*, **40,** p162-184.

Turner, W.E.S., 1956c, Studies in ancient glasses and glass making processes. Part V. Raw materials and melting processes, *J. Soc. Glass Technology*, **40**, p277-300.

Turner, W.E.S. and Rookesby H.P., 1959, A study of the opalising agents in ancient glass throughout three thousand four hundred years, Part I, *Glastechnische Berichte*, **32K**, Part VIII, special supplement, p17-28.

Turner, W.E.S., 1959, Studies in ancient glasses and glass making processes. Part VI. The composition and physical characteristics of the glasses of the Portland vase, *J. Soc. Glass Technology*, **43**.

Vandiver, P., 1982. Technological change in Egyptian faience, in *Archaeological ceramics* (eds. Olin, J.S. and Franklin, A.D.) Smithsonian, Washington, p167-179.

Vandiver, P., 1983a, Egyptian faience technology, Appendix A in *Ancient Egyptian Faience* (Kaczmarczyk, A. and Hedges, R.E.M), Warminster, Aris and Philips.

Vandiver, P., 1983b, Glass technology at the mid-second-millennium BC Hurrian site of Nuzi, *Journal of Glass Studies,* **25**, p239-247.

Vandiver, P., 1998, A review and the proposal of new criteria for the production technologies of Egyptian faience, *in* Colinart, S. and Menu, M. (ed.), *La couleur dans la peinture at l'émaillage de L'Égypte ancienne,* Centro Universitario Europeo, p121-142.

Vandiver, P., Swann, C., Cranmer, D., 1991, A review of mid-second millennium BC. Egyptian glass technology at Tell el-Amarna, *in* Vandiver, P.B., Drusik, J., and Wheeler, G.S. (ed.), *Materials Research Society Symposium Proceedings,* **185.**

Weatherhead, F. and Buckley, A., 1989, Artists' pigments from Amarna, *in* Kemp, B.J. (ed), *Amarna Reports*, **5**, London.

Young, S.M.M., Budd, P., Haggerty, R. and Pollard, M., 1997, Inductively coupled plasma mass spectrometry for the analysis of ancient metals, *Archaeometry* **39 (2)**, p379-392.

Appendix B: Figures

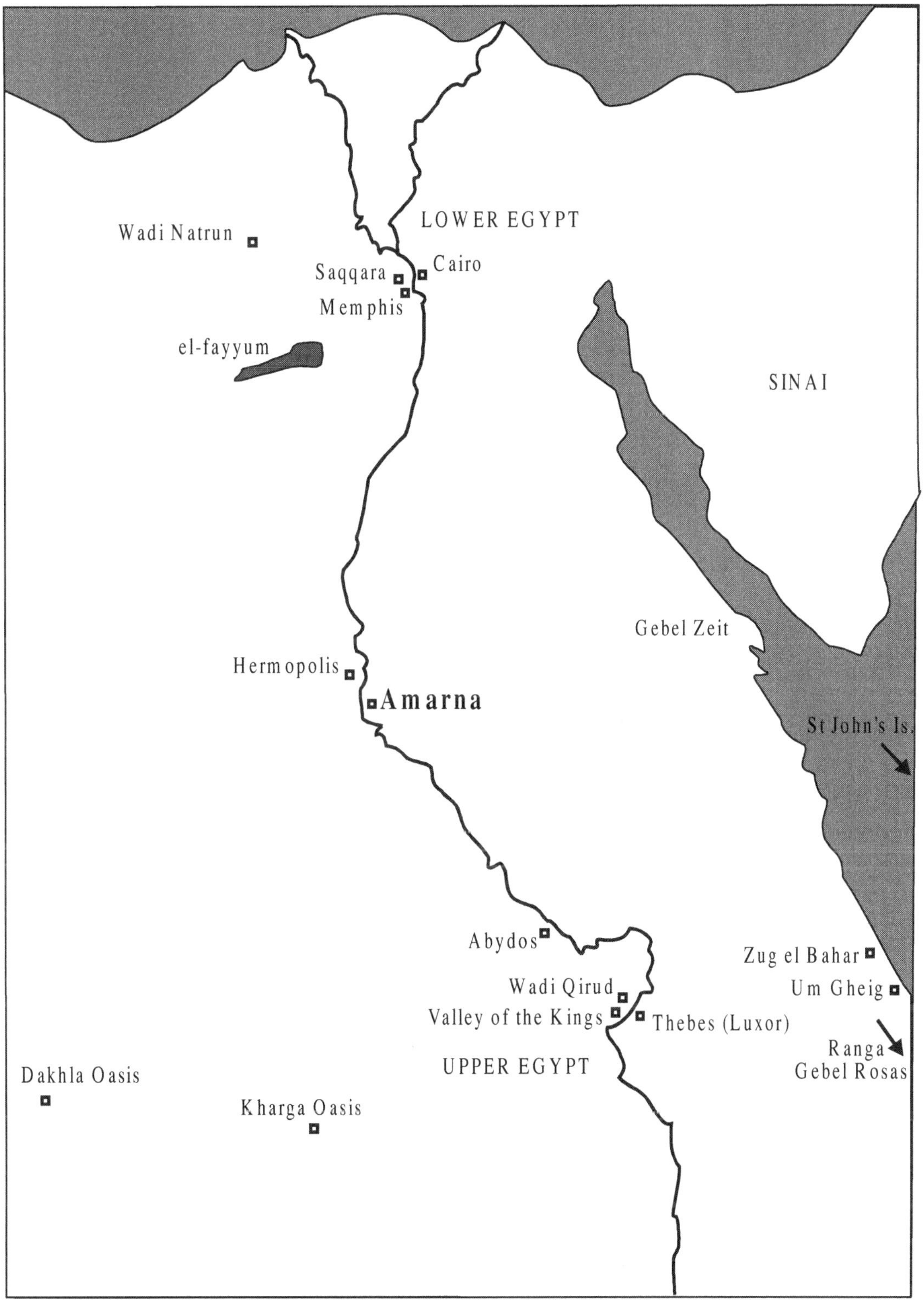

***Figure 1-1 : Map of Egypt** showing the various localities mentioned in the text of the thesis. Ranga and Gebel Rosas are on the Red Sea coast slightly south of Um Ghieg, St John's Island is in the Red Sea close to the Egyptian coast.*

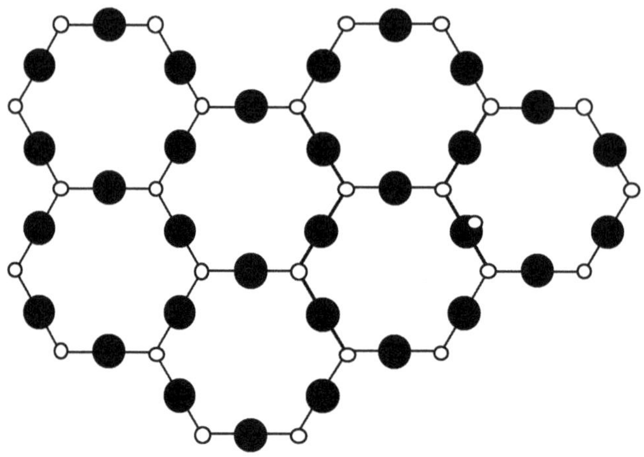

(a)

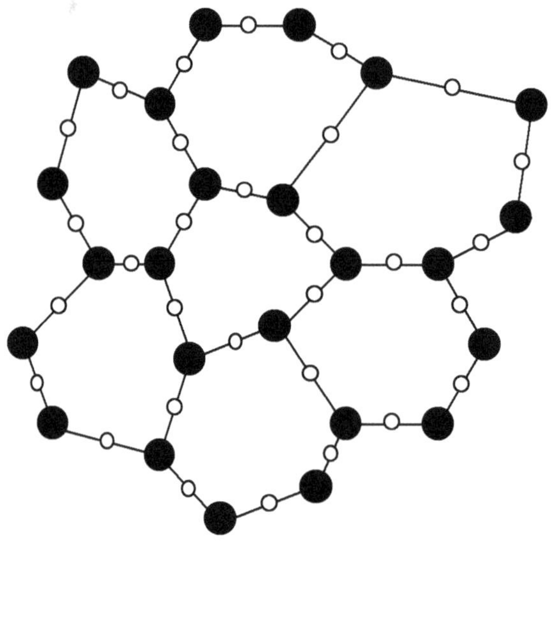

(b)

Figure 1-2 : Structures of crystal and glass. (a) The regular ordered structure of perfect crystalline quartz, (b) the disordered structure of a pure silica glass (adapted from Newton and Davison 1989, figures 1.5 and 1.6).

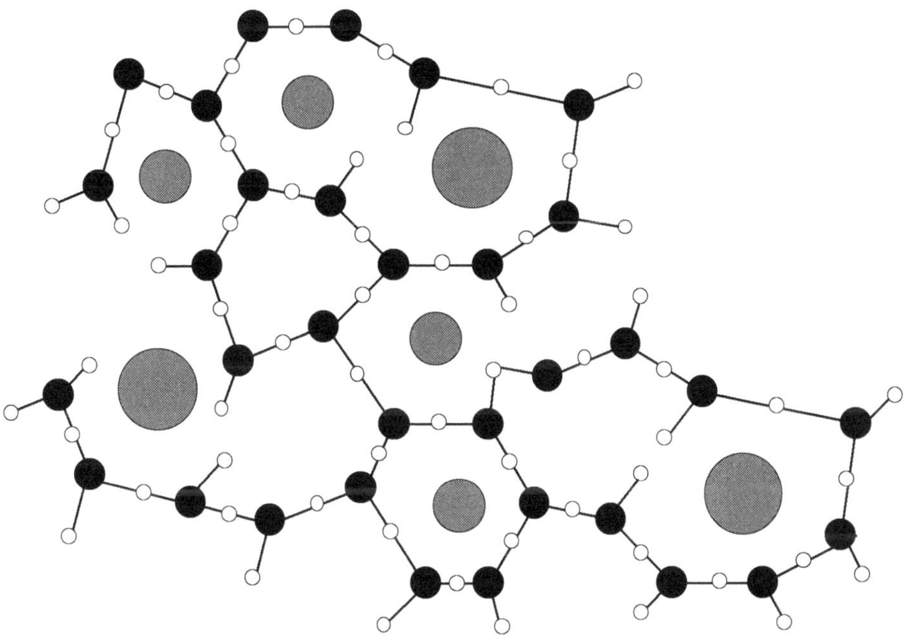

Figure 1-3 : Structure of a soda-lime-silica glass, *showing the network building silica tetrahedra forming chains and the network modifying cations in the spaces between them.* (adapted from Newton and Davison 1989, figure 1.7).

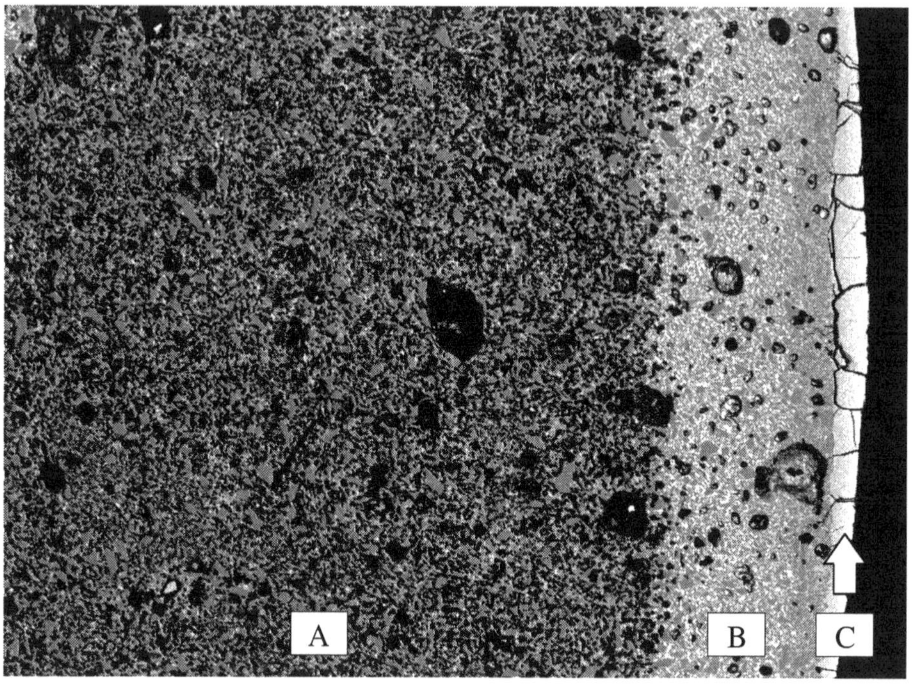

Figure 1-4 : Typical faience structure. *An SEM photomicrograph of a section through a faience fragment showing the core (A), interaction zones (B) and glaze (C). Image is approximately 200mm across.*

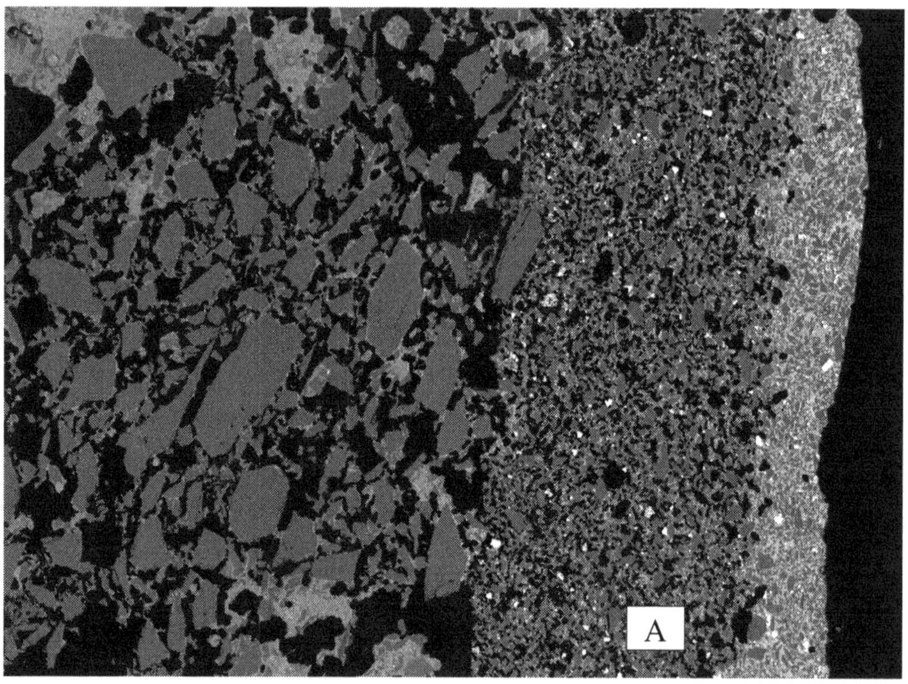

Figure 1-5: Variant A structure in a faience tile, *showing a Variant A slip like layer (A) applied to the top of the core.*

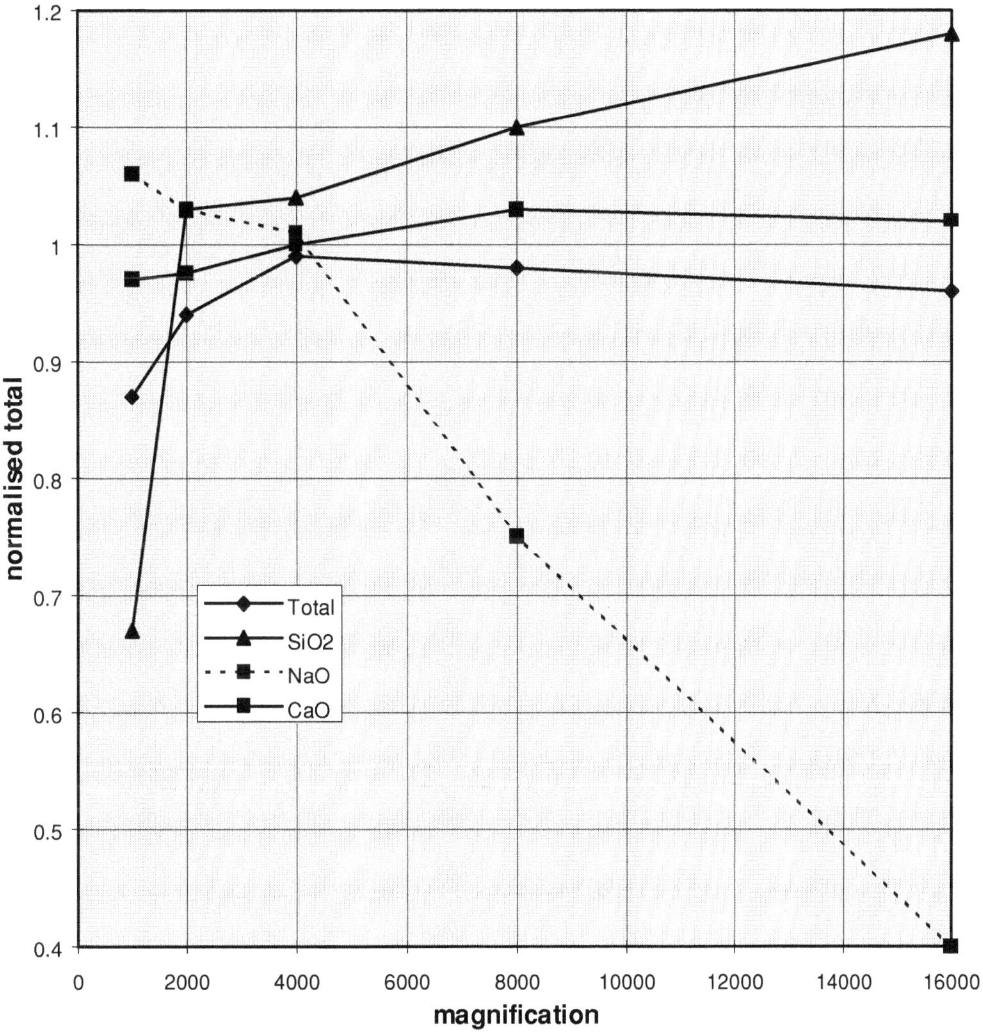

Figure 2-1 : The effect of magnification on WDS totals. *The ratio of the analysed percentage of each element at different magnifications to that published analysis for the Corning A standard (0.8 on the above graph therefore equates to an analysis that has only 80% of the published amount for the particular element*

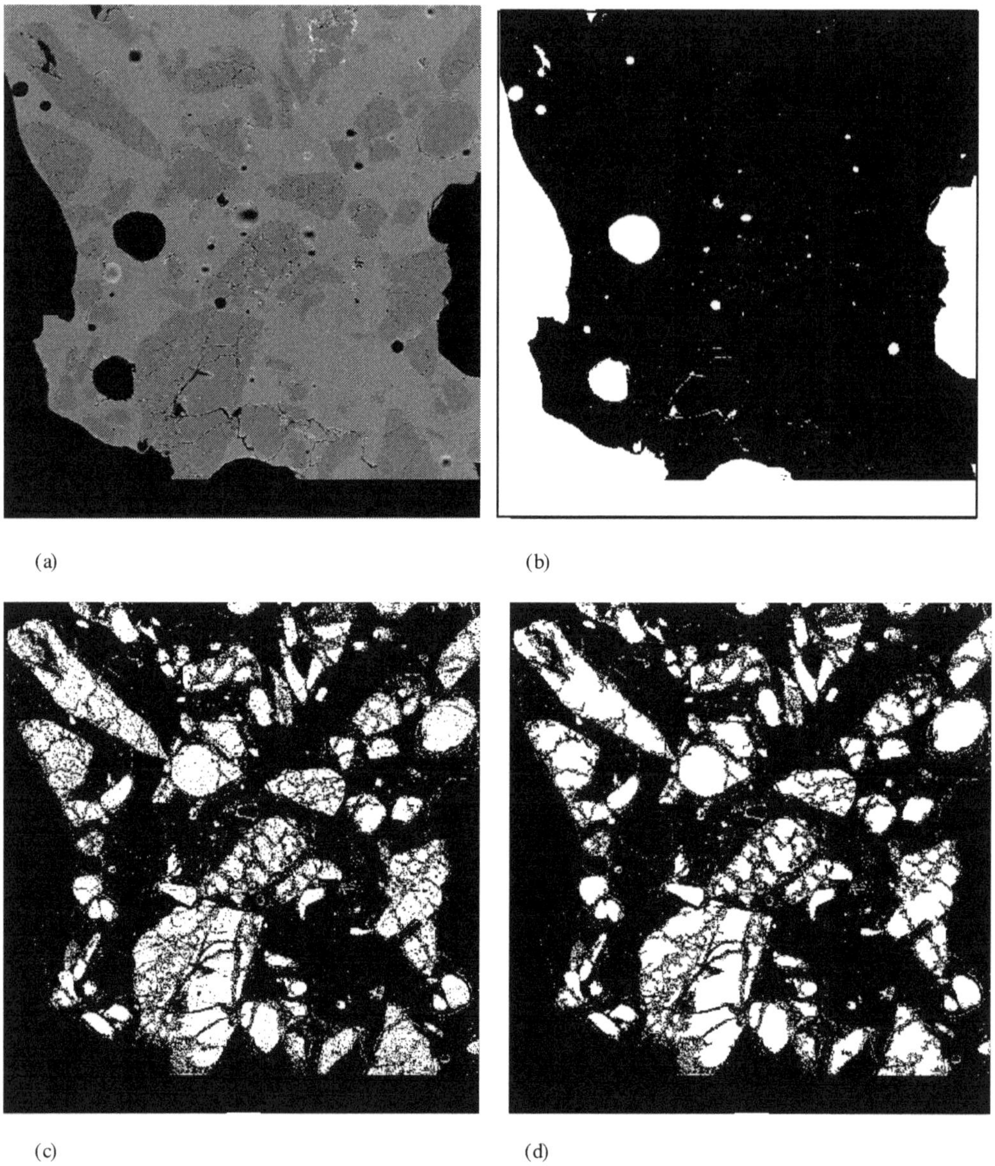

Figure 2-2 : Image analysis of frits. *Four images of frit AM34. (a) is the original SEM image, (b) the amount of void space plus headers and footers to the image, (c) is the amount of quartz straight from VISILOG, (d) is the amount of quartz using (c) plus the HOLE_FILL algorithm.*

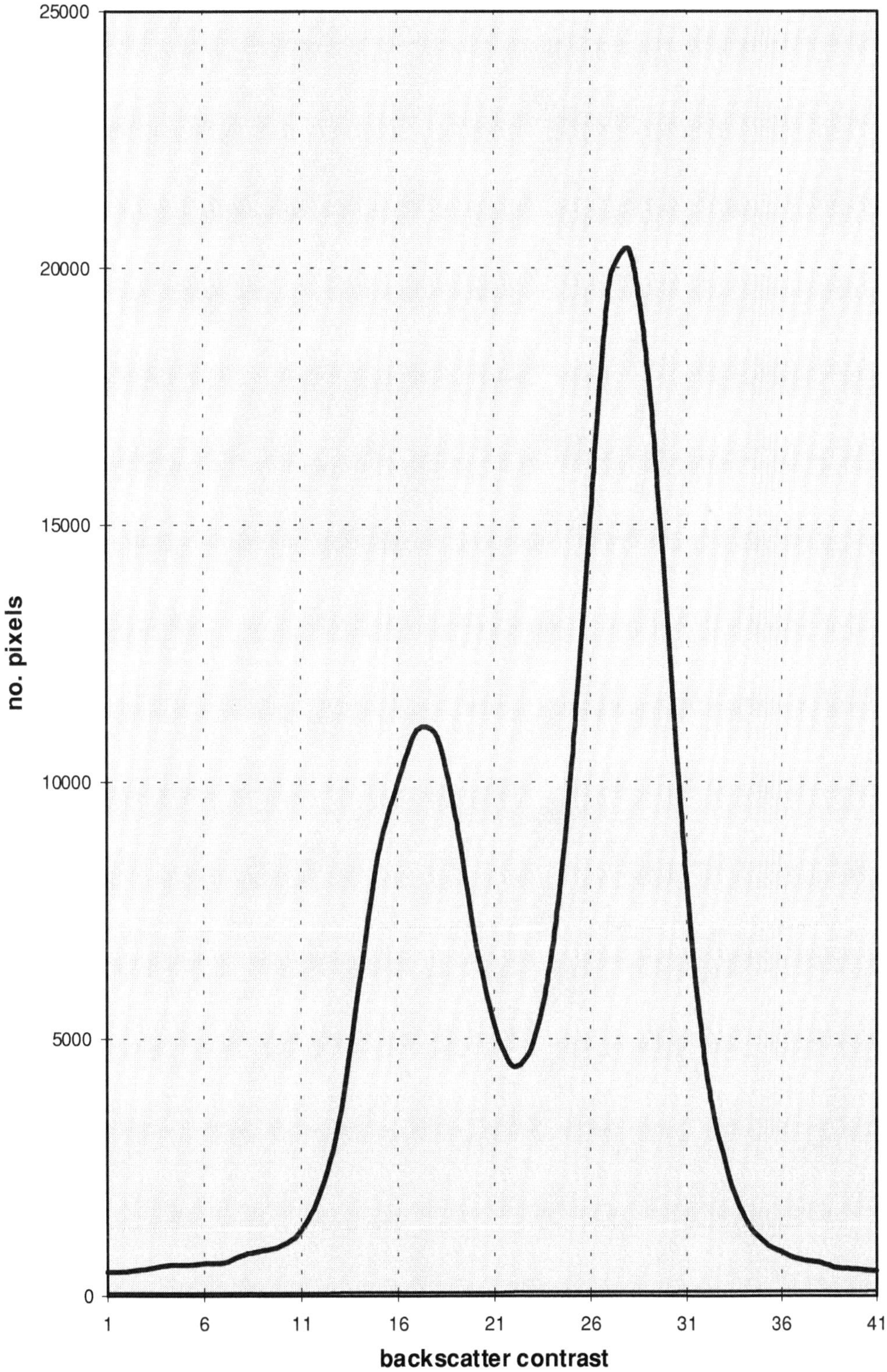

Figure 2-3: Visilog graph of the brightnesses from the image analysis. *The peak on the left represents the quartz backscatter brightness, the peak on the right the glass. Brightness 0, representing void space, is not shown on the graph.*

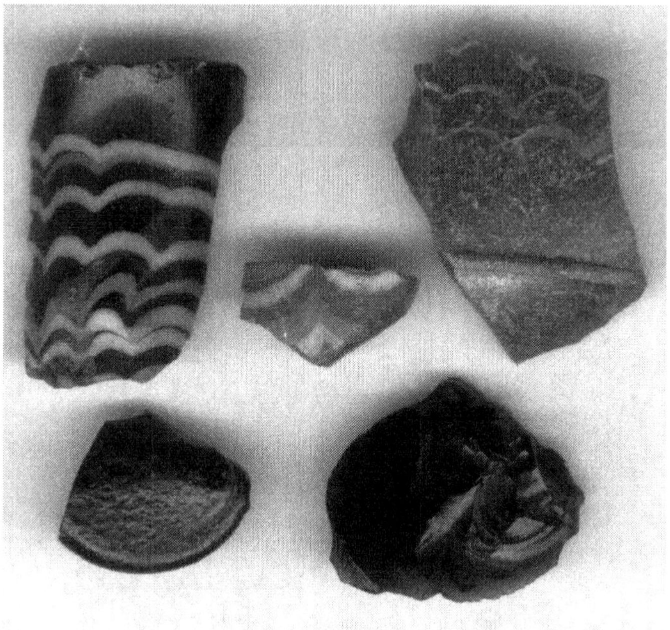

Figure 3-1 : Patterns of glasses: *From the top left, E5613/1 deep blue body with white and yellow trailed and marvered decoration, E5613/6 small turquoise fragment again with white and yellow trailed and marvered decoration, E5613/7 weathered blue vessel with yellow trailing. Bottom left, E5042 blue glass spill and bottom right E5051 purple glass spill. E5613/6 is 15mm across. See Plate 1 for colour image.*

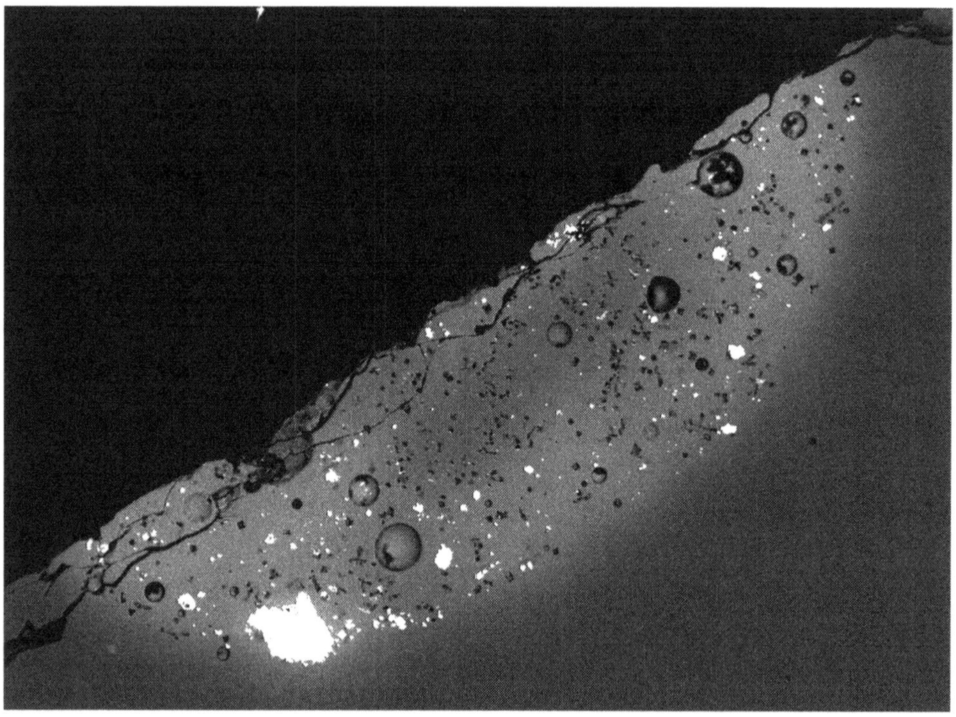

Figure 3-2 : Opacifiers and their effect on microstructures. *A rod of yellow glass has been applied to and marvered into a body of blue glass. Here, in cross-section, the yellow glass appears as a light grey half-ellipse embedded into the mid grey body glass. Throughout the yellow glass are bright particles of lead antimonate, the largest at the bottom of the image being 0.05mm across. Sample E5613/1, image is 0.6mm across*

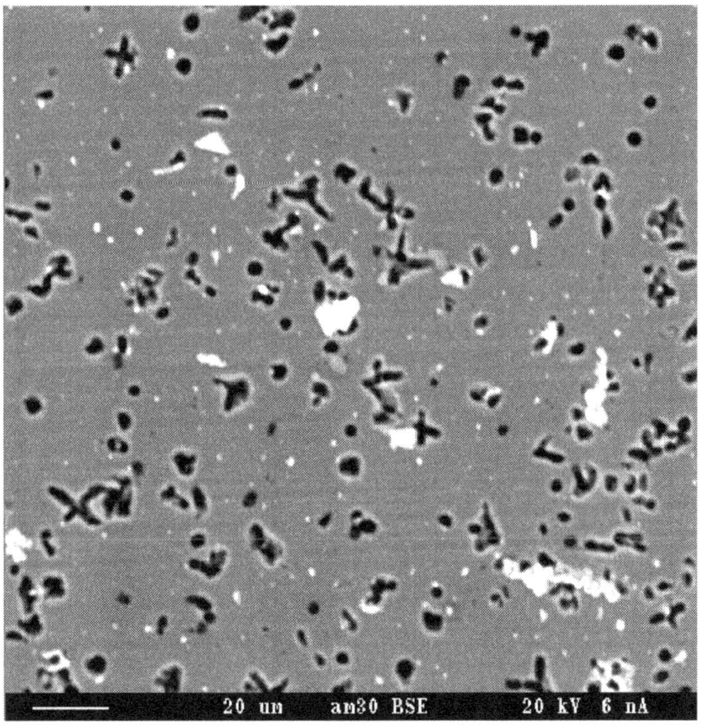

Figure 3-3 : Calcium antimonate opacifier. *A high magnification photomicrograph of glass rod 406, showing the particles of calcium antimonate, appearing in bright backscatter contrast. Also evident are the characteristically shaped void spaces.*

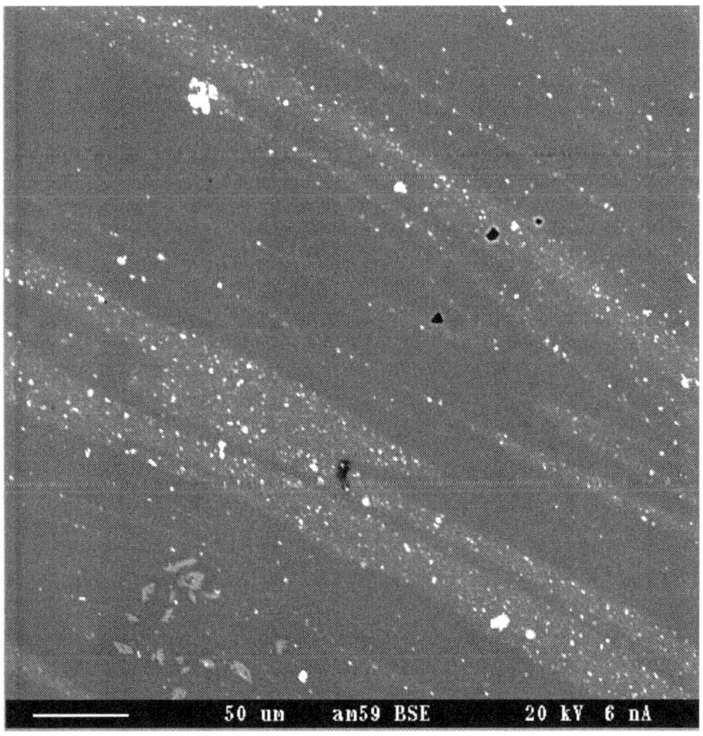

Figure 3-4 : Incomplete mixing of opacifiers. *The high magnification photomicrograph of the body of glass vessel E5613/6 above shows evidence for the late mixing of calcium antimonate, leaving streaks rich in bright particles of opacifier poorly mixed in the darker backscatter of the glass. Small grains of wollastonite are also visible at the bottom left.*

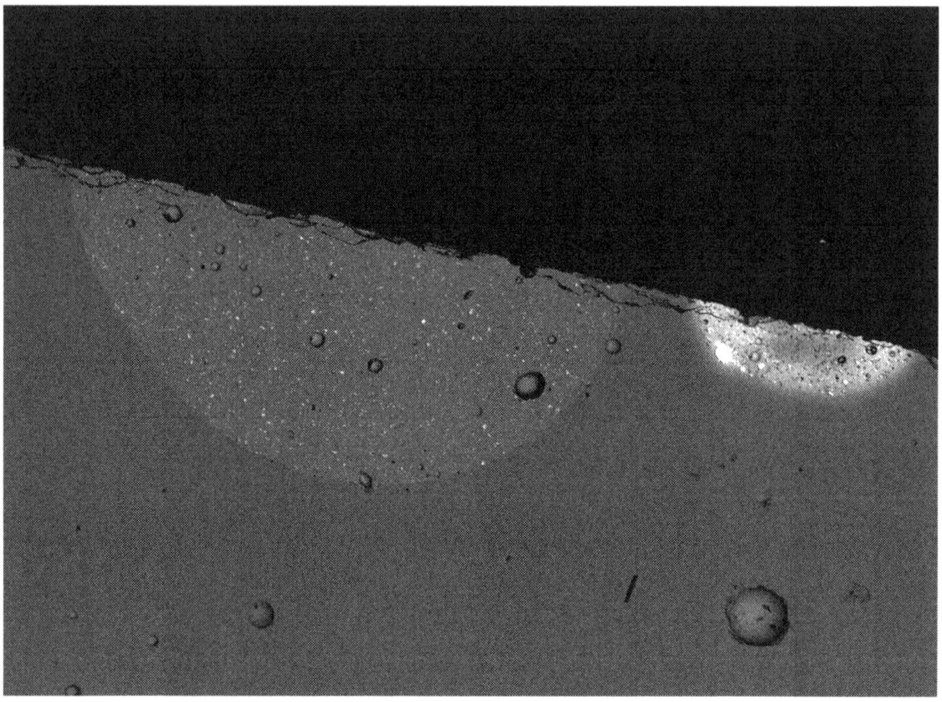

Figure 3-5 : Trailing and marvering. *E5613/1, showing a cross-section through the wall of a vessel which has had two glass rods trailed onto it and then marvered in. The larger and darker of the two (left) is opaque white, and the smaller brighter one (right) is opaque yellow. Image is 2.4mm across.*

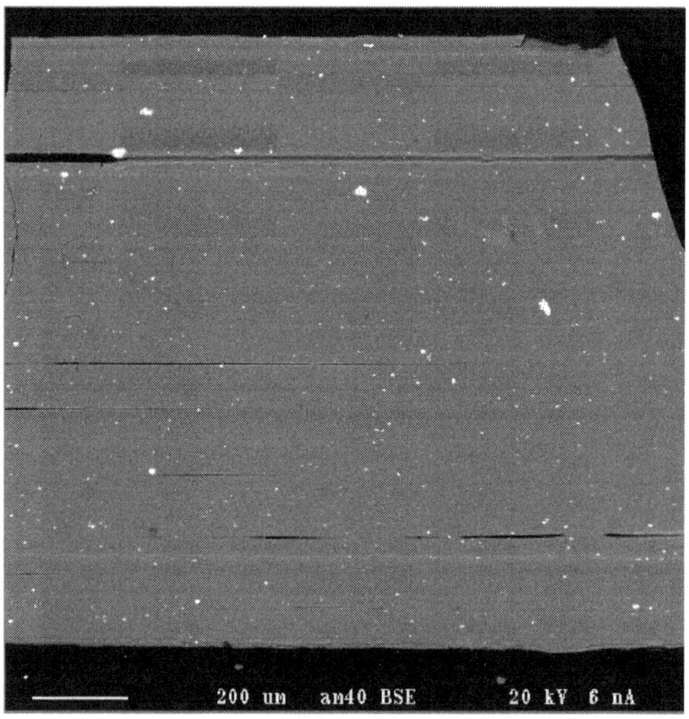

Figure 3-6 : Drawing glass rods. *Green glass rod 593a, showing the striations that develop when a glass rod is drawn. The vesicles present in the glass during manufacture, are drawn out into long thin voids. A large one is present toward the top of the figure, and several smaller ones in the middle and at the bottom. Particles of the lead antimonate opacifier are also visible in the glass. Image is approximately 1.5mm across.*

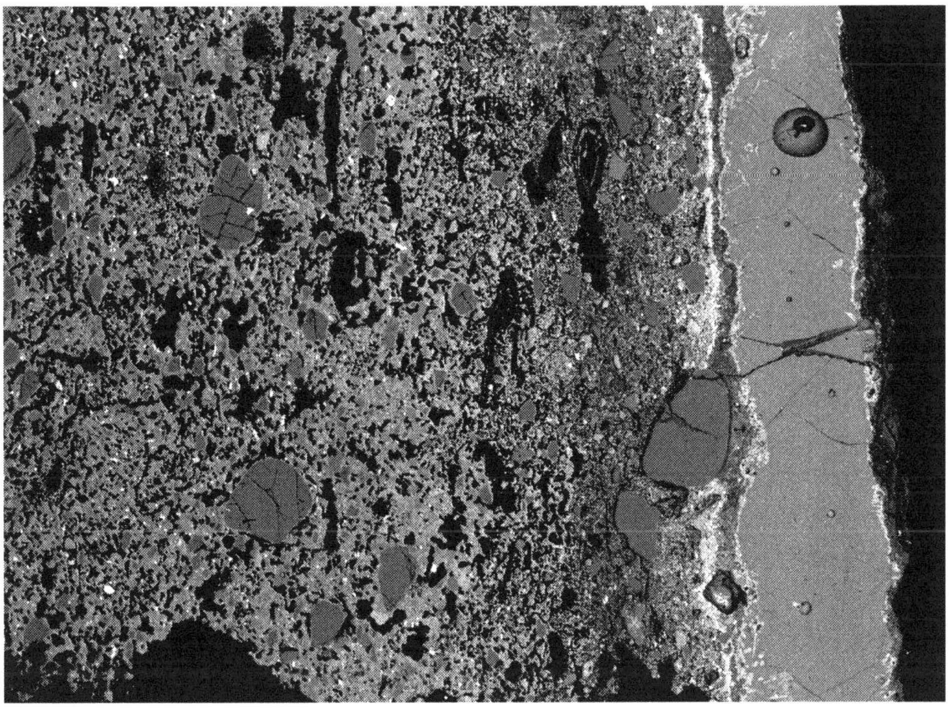

Figure 3-7 : Contact microstructures: *Typical contact between a glass and the white slip of a ceramic, showing a thin layer of white accicular grains of wollastonite growing into the glass layer (mid grey to the right). Glass layer is approximately 300mm thick, white slip is 0.5-0.6mm thick and lies immediately to the left of the glass, grading in to normal Nile silt ceramic at the far left. Sample AM22, total image is 2.4mm across*

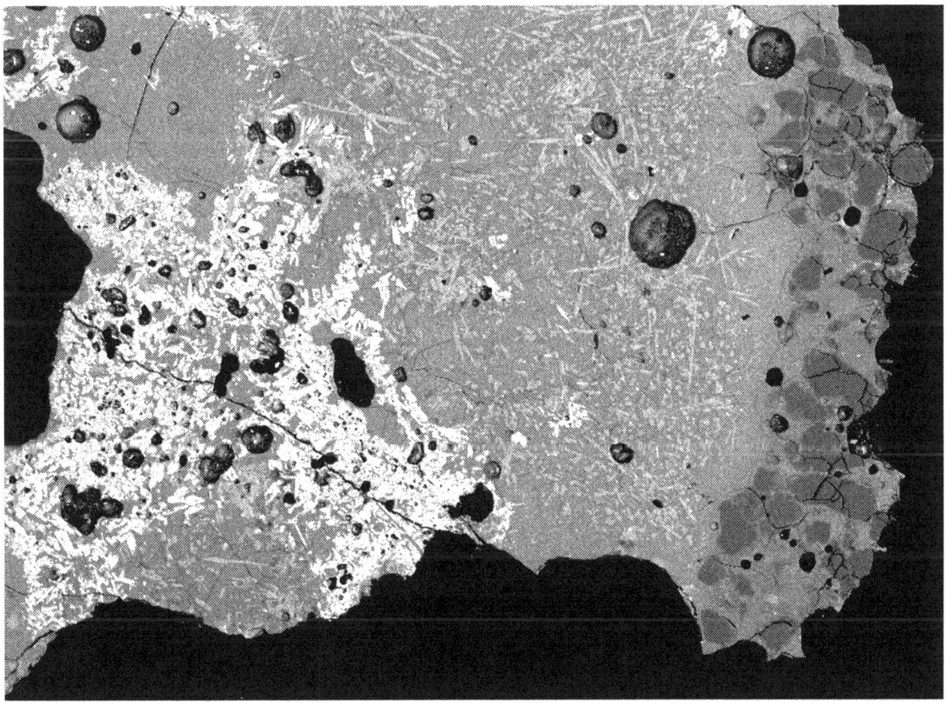

Figure 3-8 : Contamination of the glass by the presence of a lime rich slip. *Sample AM30 above shows a high degree of contamination in a glass caused by the proximity of a ceramic (right of image). In the centre of the image is the mid grey glass with many thin crystals of diopside (CaMgSi2O6)in a lighter grey. To the left a large area of partially dissolved lime can be seen, surrounded by fringing wollastonite (both white). Total image is 2.4mm across.*

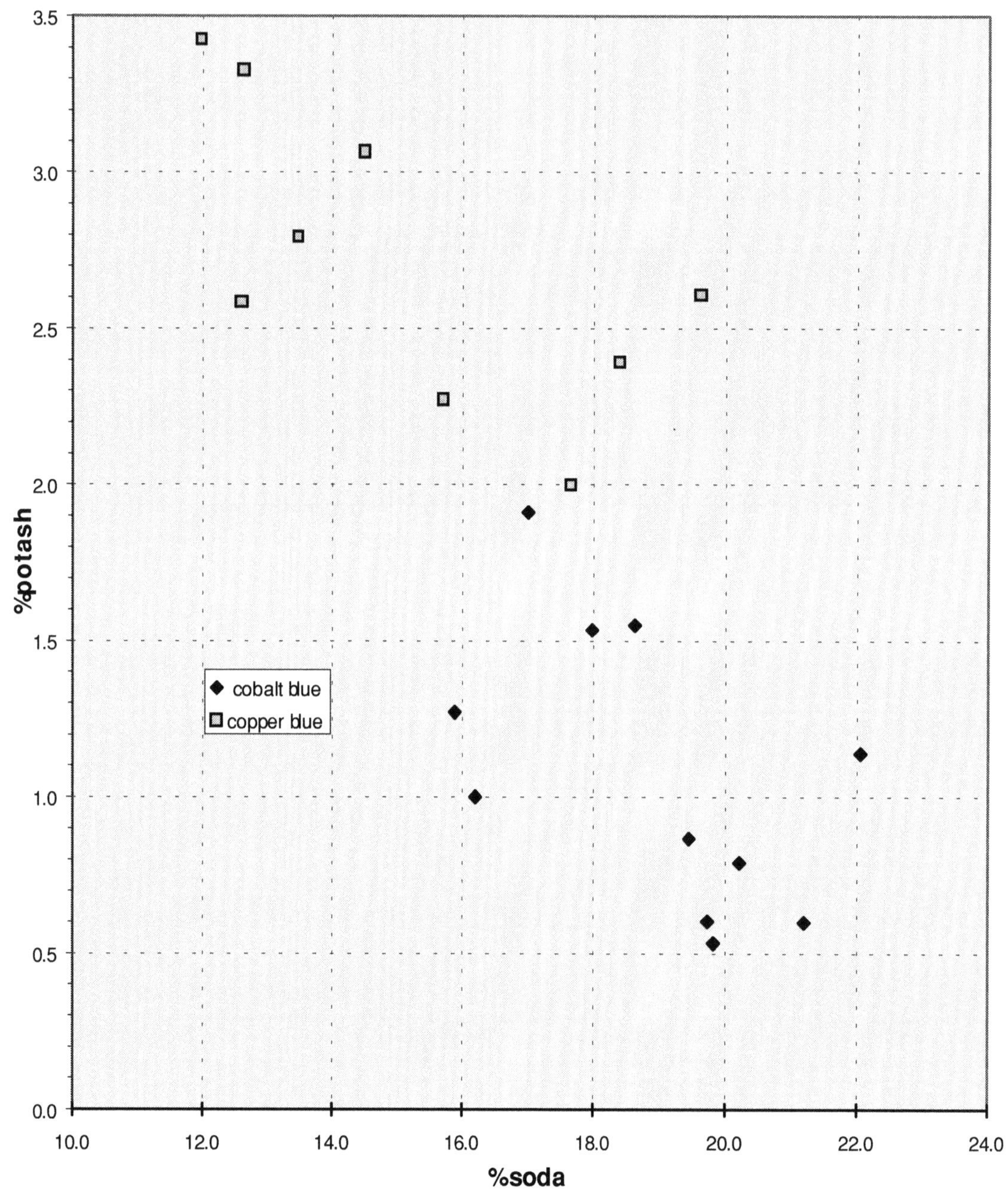

Figure 3-9 : Soda against potash for the glasses. *Plotting the soda against potash for the blue glasses shows a clear split, with copper glasses having relatively high potash and low soda, whereas cobalt glasses have low potash and high soda. It is clear in both cases, however, that soda is still the dominant alkali.*

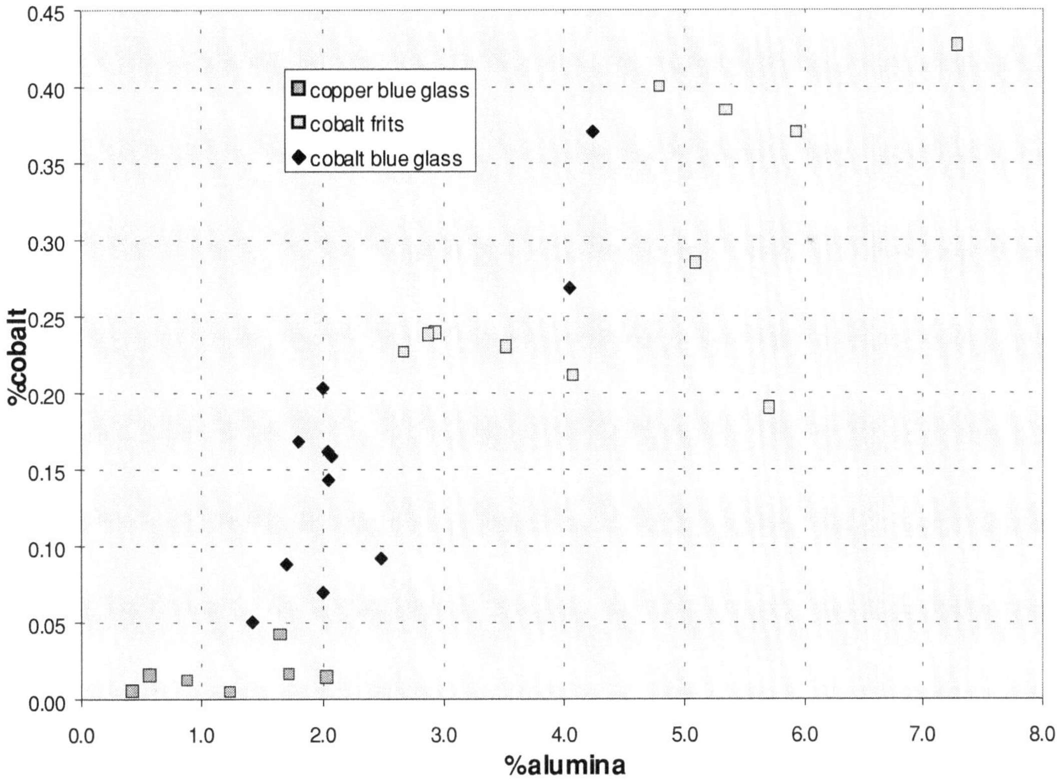

Figure 3-10 : Cobalt against alumina. *Plotted are the glasses coloured by copper and those coloured by cobalt. Also plotted to make the relationship of the two elements clearer is the data for the same elements for the glass matrix of the cobalt coloured frits.*

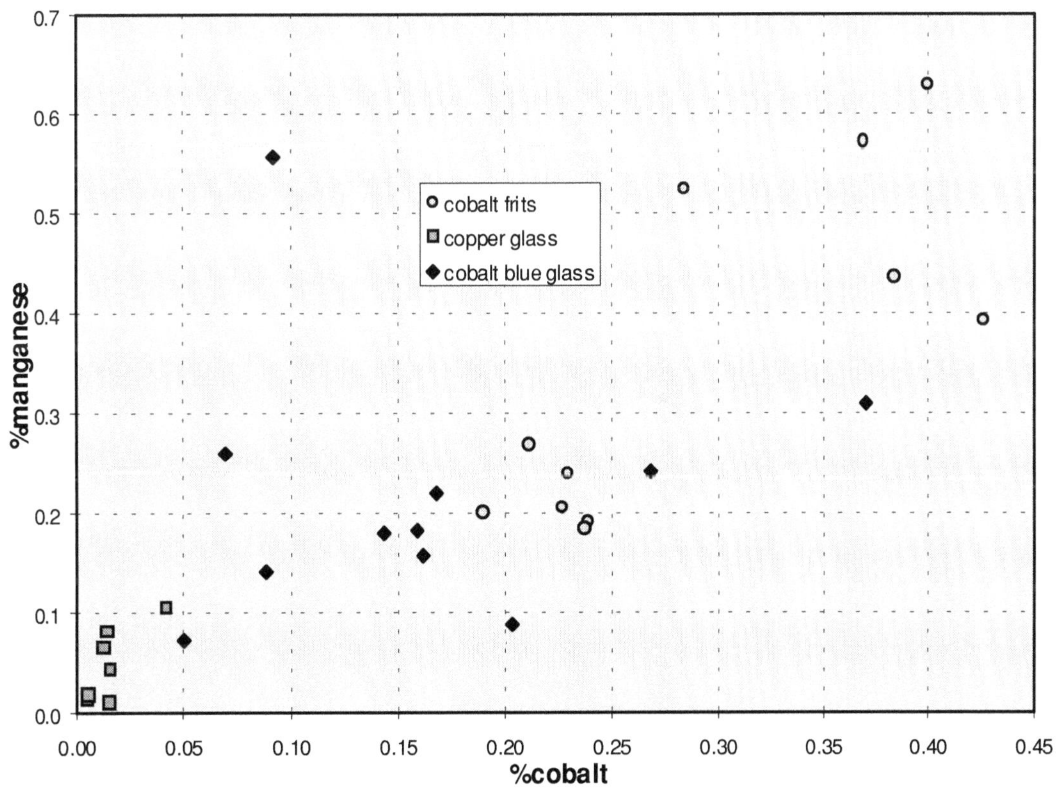

Figure 3-11 : Cobalt against manganese. *Plotted are the glasses coloured by copper and those coloured by cobalt. Also plotted to make the relationship of the two elements clearer is the data for the same elements for the glass matrix of the cobalt coloured frits.*

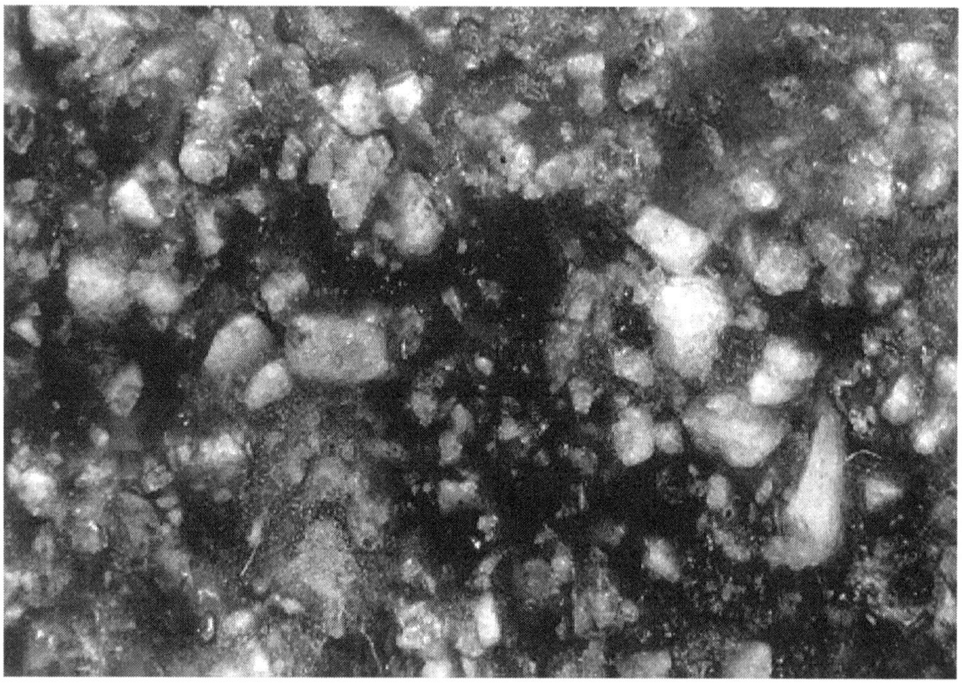

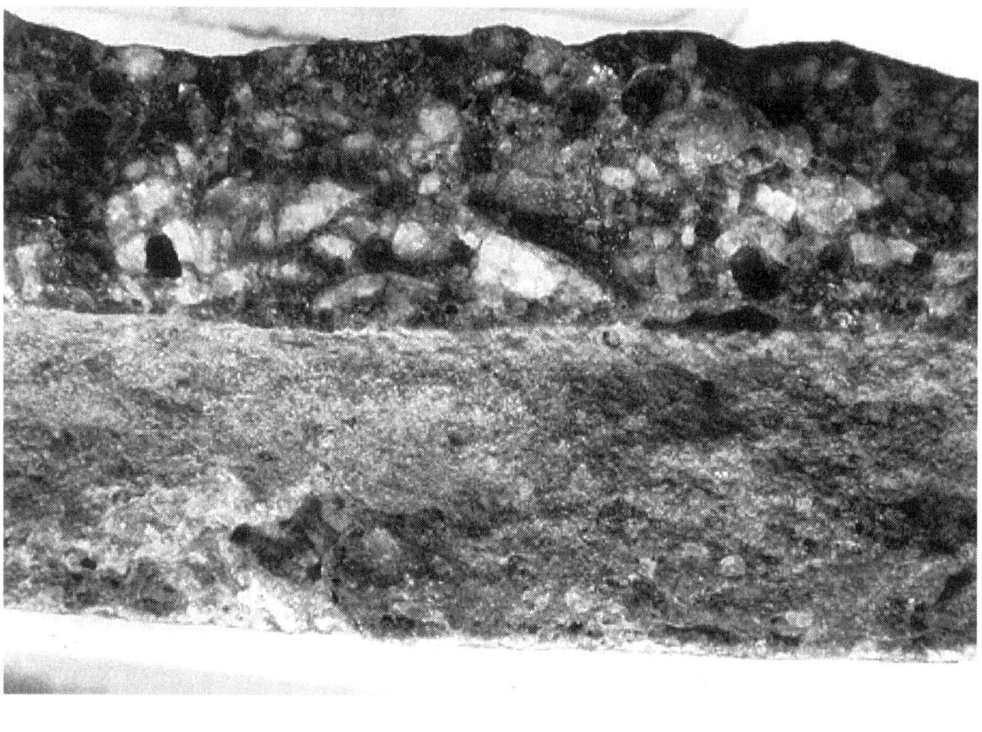

Figure 3-12 : Typical frit. *. Two low magnification images of the frit lying in fritting pan 21977, showing the white quartz grains in a blue matrix, (top) from above illustrating the incomplete covering revealing the ceramic of the fritting pan below (image 15mm across), (bottom) in cross section, showing that the thickness of the pan to be about 8.0mm, witha thin yellow slip on the upper surface, and an irregular layer of frit above, here again 8.0mm thick. For a colour image see Plate 2.*

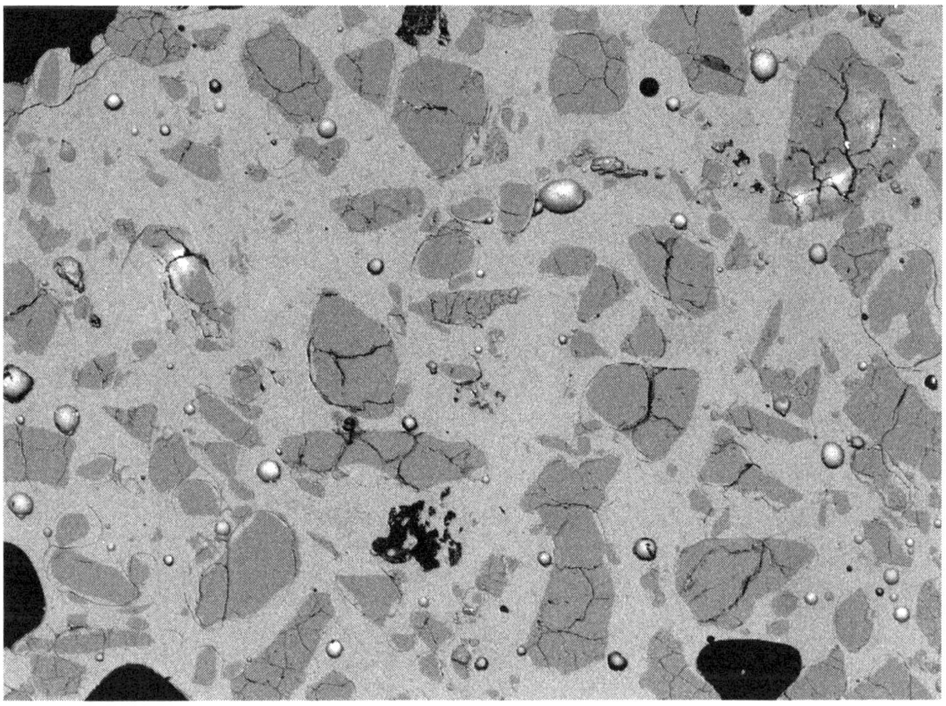

Figure 3-13 : Typical frit. *An SEM image of AM32 showing the typical microstructure of the majority of the frits, with large rounded grains of quartz in a homogeneous glass matrix. No other phases are present. Image is 3.0mm across.*

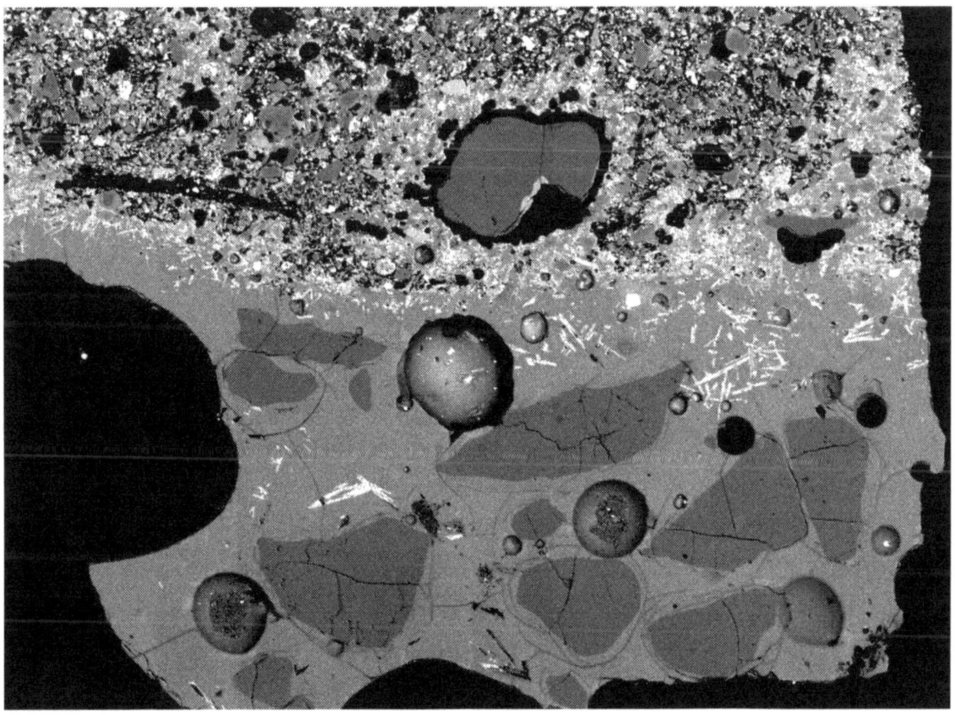

Figure 3-14 : Secondary growth in frits. *AM31 preserves the boundary between the frit and the lime rich white slip of a ceramic. The white grains growing from the slip into the frit are wollastonite grains. As can be seen, they do not penetrate far into the slip. Image is 1.5mm across.*

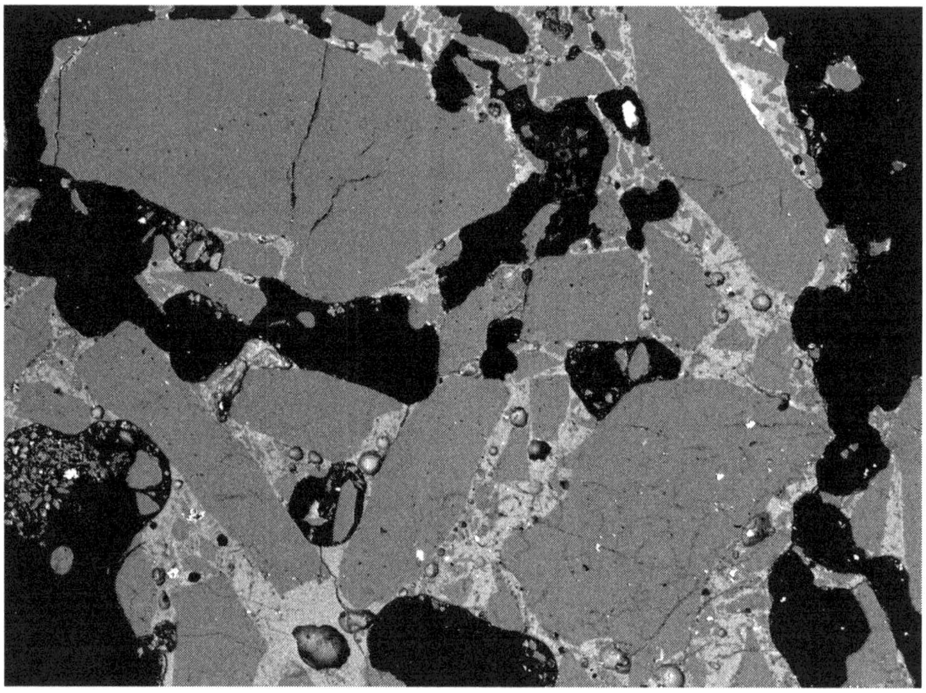

Figure 3-15 : Angular quartz grains. *Frit 25038, the copper coloured frit, which shows a high degree of angularity, one extreme of the degree of rounding seen in the frits. Here, large almost unaltered and unshattered quartz grains are cemented by only small quantities of glass matrix. Image is 1.5mm across.*

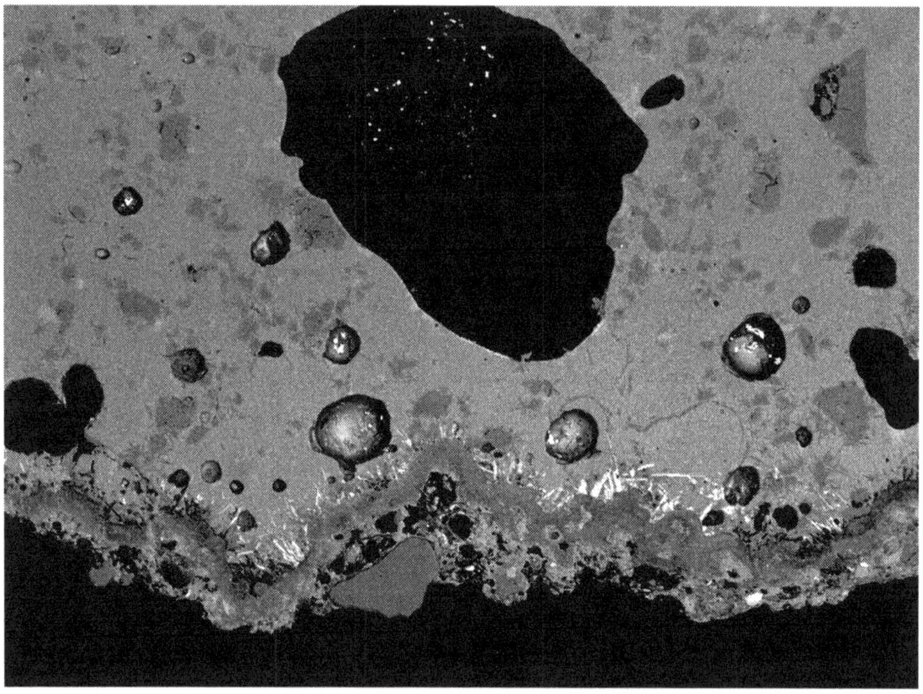

Figure 3-16 : Rounded quartz grains. *AM28 shows rounded quartz grains with highly reacted grain boundaries, along with a reaction zone to a lime rich slip (at the base of the picture). Image is 1.5mm across.*

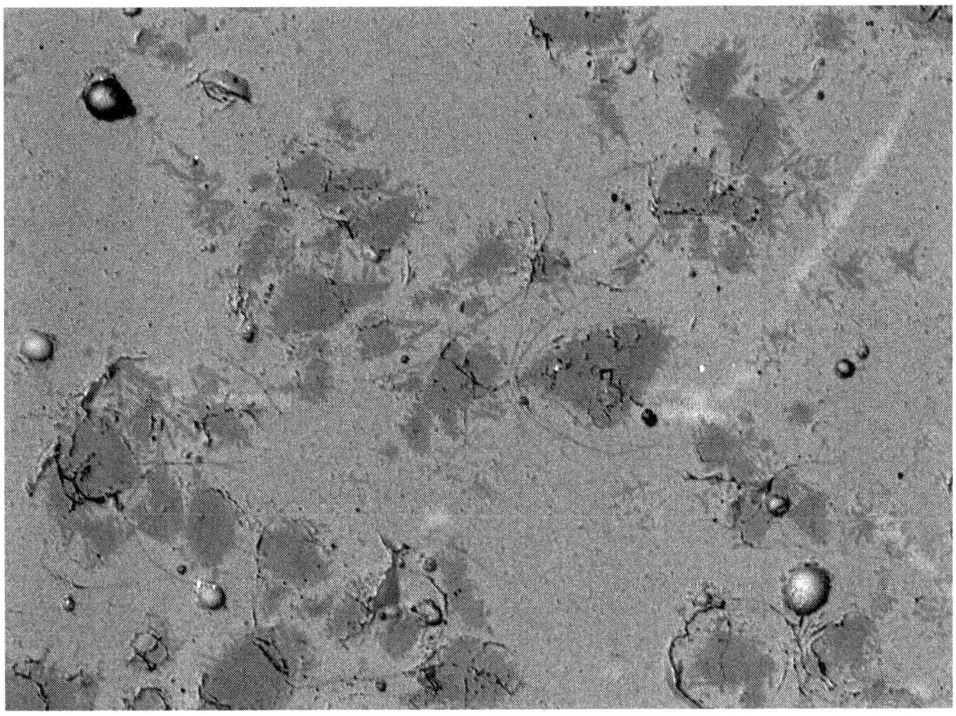

Figure 3-17 : Cristobalite and beta to alpha quartz transitions. *A high magnification photomicrograph of frit 403 showing the highly altered quartz grains that can be seen in those frits assessed as being at the top end of the temperature scale. A number of the grains show fringes of cristobalite, particularly well developed in the grains in the centre of the picture. The grains at the top right show the perimeter cracking indicative of the b to a quart transition. 0.5mm across.*

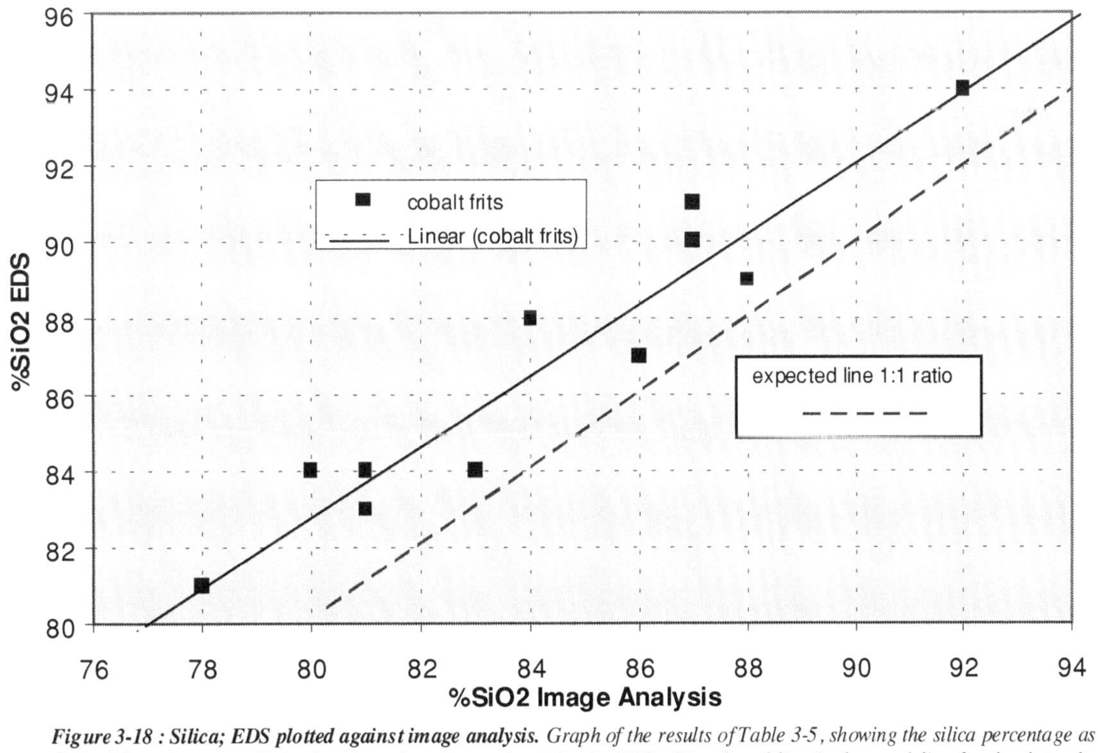

Figure 3-18 : Silica; EDS plotted against image analysis. *Graph of the results of Table 3-5, showing the silica percentage as derived from image analysis plotted against that analysed by the EDS. Fine dotted line is the trend line for the data, the dashed line is that expected if the EDS and image analysis calculated the same silica percentage.*

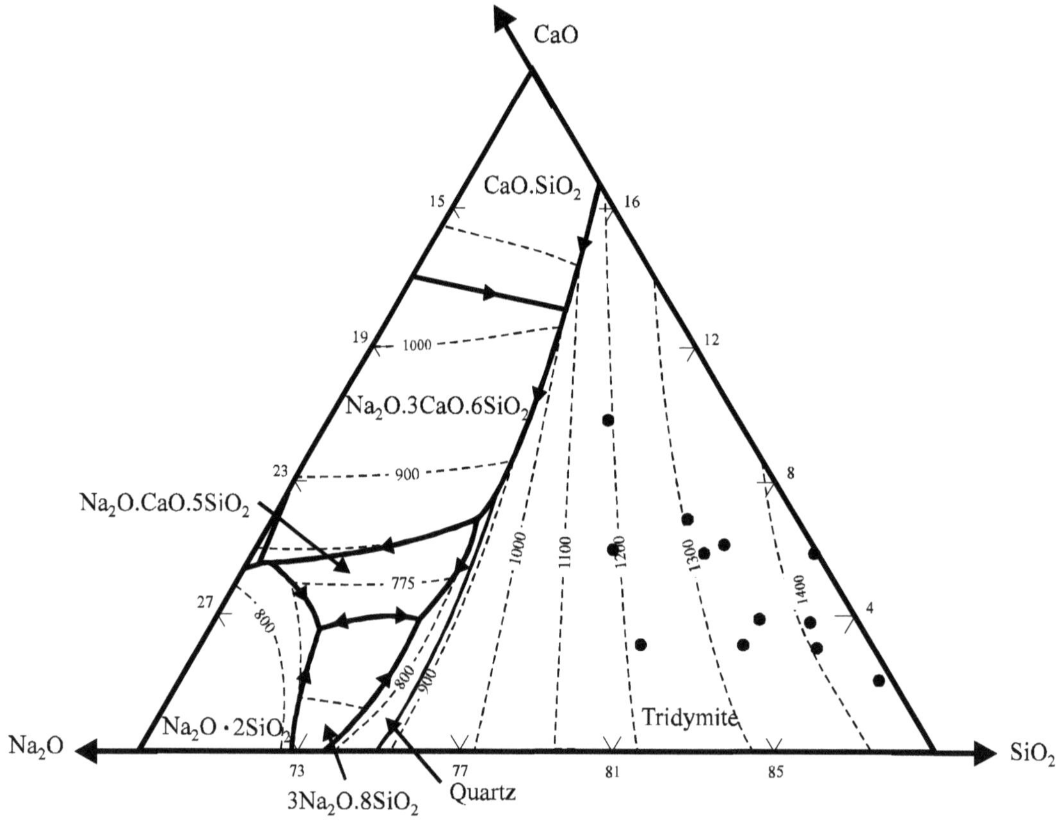

Figure 3-19 : Ternary plot of the region 69-89% SiO2 11-31% Na2O and 0-10% CaO for the glass matrices of the frits. *Eutectics and peritectics for the region are shown. Plotted by weight % from the three elements, normalised to 100%. .*

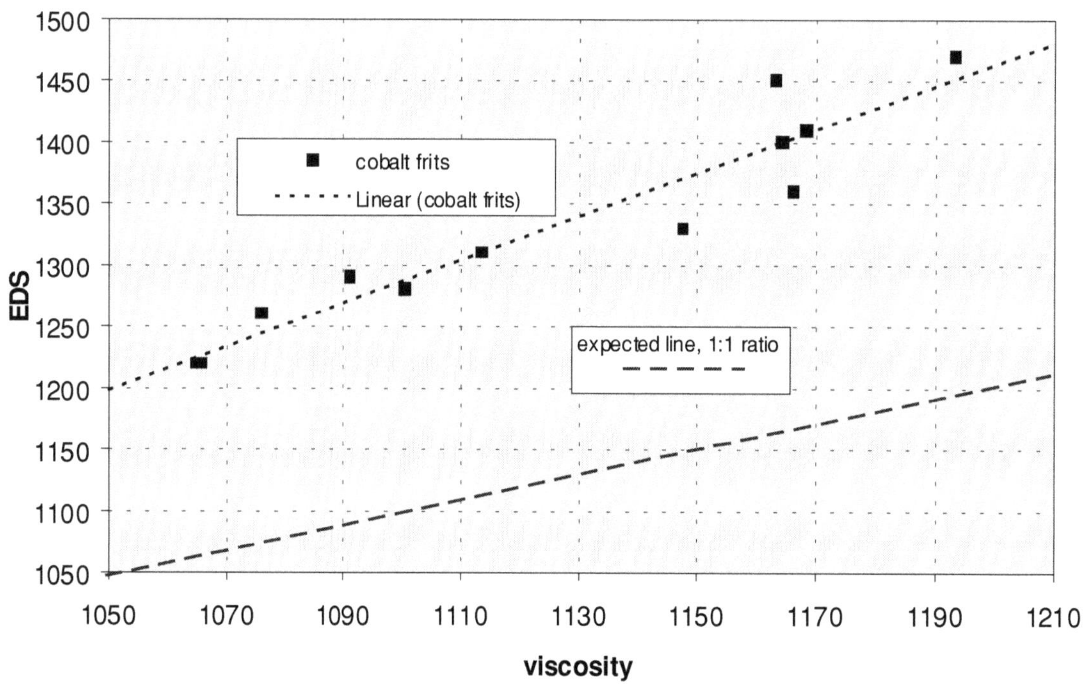

Figure 3-20 : Temperature comparison. *The temperature derived from Lakatos et al (1972) and the ternary diagram (Figure 3-18) shown in Table 3-6 are plotted. The blue dotted line is the trend line for the data, the red dotted line the expected line that the data would plot on if the two values were identical.*

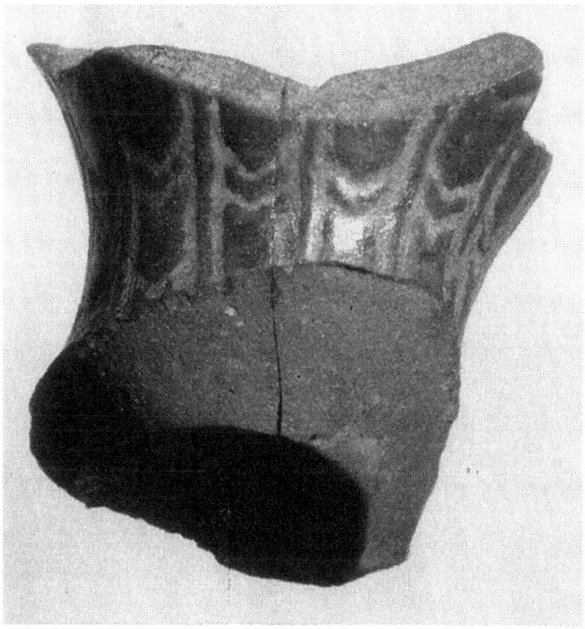

Figure 3-21 : Variant D. *Low magnification photograph showing a close up of the core of a faience vessel made from Variant D faience, demonstrating the pale colour at its top that is typical of the variant when weathered , and the deep blue typical of a fresh break at the bottom.. Sample Berlin 11/674 from House N48.16 at Amarna.41mm high. See Plate 3 for a colour version.*

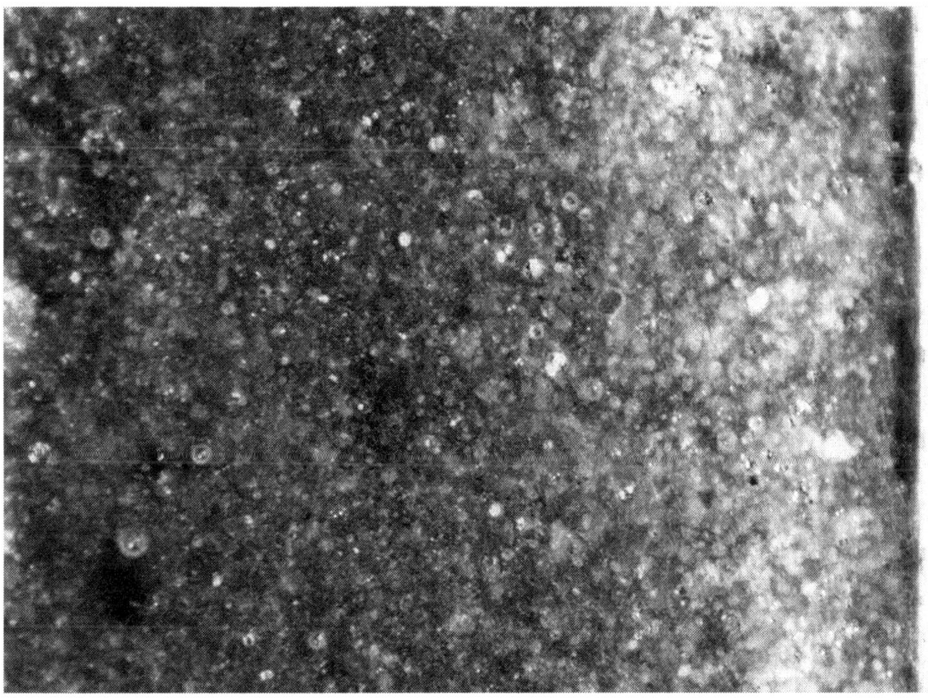

Figure 3-22 : Variant D core. *Low magnification photograph of a cross section through a core, showing an inlay with a deep blue colour core and the green glaze (to the right). Sample is 3556, 5mm across*

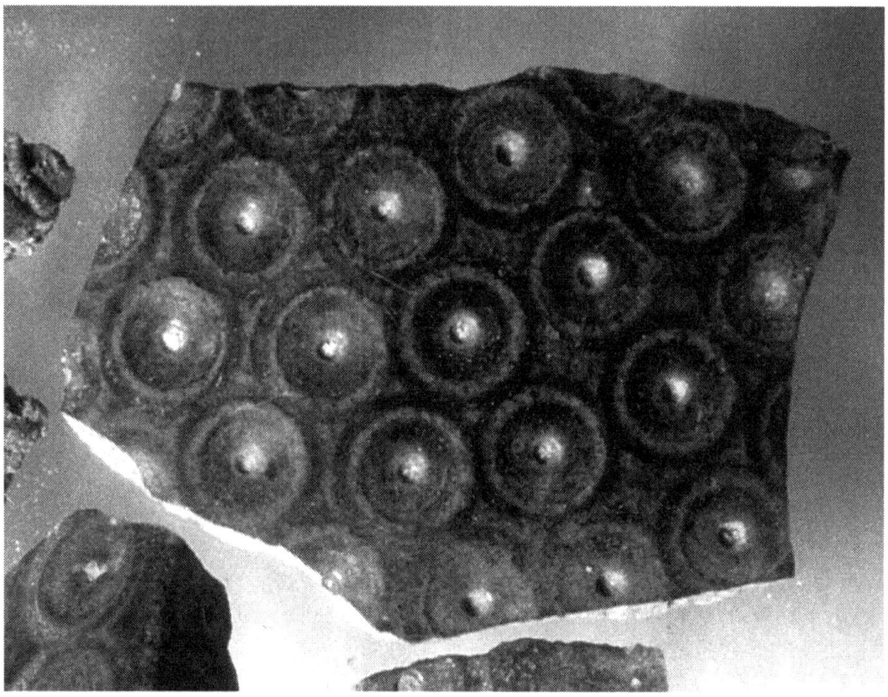

Figure 3-23 : Surface working on a Khepresh crown. *This crown has been drilled will small holes at the centre of the rosettes after the glaze has cooled. this pattern of cold working is typical of blue crowns at Amarna. Berlin 11/12/764 from house P49.6 at Amarna. Fragment 54mm across.*

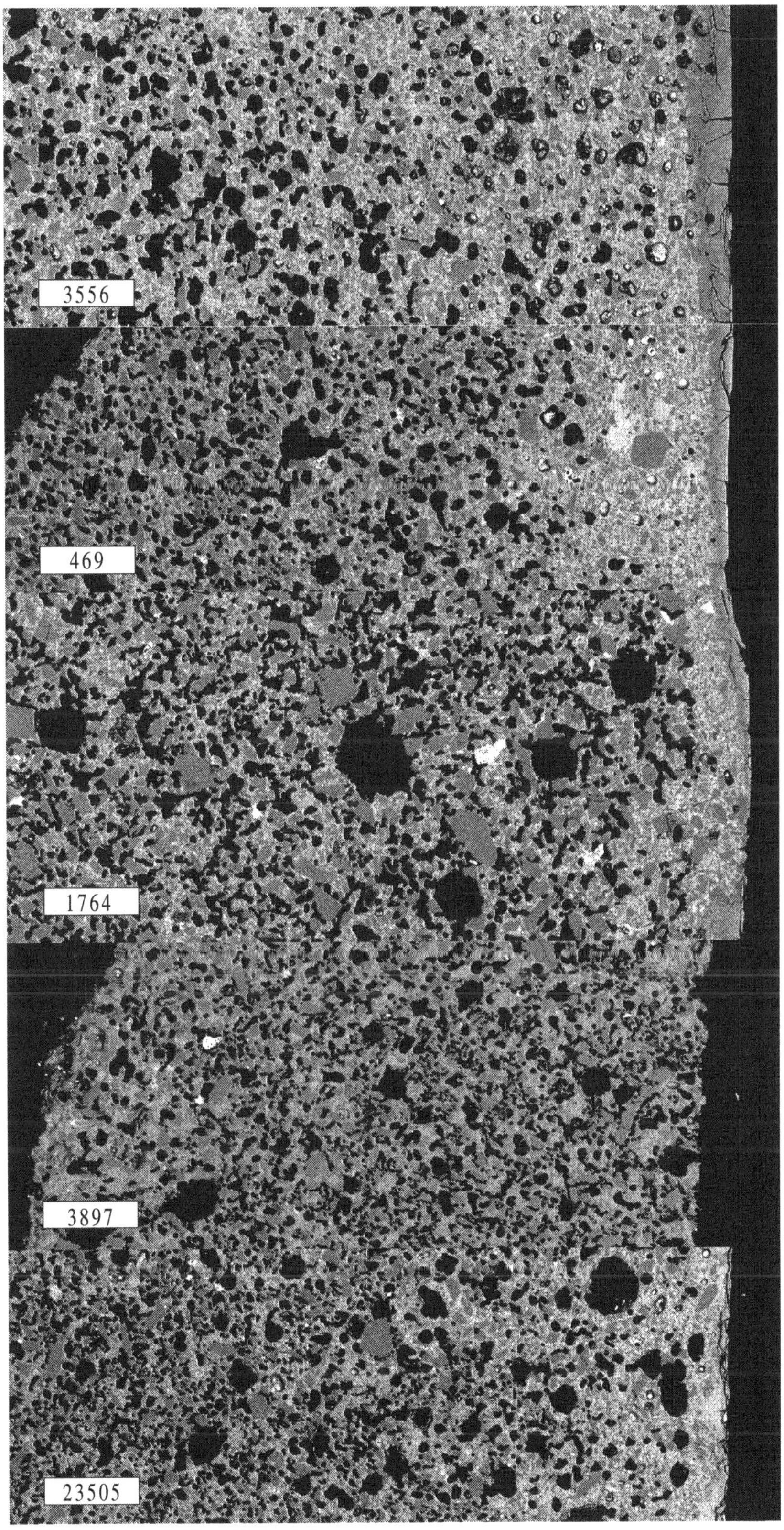

Figure 3-24 : Variant D faience; *Five SEM images of cross sections through the more vitreous of the Variant D faience samples, ordered by decreasing amount of glass in the cores. All images 2.5mm across. From the top down, 3556, 469, 1764, 3897 and 23505*

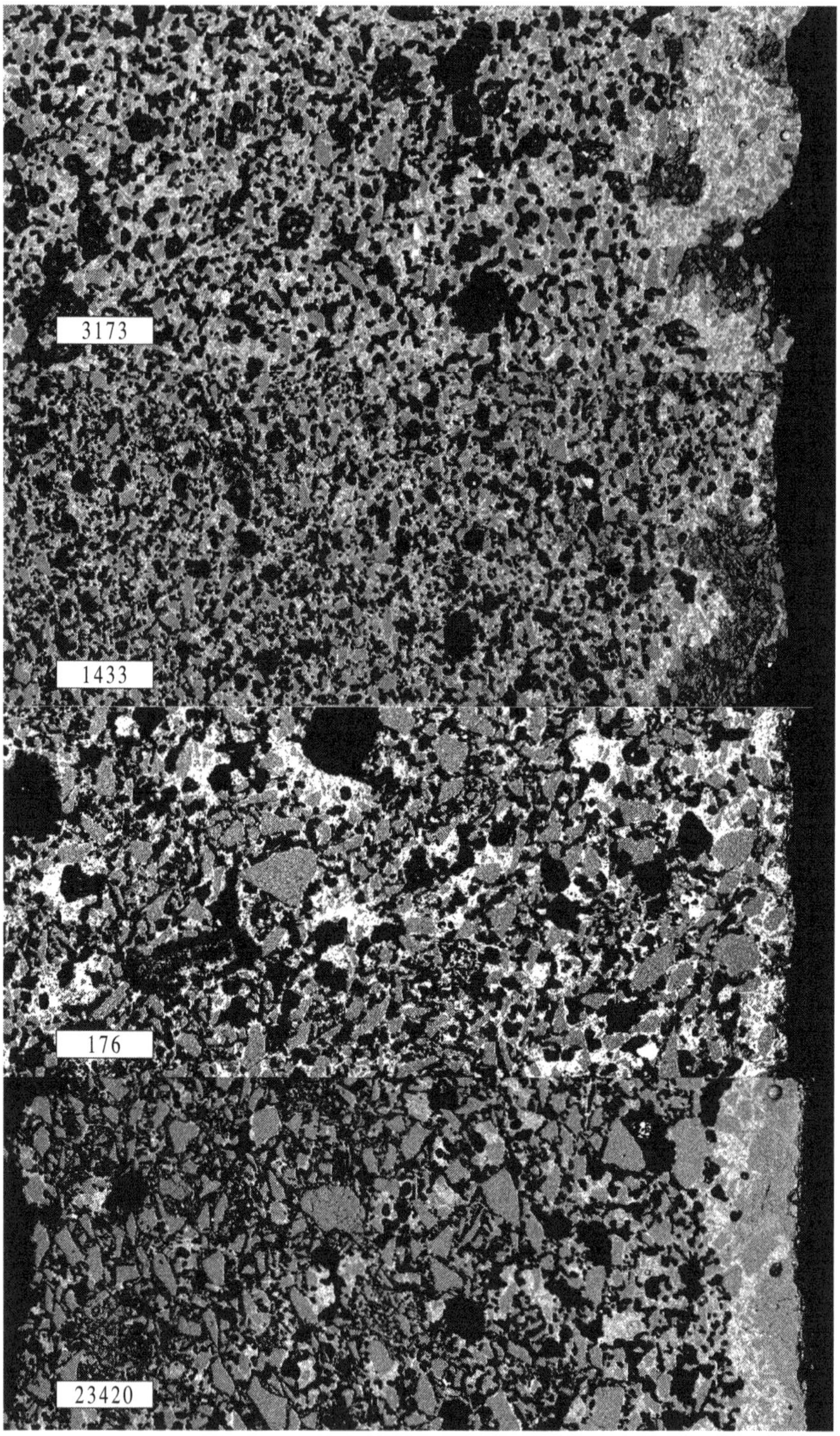

Figure 3-25 : Variant D faience; *Four SEM images of cross sections through the less vitreous of the Variant D faience samples, ordered by decreasing amount of glass in the cores. All images 2.5mm across. From the top down, 3173, 1433, 176 and 23420.*

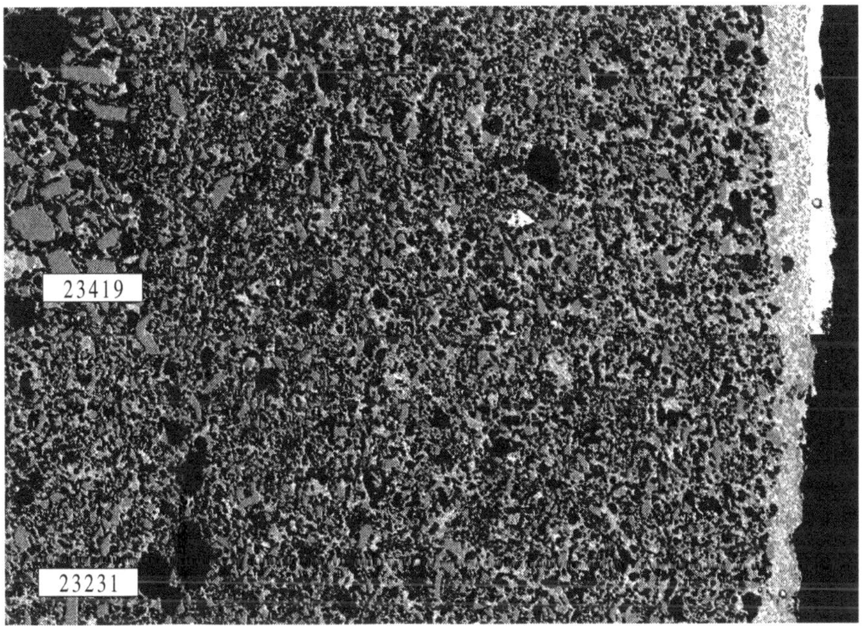

Figure 3-26 : Variant D faience with Variant; *SEM images of cross sections through two pieces of Variant D faience, each with the fine grained slip layer classified 'Variant A' by Lucas (coarse grained core just off the image to the left on 23231). Both images 2.5mm across. From the top down, 23419 and 23231.*

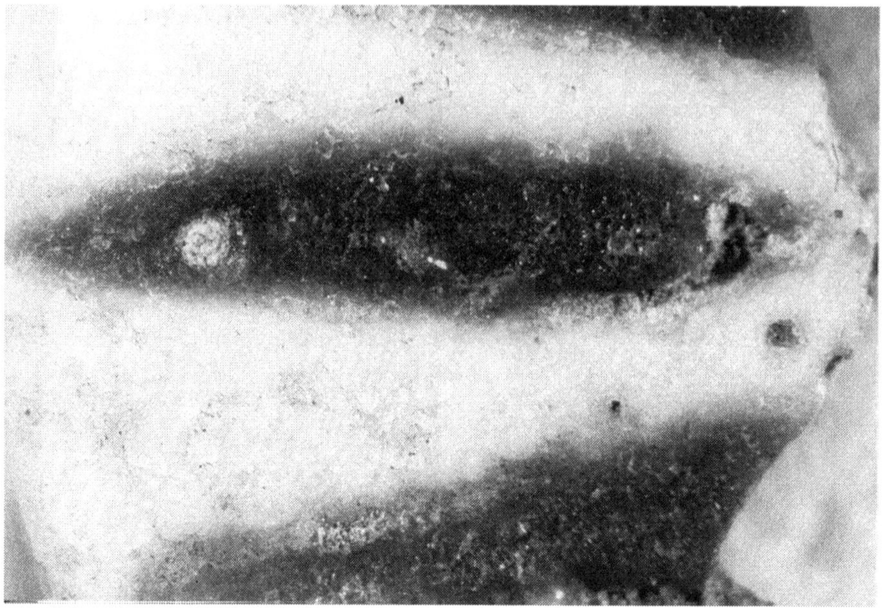

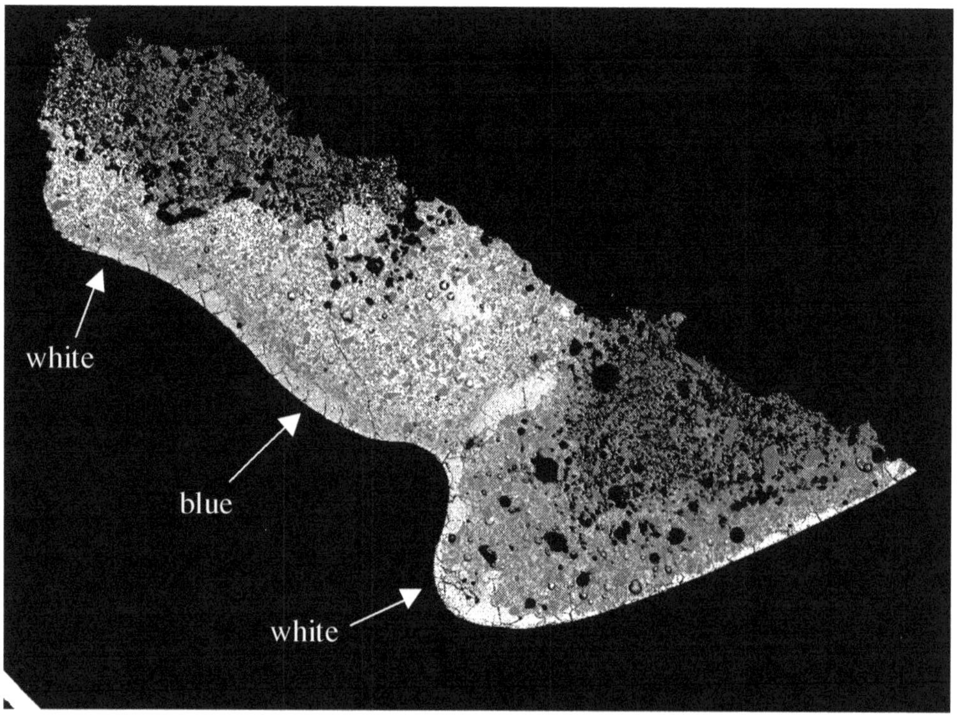

Figure 3-27 : White faience inlayed bead E5053. *Top shows the bead under low magnification reflected light showing the inlays of blue and red leaves or petals on the white background. Bottom is a photomicrograph of a cross section of the bead showing three different areas. The top and bottom of the bead as marked are the white faience, whereas the middle area with the highly vitreous microstructure is blue*

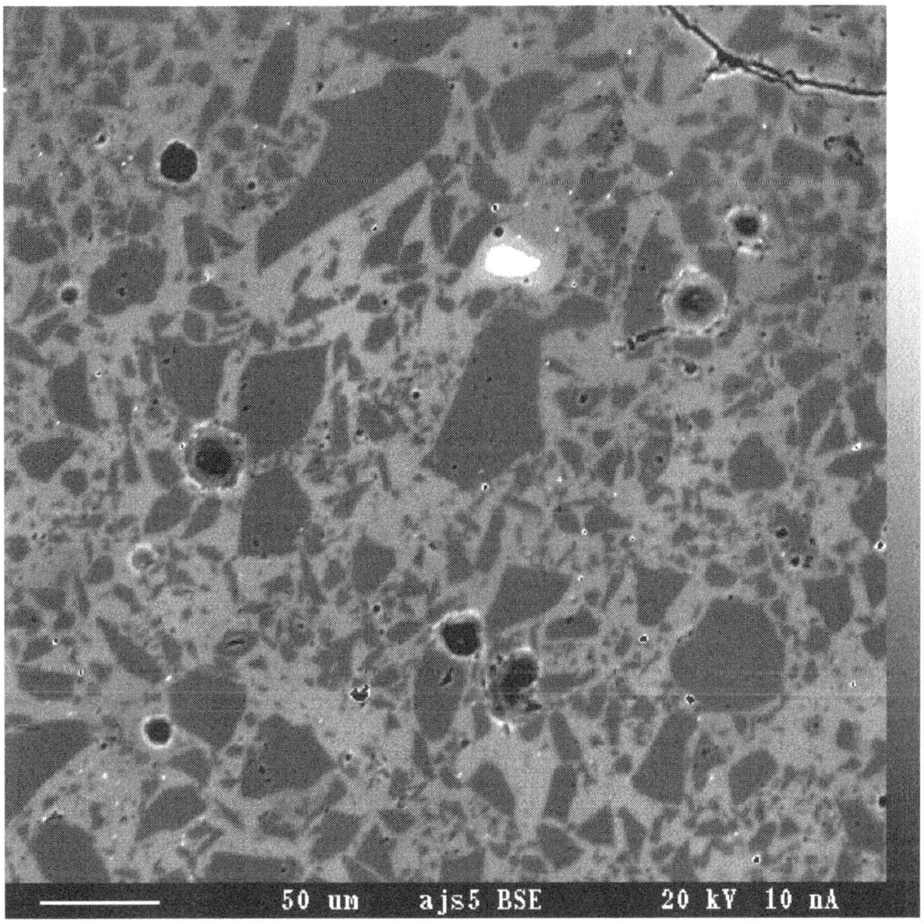

Figure 3-28 : Close up of the highly vitreous blue inlay to E5053, *showing the angular quartz grains (mid grey) in a glass matrix (light grey).*

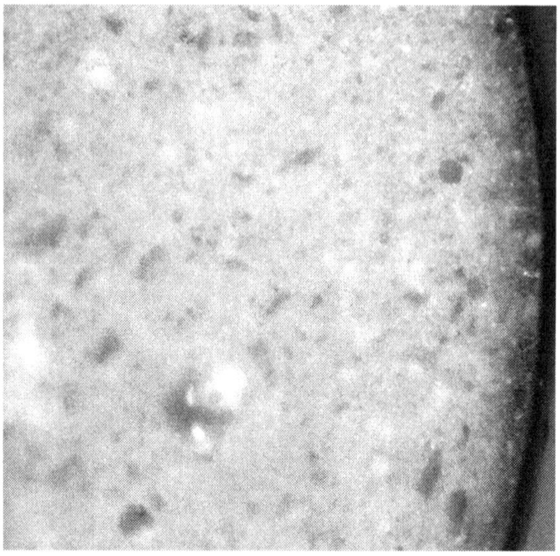

Figure 3-29 : Yellow faience bead E5032/4 *in polished cross section showing the core and the thin yellow translucent glaze (image is 1.5mm across). See Plate 4 for a colour image.*

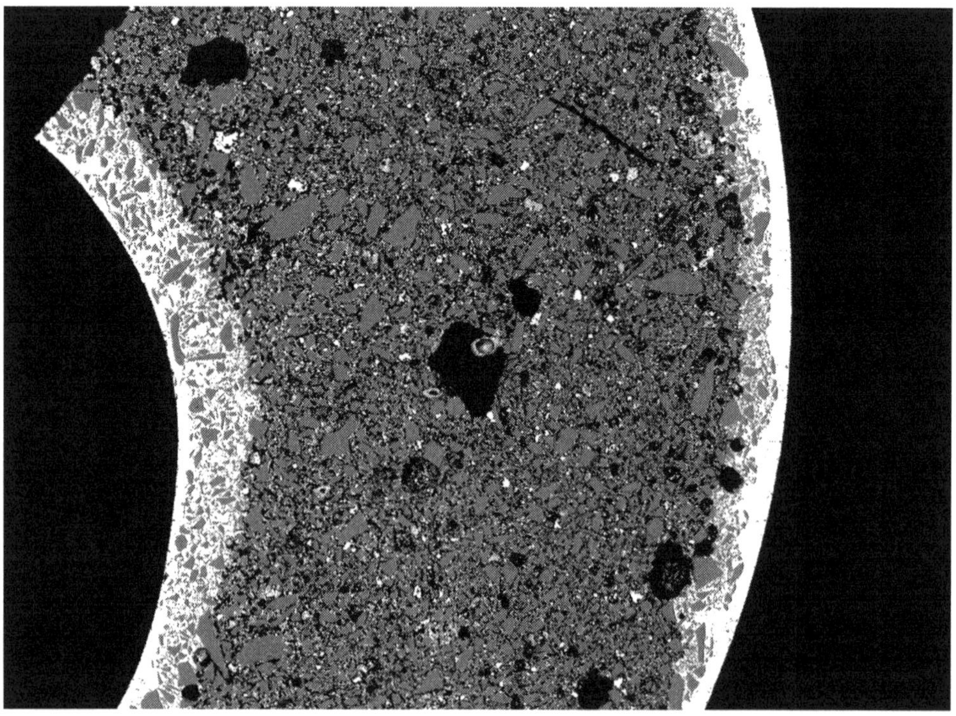

Figure 3-30 : A photomicrograph of yellow faience bead E5032/4 *showing the bright contrast of the glaze, interaction layer and particles within the core that is indicative of concentrations of high atomic number elements, in this case PbO. (image is 3.5mm across)*

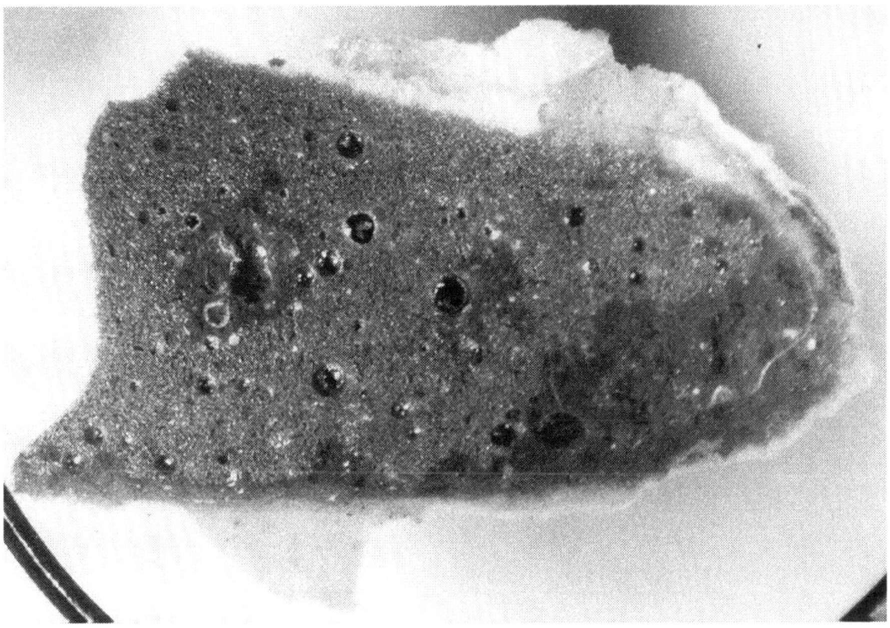

Figure 3-31 : Red faience inlay 770, *showing a polished cross section of the red faience illustrating the the core and the thin greenish glaze. See Plate 5 for colour image.*

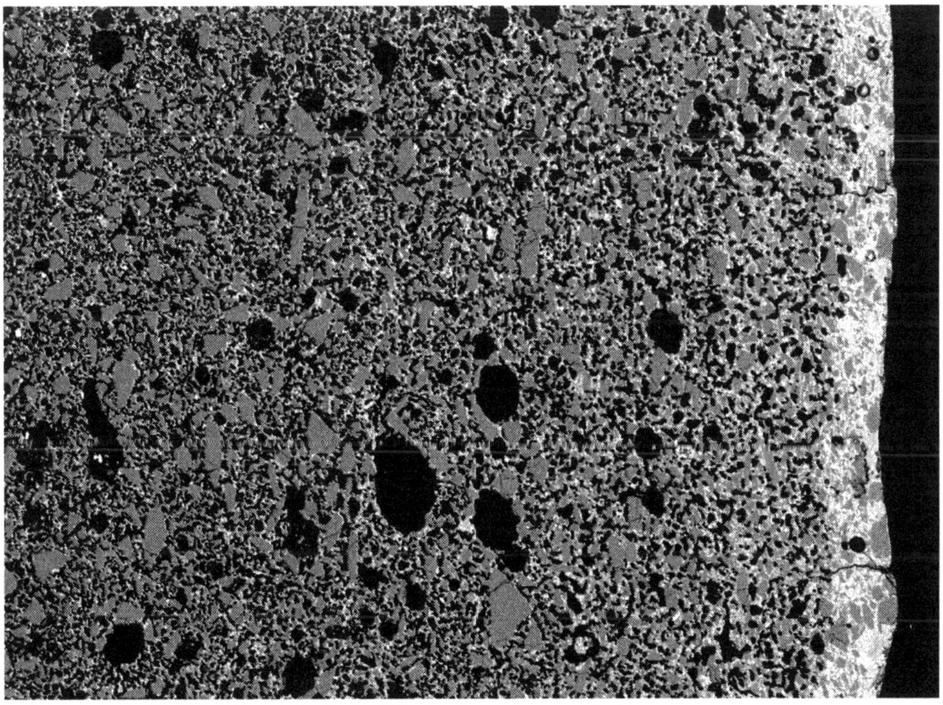

Figure 3-32 : Red faience inlay 770, *a photomicrograph showing the "Variant A" slip layer which is 3mm thick and lies above the coarse core which is off shot to the left. The distinct interaction layer is clearly visible, although the glaze is not well preserved in this area. Image is 2.5mm across.*

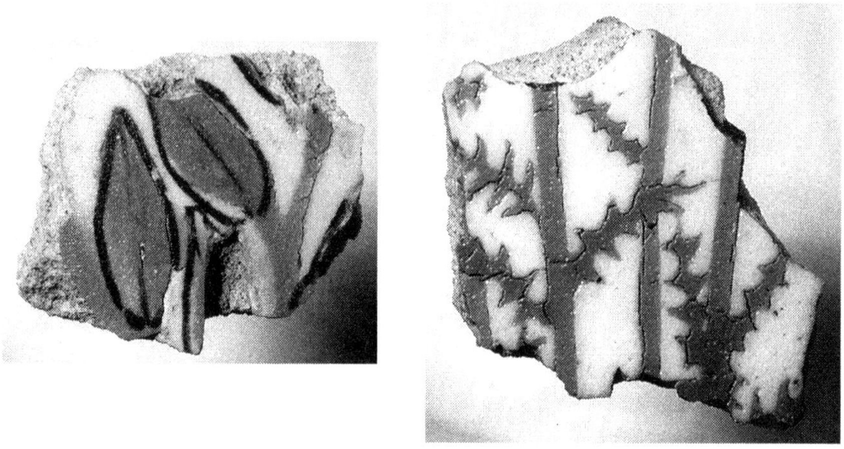

Figure 3-33 : Painting and inlaying in tiles. Left, *tile Berlin 11/12/876, a white tile inlayed with a green plant, with the leaves outlined with paint in black/brown (tile 47mm across).* ***Right,*** *tile Berlin 13/14/372, a white tile inlayed with a green plant, without any paint outlining. (tile 64mm across)*

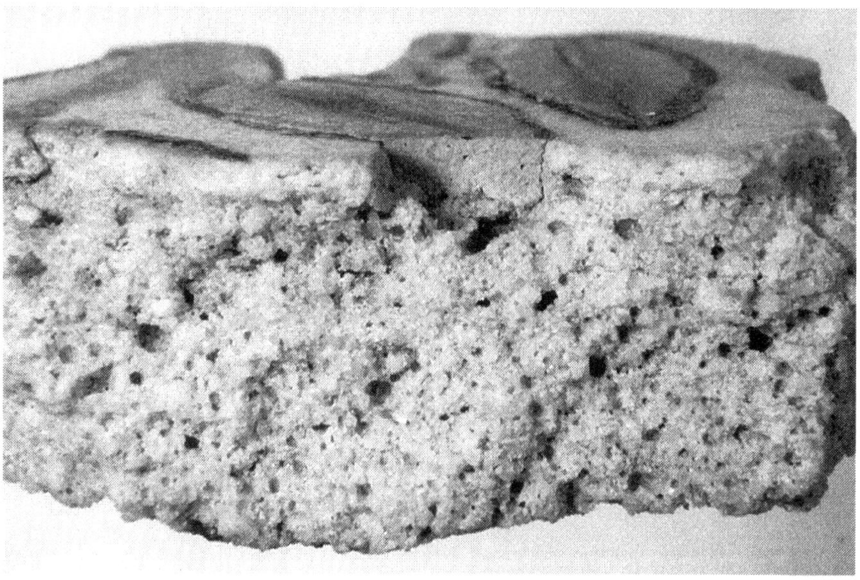

Figure 3-34 : Painted and inlayed tile. *Close up under low magnification of the broken edge of a tile showing inlayed elements. Tile Berlin 11/12/876, 47mm across.*

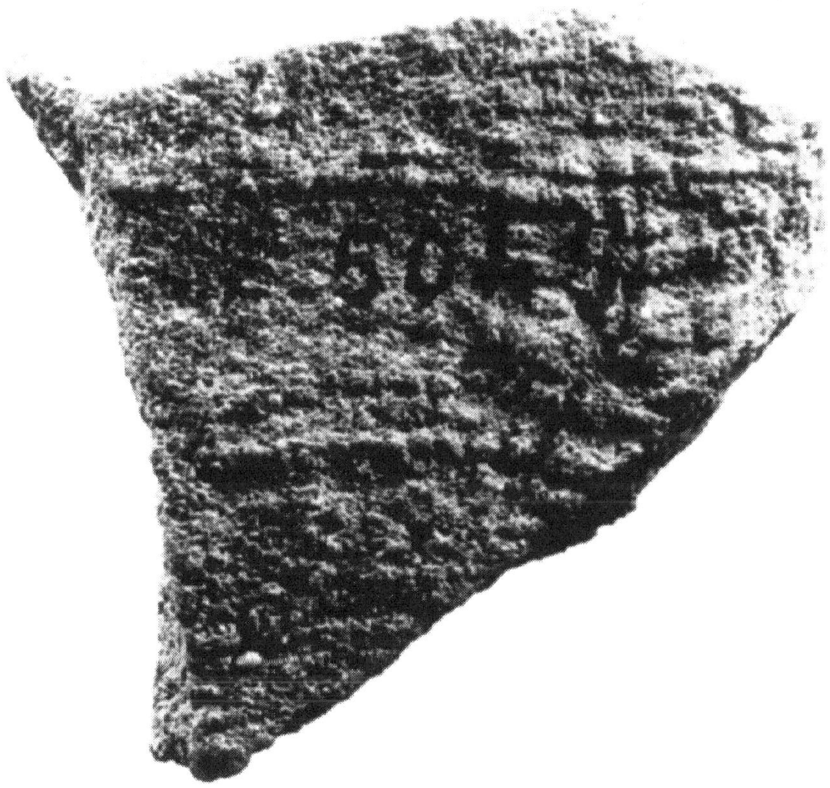

Figure 3-35 : Underside of tile E5045 : *In common with many of the tiles from Amarna, E5045 shows patterned ridges on its underside that show that it was dried on a textile or reed mat. Sample is 36mm across.*

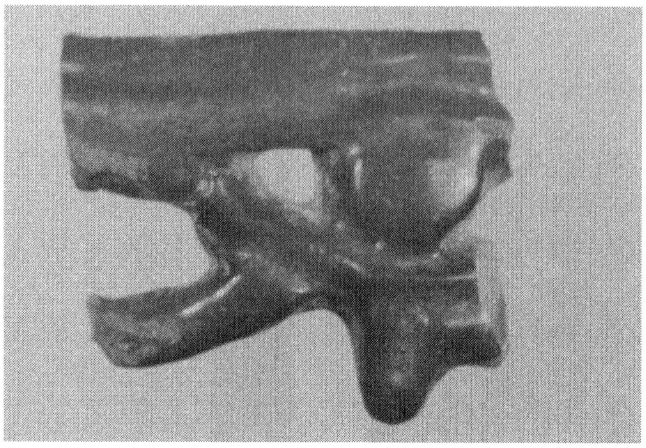

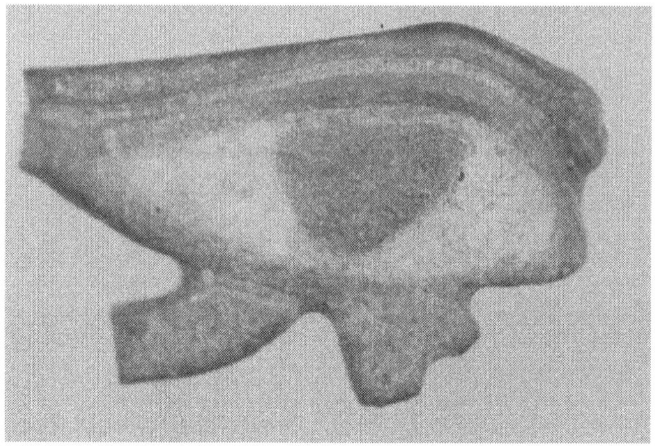

Figure 3-36 : Two faience finger rings. Wadjet-*eye finger ring bezels similar to the broken items from which the samples for the study were taken. On the left is an example in turquoise blue, typical of a colouring by copper, and on the right is a dark blue example, typical of a cobalt colorant, and inlayed in white and slightly weathered. Both examples 18th dynasty and from the Liverpool University Museum, E5053.*

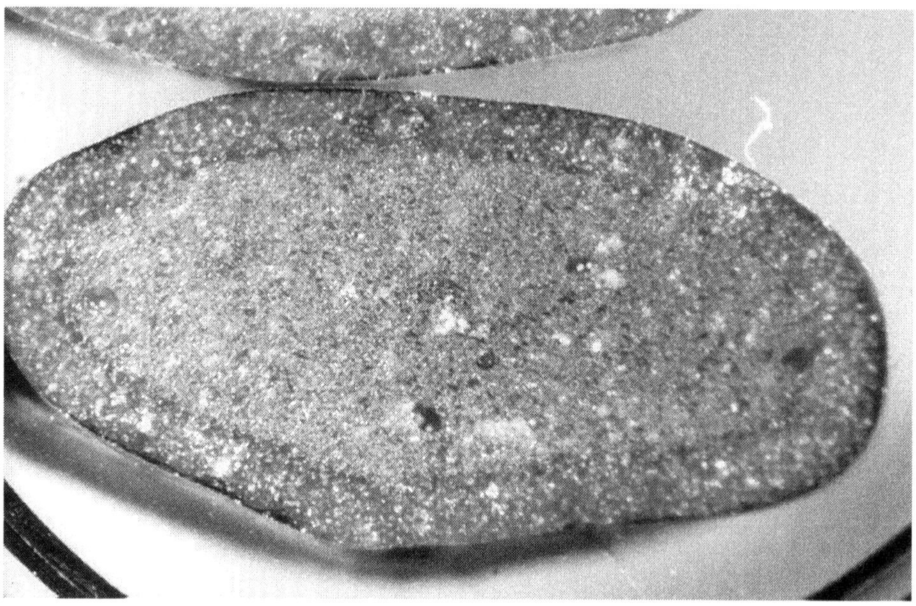

Figure 3-37 : Cross sections of ring shanks showing their shapes and colouring. *Flattened ring shank E5053/3, 4mm along the long axis. Flattening is perpendicular to the plane of the ring.*

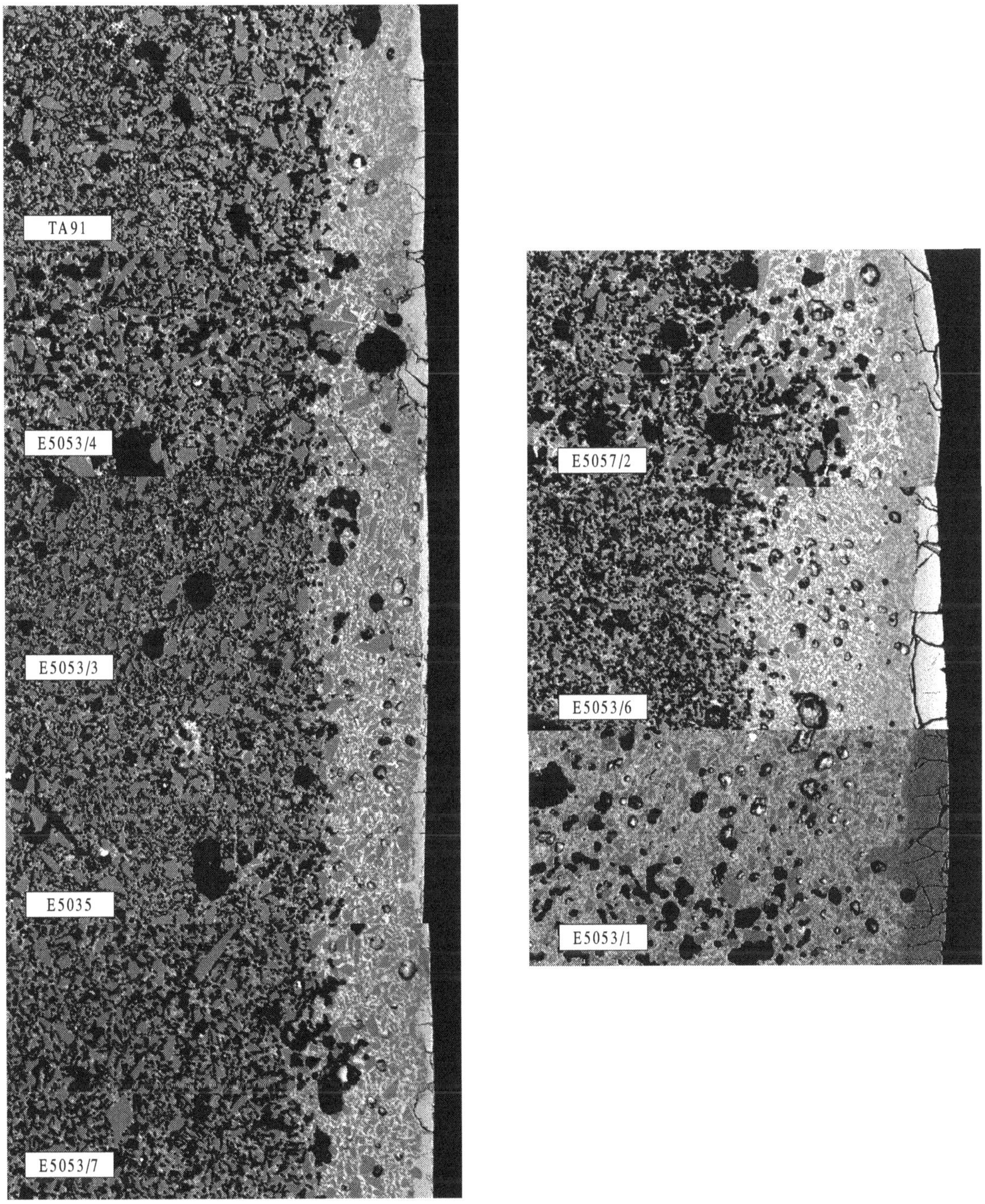

***Figure 3-38 : Microstructure of typical ring shanks.** Left are photomicrographs of five examples of typical sections through ring shanks, demonstrating the consistency seen in the microstructures. Right are photomicrographs of three other examples of sections through ring shanks, showing some variability in the microstructures. 5053/1 is one of the dark blue examples. All images are at the same magnification and 1.5mm across. In the left column, from the top, TA91, E5053/4, E5053/3, E5035 and E5053/7. In the right column, from the top, E5057/2, E5053/6 and E5053/1*

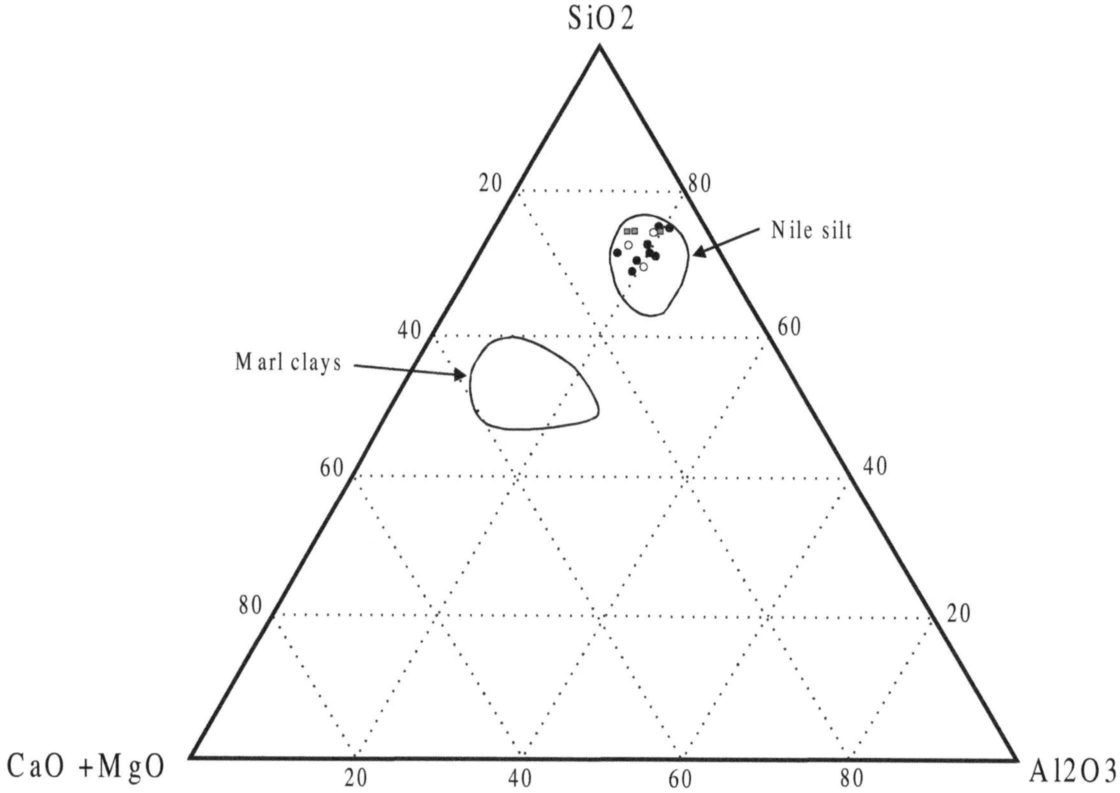

Figure 4-1 : Triangular diagram of SiO2-Al2O3-CaO+MgO*, showing the plots of the ceramics associated with vitreous materials (black circles), the unfired Nile silts (white circles) and the kiln wallfragments (grey squares). The fields for Nile silt and marl clay are from Noll (1981).*

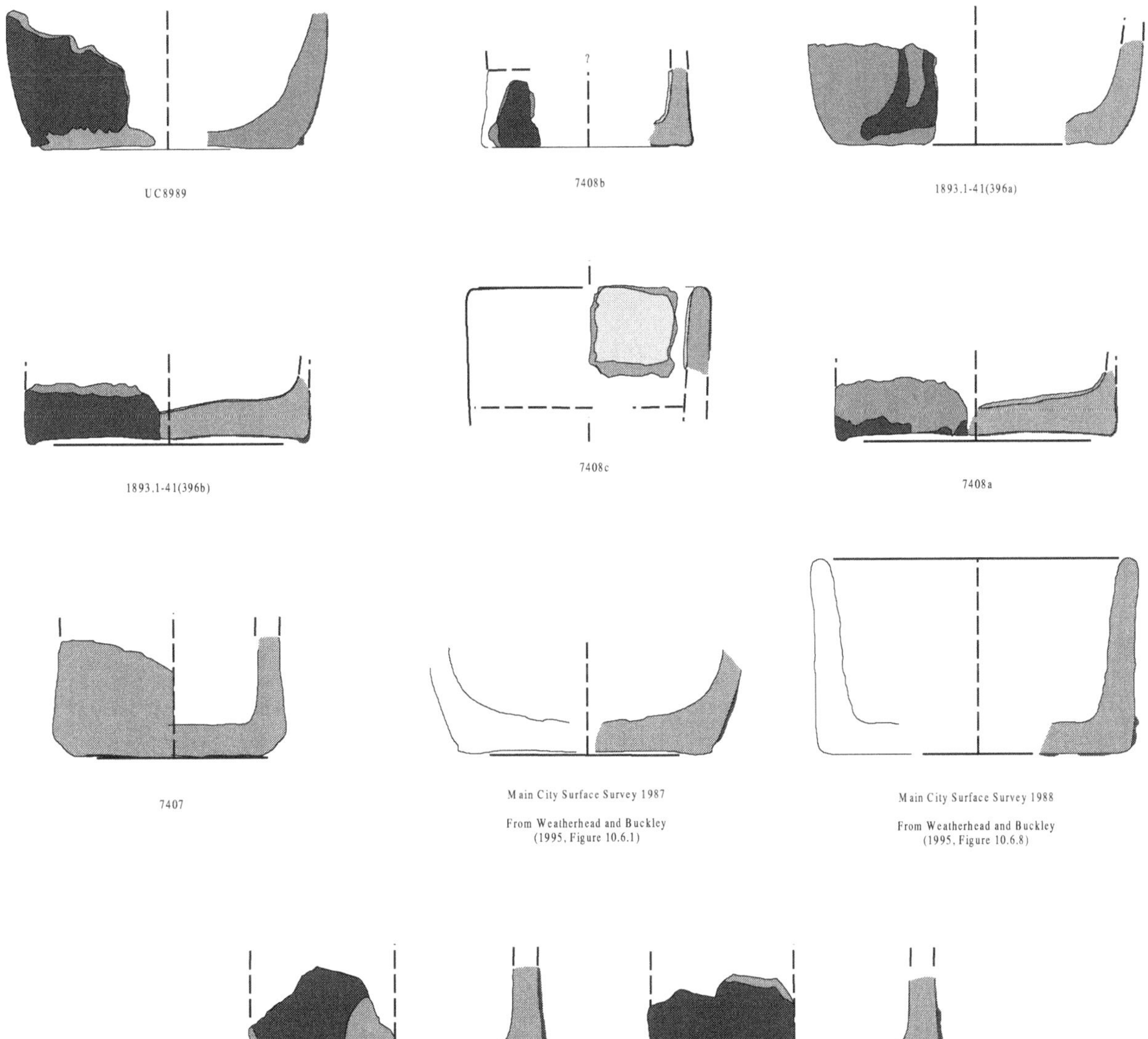

Figure 4-2a : Cylindrical vessels from Amarna with external glass only, *schematically showing the position of white slip (light grey), glass (dark grey).*

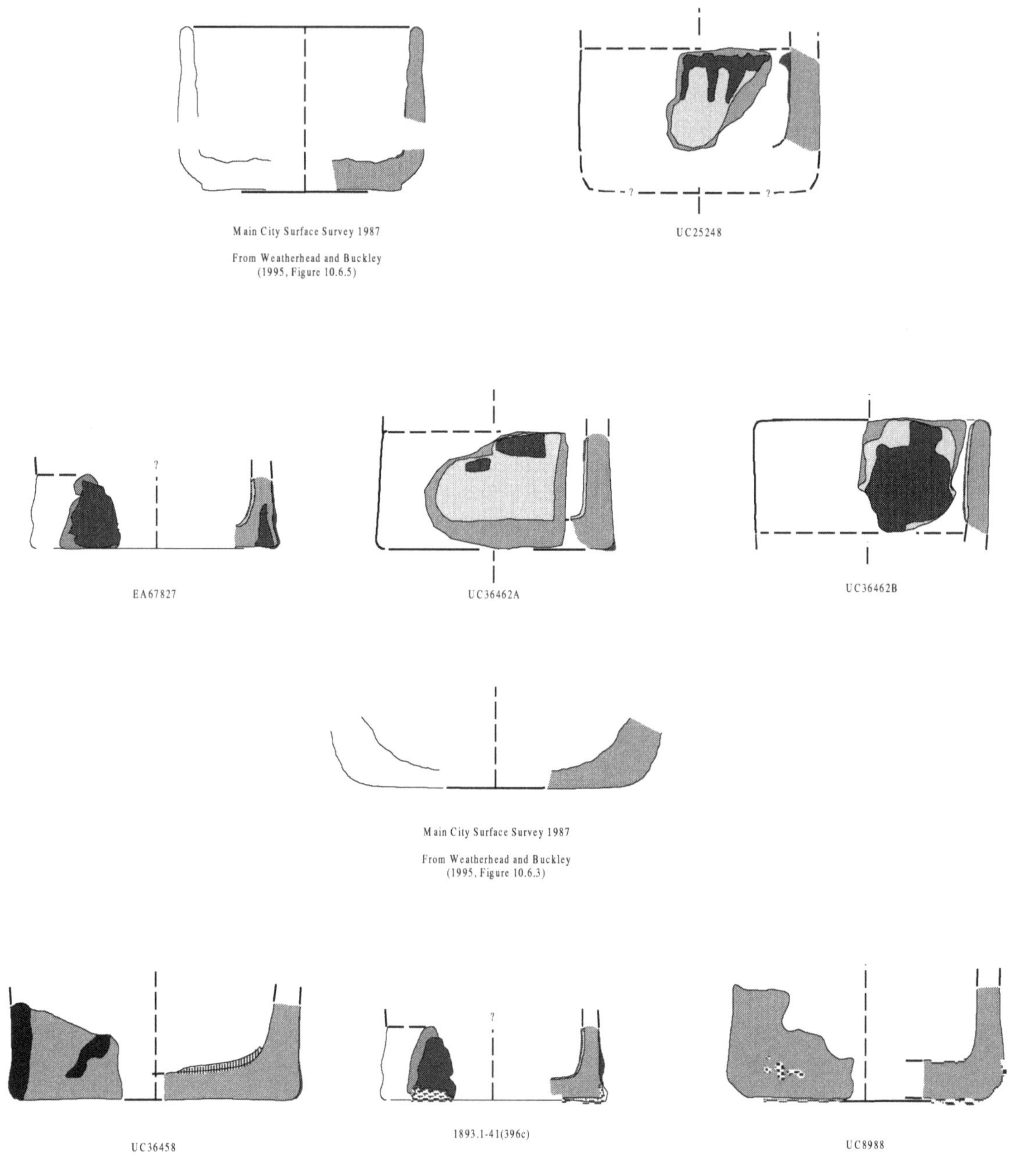

***Figure 4-2b : Cylindrical vessels from Amarna with other combinations of glass and frit**, top row - internal glass only, second row - internal and external glass, third row - no vitreous materials, bottom row - two different vitreous materials. Showing schematically the position of white slip (light grey), glass (dark grey), frit (dots) and Egyptian blue (stripes).*

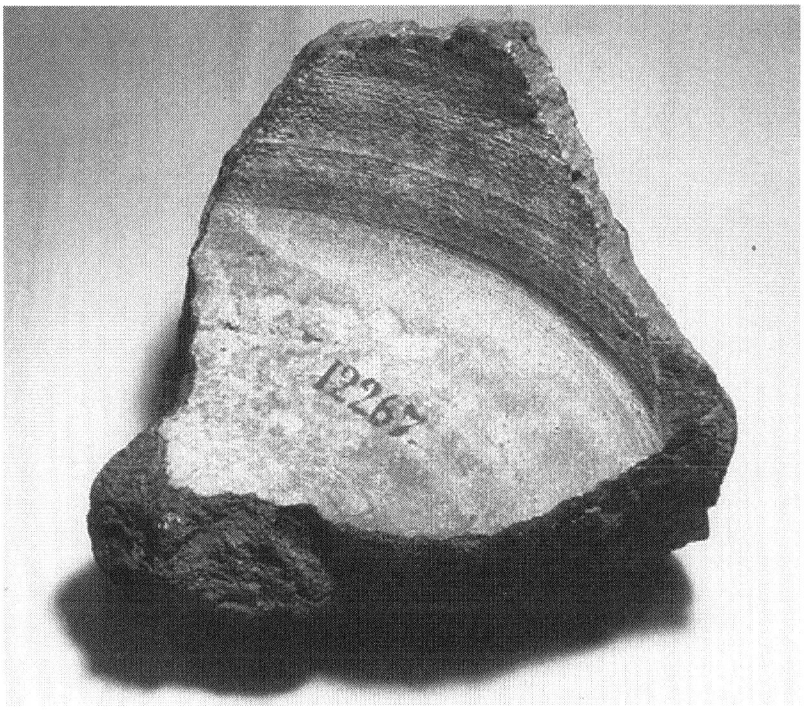

***Figure 4-3 : Cylindrical vessel**: Berlin 12267a, a typical example of a cylindrical vessel showing the internal white slip and vertical wall. Fragment is 120mm across.*

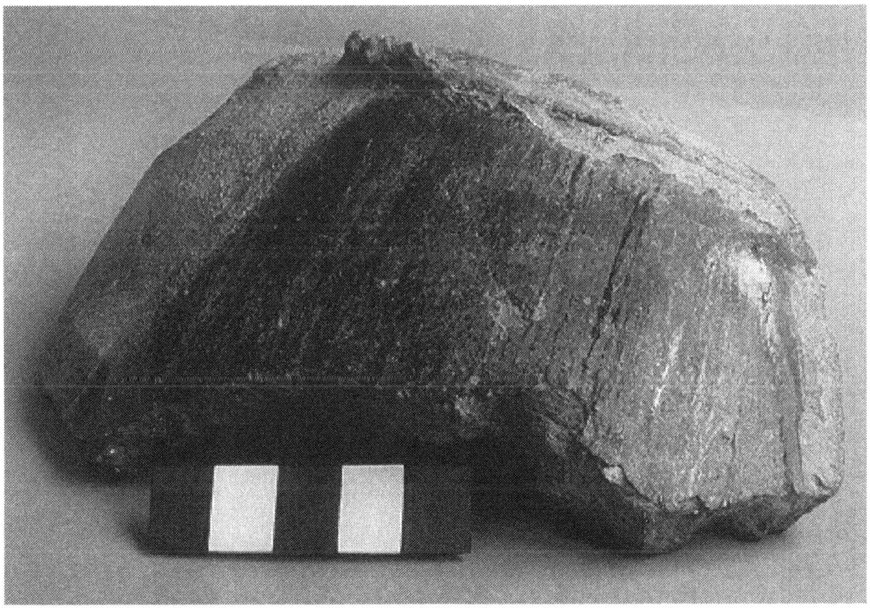

***Figure 4-4 : External glass on a cylindrical vessel.** Photograph of 8989 showing coverage of glass over the external wall of the vessel. Copyright of the Petrie Museum of Egyptian Archaeology UC 8989.*

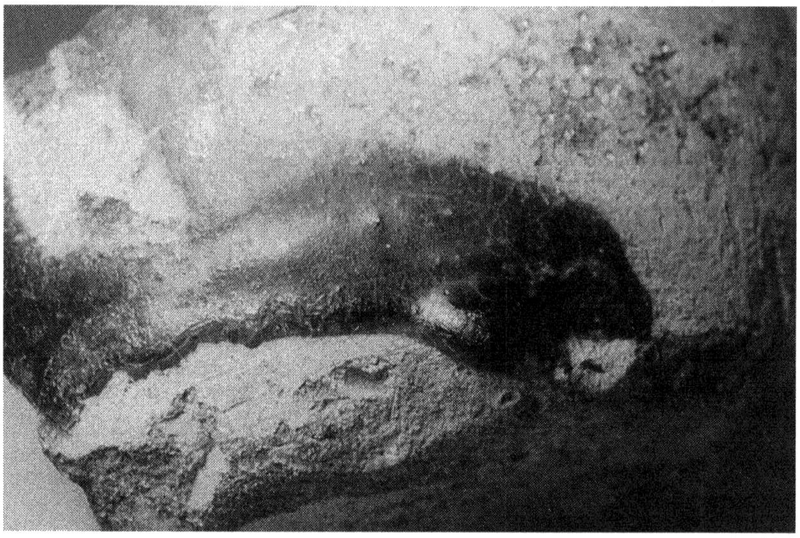

Figure 4-5 : Cylindrical vessel with drip: *396a, a typical example of a cylindrical vessel showing glass teared externally.*

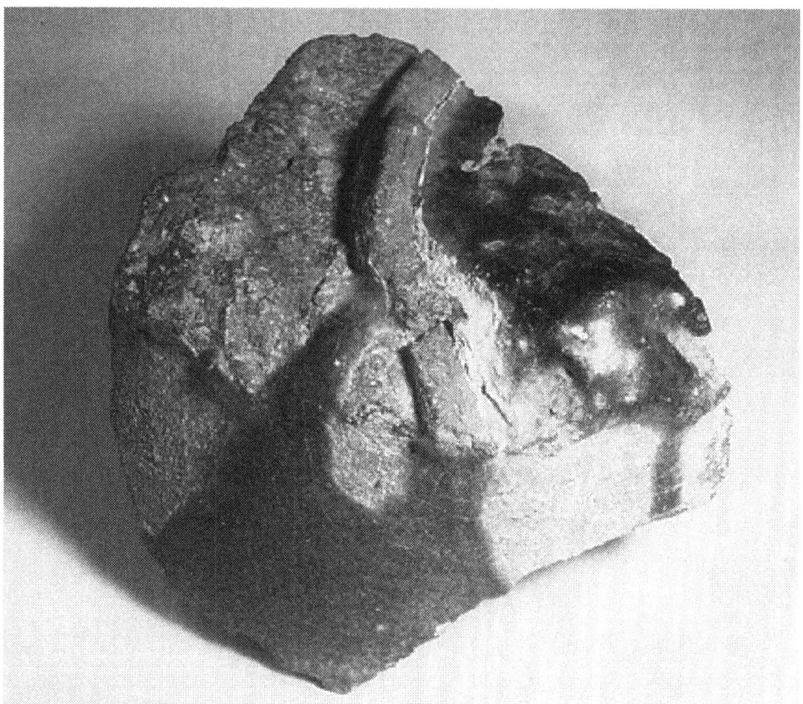

Figure 4-6 : Base of cylindrical vessel 12276a, *with glass running down and an adhering rim of a second cylindrical vessel attached to the underside of the base.*

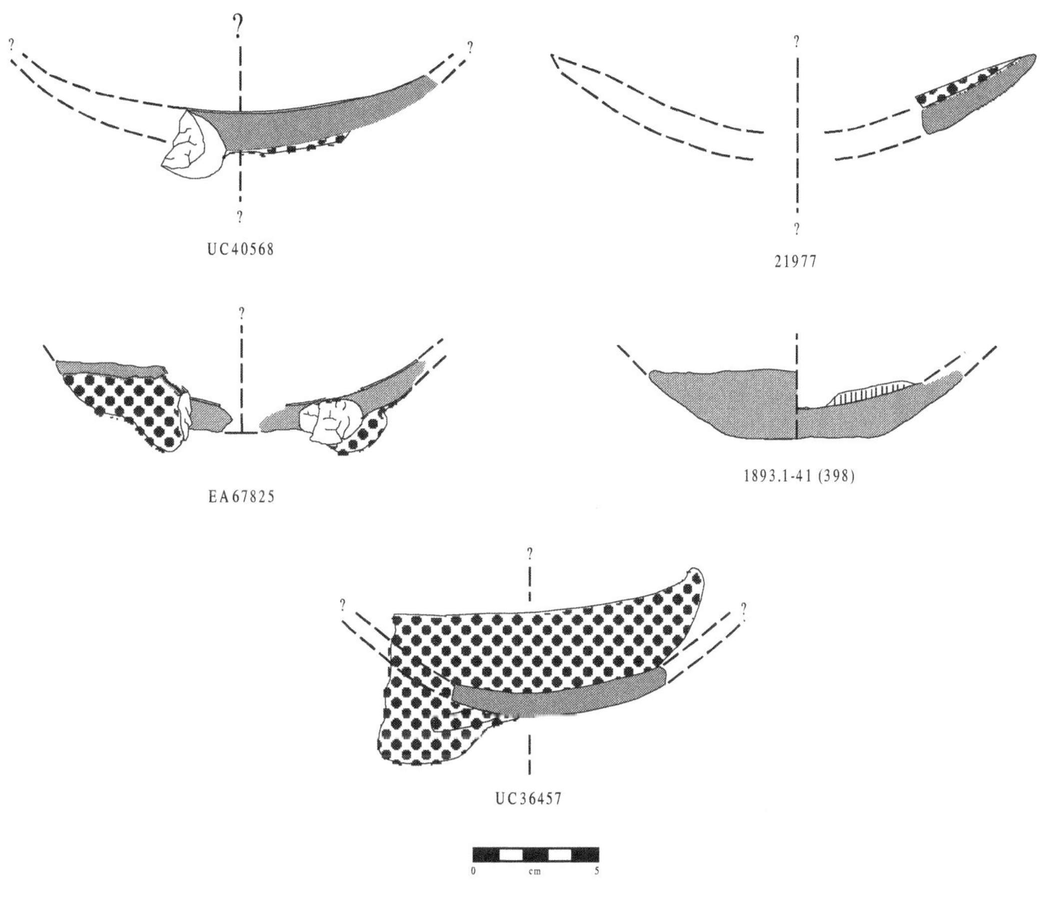

***Figure 4-7 : Fritting pans from Amarna**, schematically showing the position of white slip (light grey), frit (dots) and Egyptian blue (stripes).*

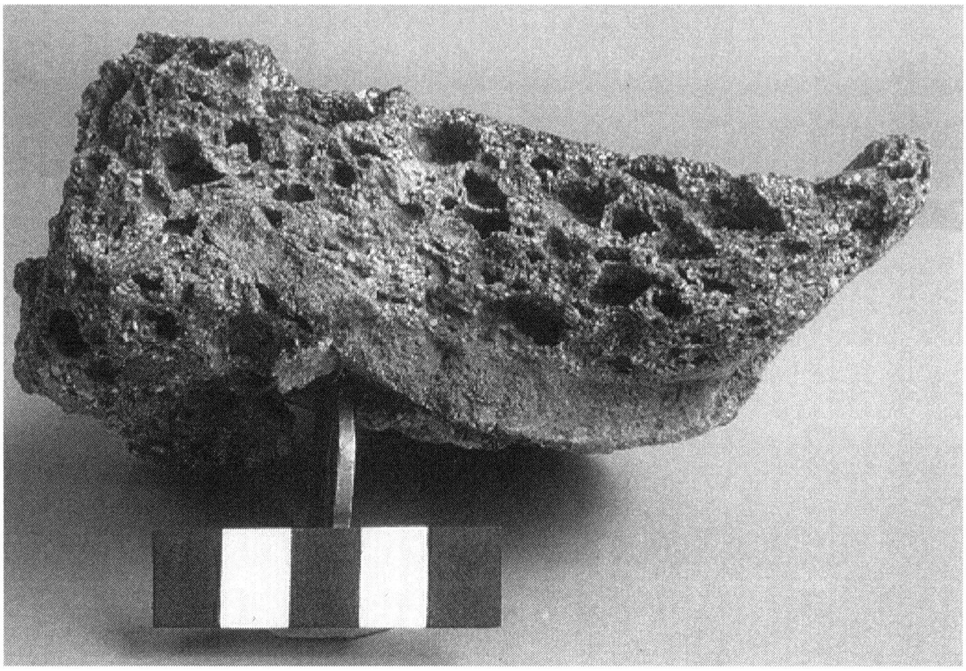

Figure 4-8 : Fritting pan. *Photograph of the largest surviving example of a fritting pan, UC 36457, showing the fragmented pan with large amounts of internal frit. Copyright of the Petrie Museum of Egyptian Archaeology, UC 36457.*

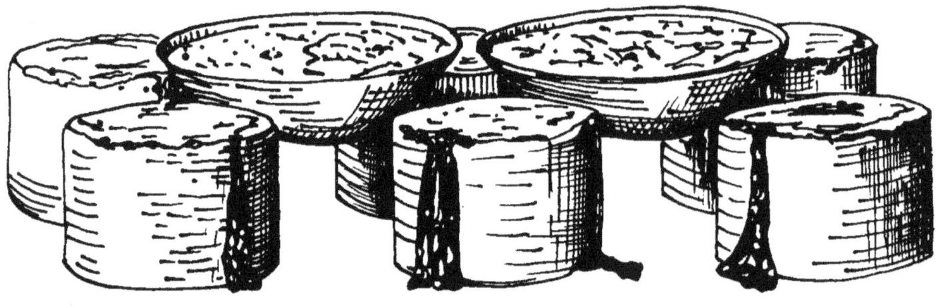

Figure 4-9 : Petrie's proposed reconstruction *of the orientation of the fritting pans and cylindrical vessels in the kilns (Petrie 1894). The caption to the drawing reads, "Fritting Pans, supported in the furnace on jars inverted, down which the glaze runs. W.M.F.P."*

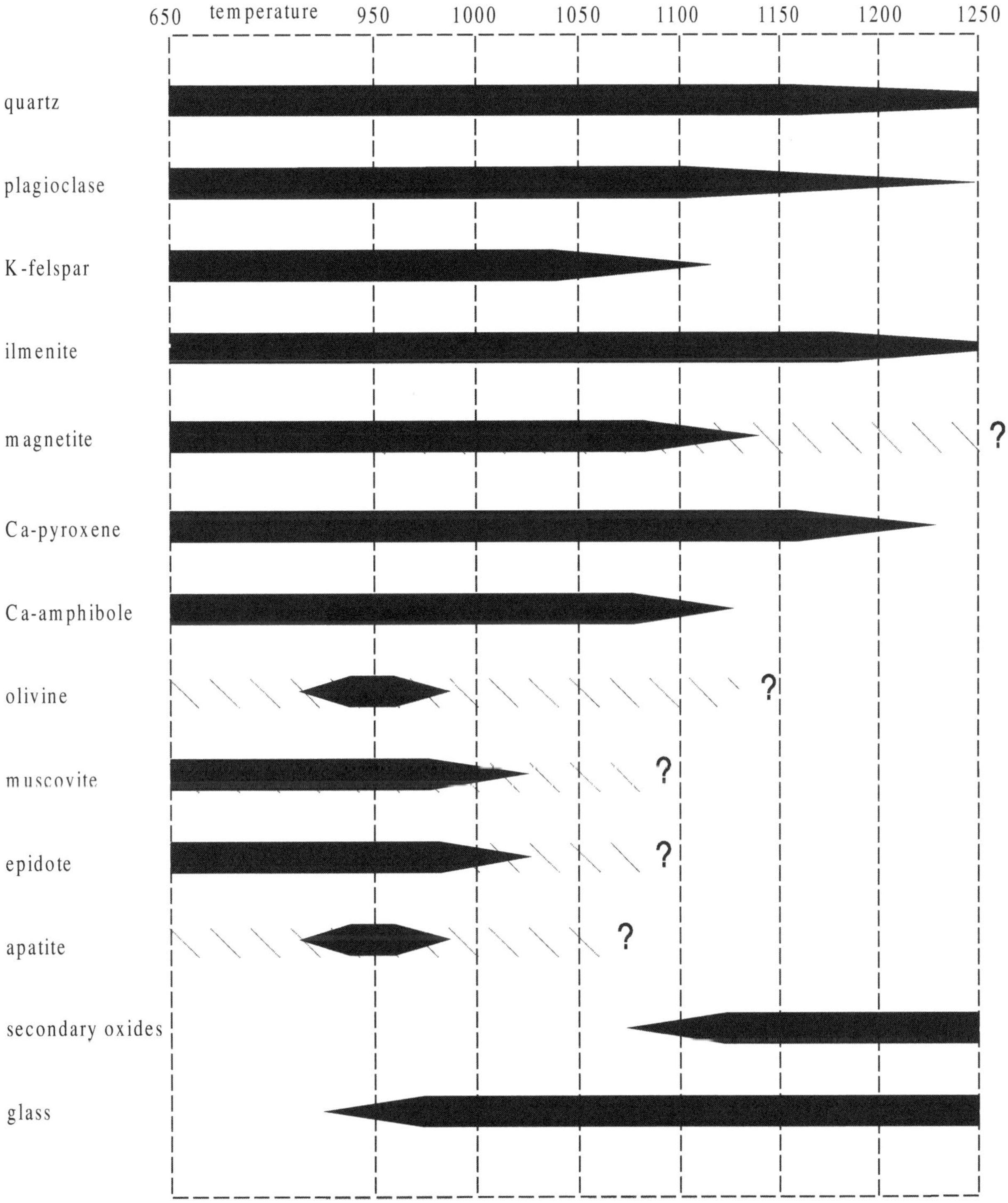

Figure 4-10 : Summary of the range of survival of the minerals in the Nile silt *at various temperatures based on the results of observation in the SEM of the experimentally refired AM45 samples.*

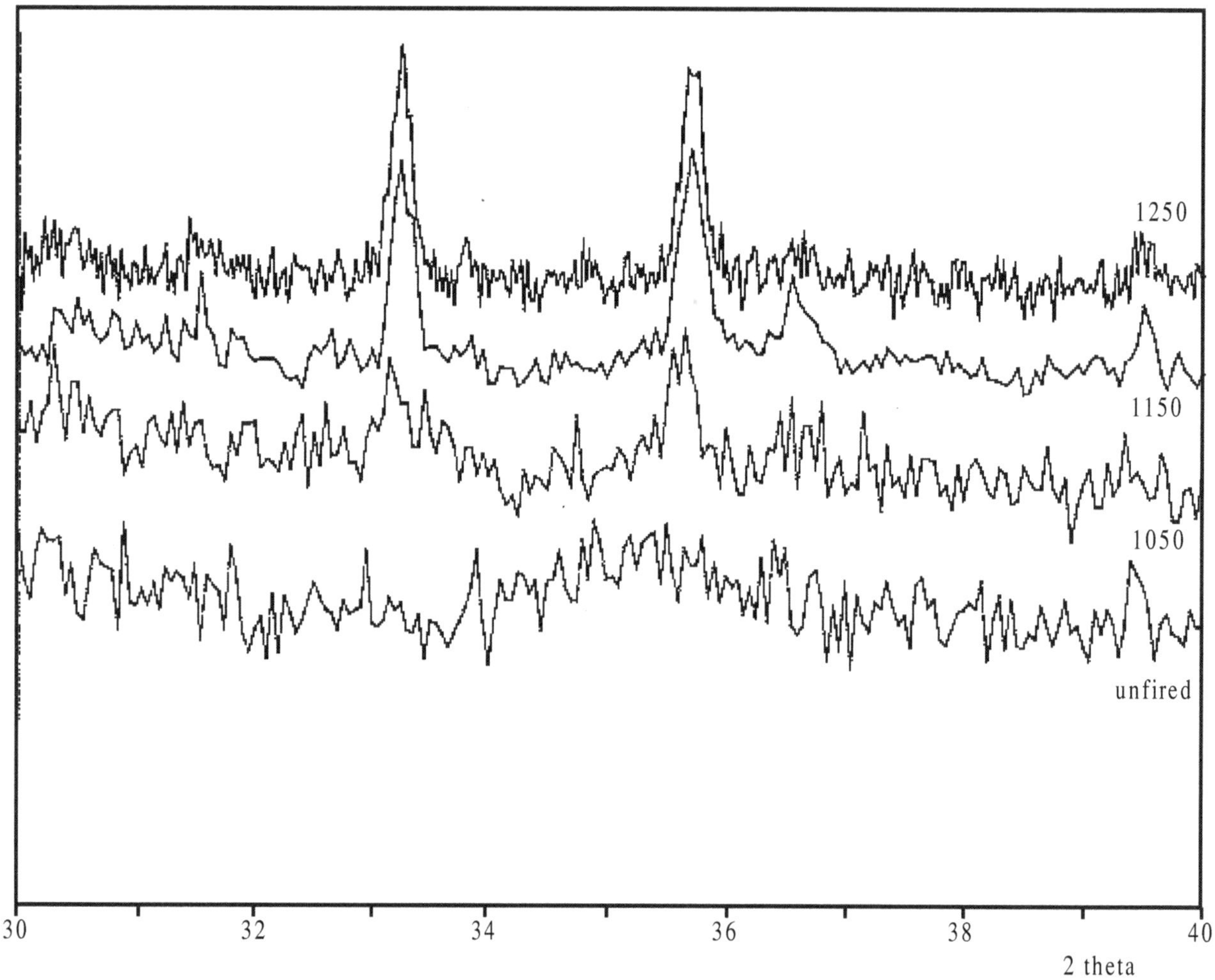

Figure 4-11 : Original XRD traces *for the experimentally refired Nile silt AM45 for three temperatures and the unfired original shown on the same vertical scale (but displaced), showing the development with temperatures of the principal hematite peaks.*

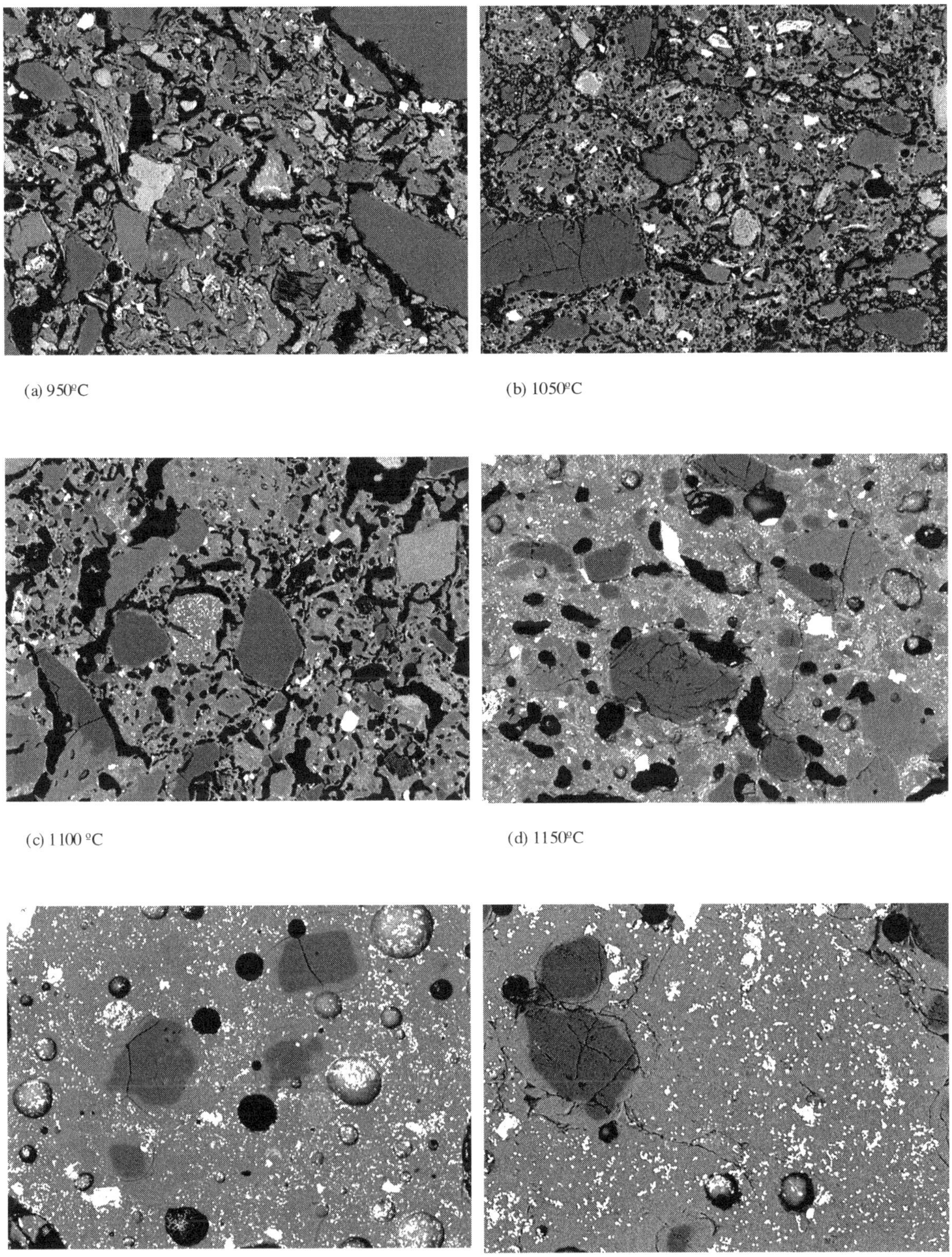

(a) 950ºC

(b) 1050ºC

(c) 1100 ºC

(d) 1150ºC

(e) 1200 ºC

(f) 1250ºC

Figure 4-12 : Nile silt AM45 experimentally refired *to various temperatures all images 0.33mm across.*

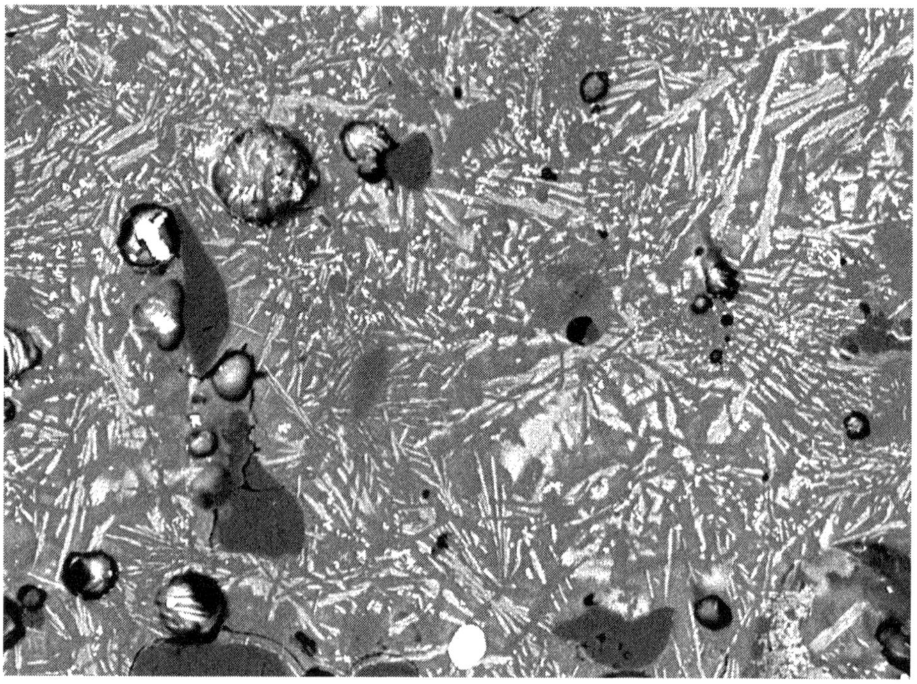

Figure 4-13 : Vitrified clay AM 24 *showing extreme development of secondary grains of diopside.*

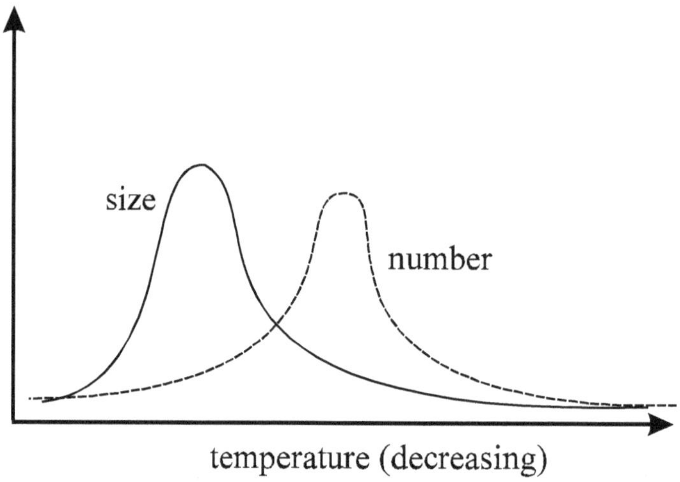

Figure 4-14 : Secondary mineralisation. *Schematic graph showing amount of crystal growth (solid line) and number of crystals (dashed line) in relation to the temperature. If the temperature falls slowly, small number of large crystals are formed, a little faster and large numbers of small crystals form, whilst if the temperature cools very rapidly through both these zones, no crystals form at all.*

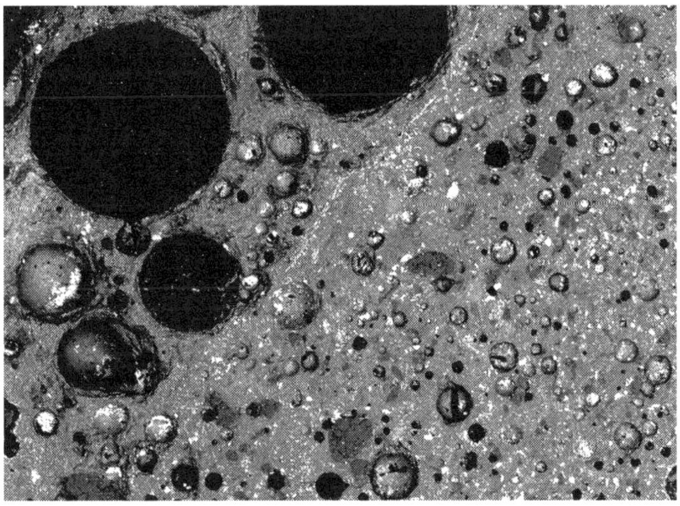

(a) high temperature inner wall

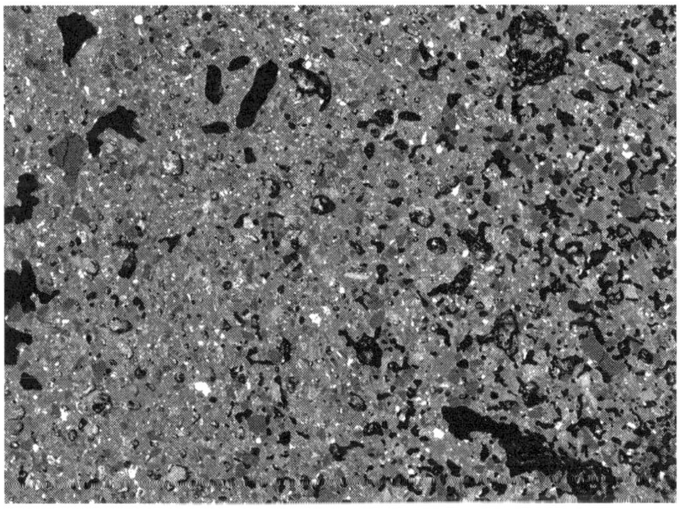

(b) change from extensive vitrification to continuous vitrification (middle area)

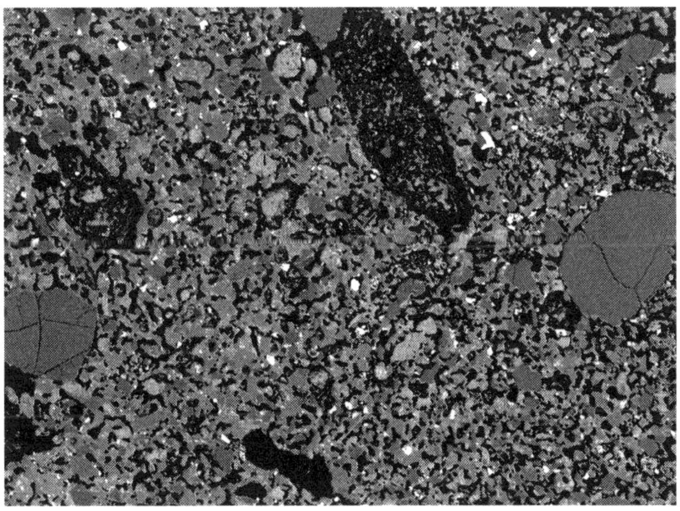

(c) low temperature body.

Figure 4-15 : Kiln wall fragment AM42c *showing the different microstructures at (a) the surface of the kiln wall, (b) 20-23mm into the wall and (c) 33mm deep into the body of the wall.*

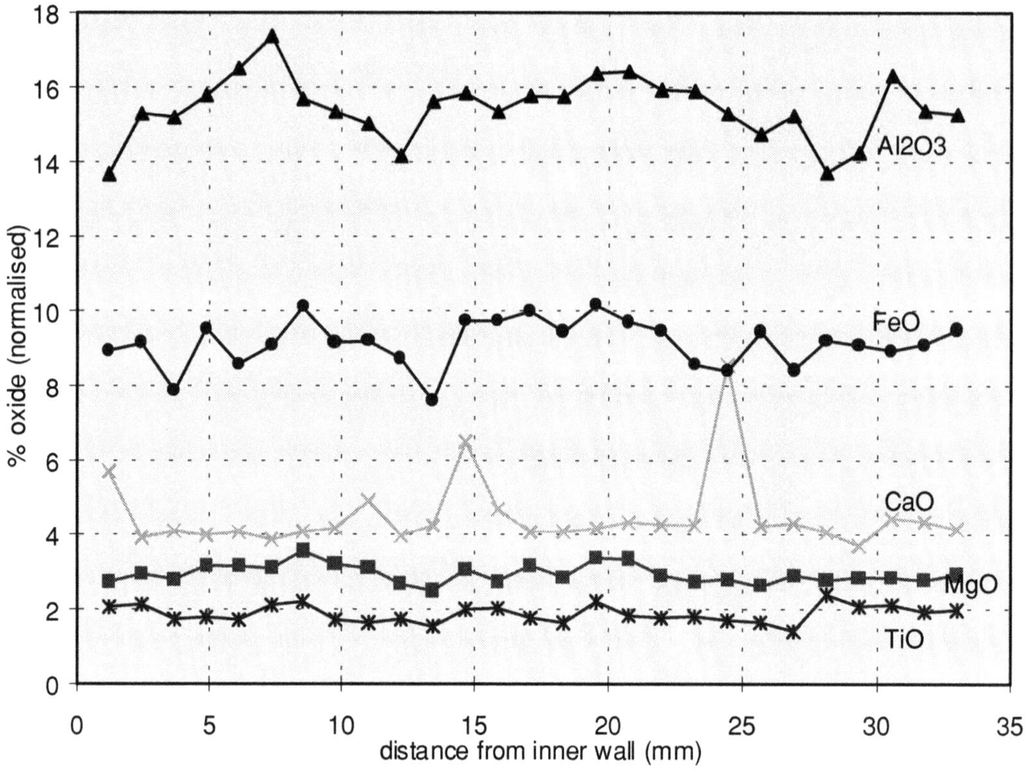

Figure 4-16 : Graph showing normalised oxide percentages, *from the wall of the kiln to 33mm into the body.*

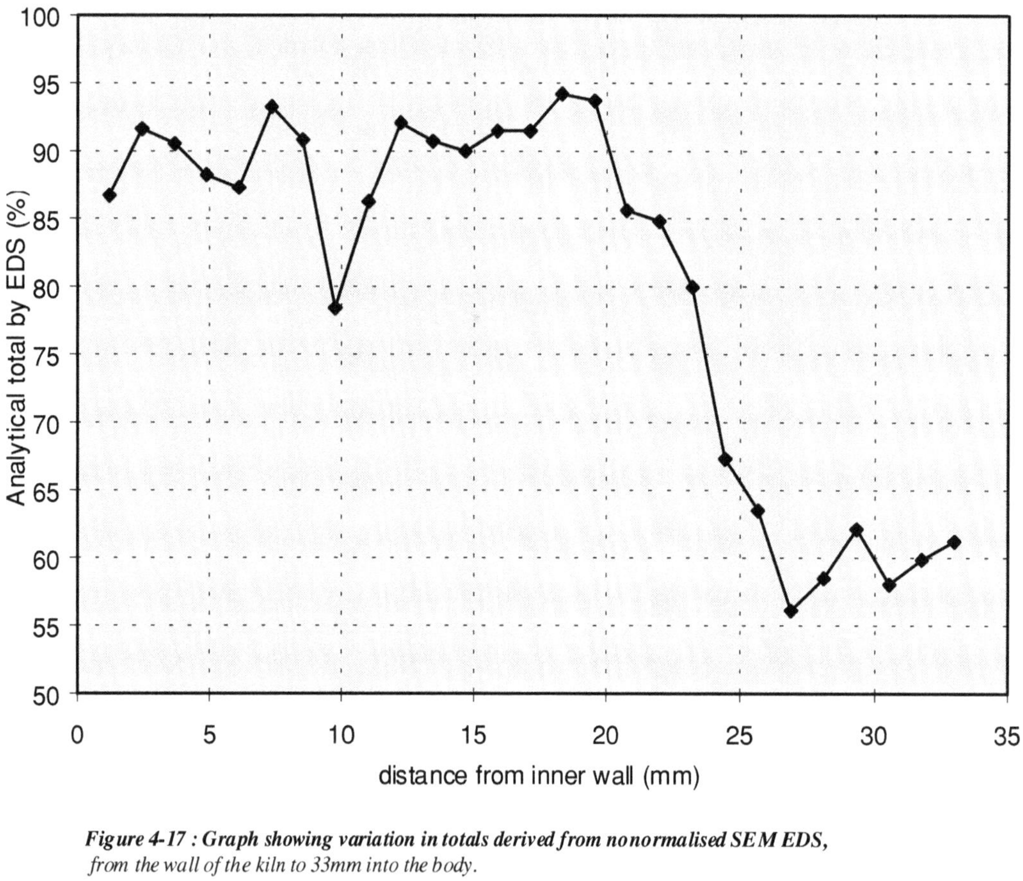

Figure 4-17 : Graph showing variation in totals derived from nonormalised SEM EDS, *from the wall of the kiln to 33mm into the body.*

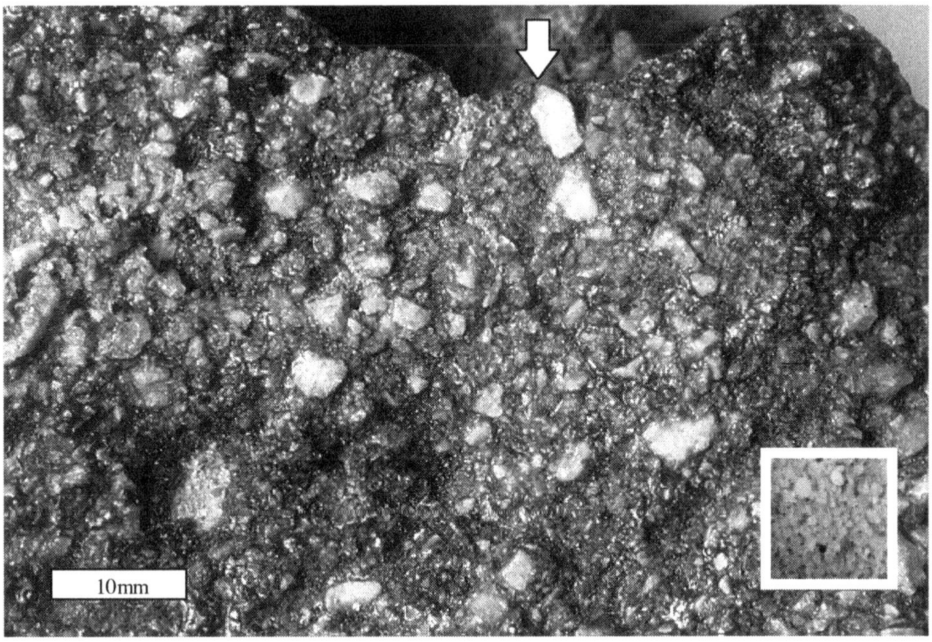

Figure 5-1 : Quartz grains in UC36457. *Close up under low power magnification of the surface of the frit in fritting pan UC36457, showing the size and angularity. AM45, the coarse sand, is shown inset at the same scale. The long axis of the grain marked is 6mm long.*

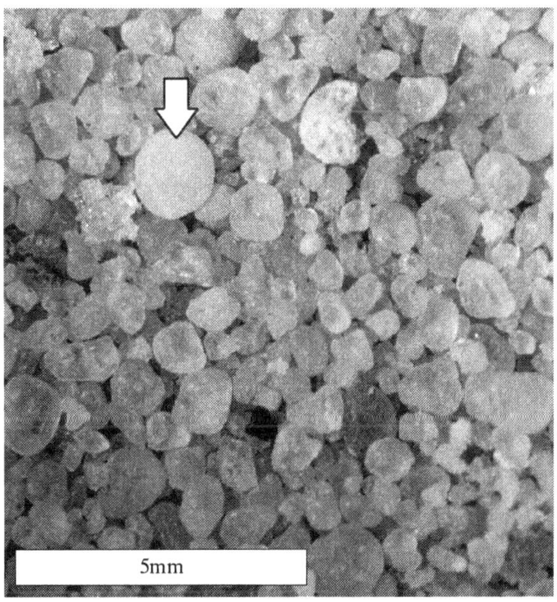

Figure 5-2 : Rounded sand grains in AM44. *Close up under low power magnification of the well sorted sand grains of AM44 from Amarna, showing an average grain size of 0.75mm and well rounded shape. The grain marked is 1mm across.*

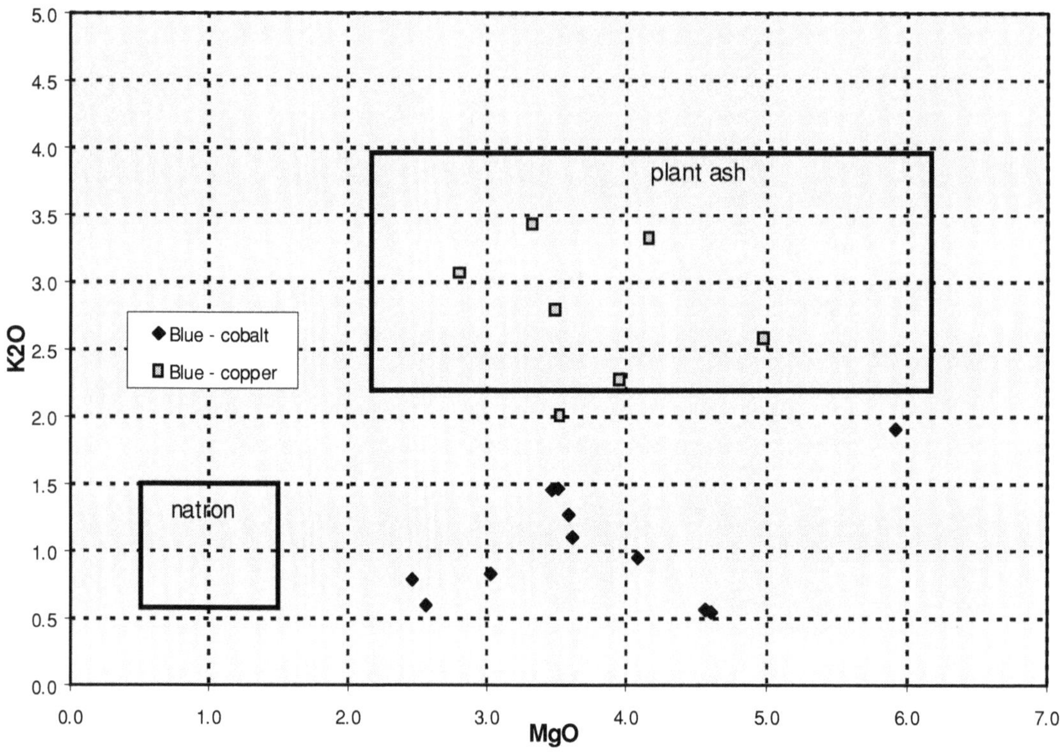

Figure 5-3 : Graph of MgO against K2O for the glasses. *Data from SEM WDS. Also plotted are the fields in which glass made with natron alkali and plant ash alkali should plot according to Lilyquist and Brill (1995).*

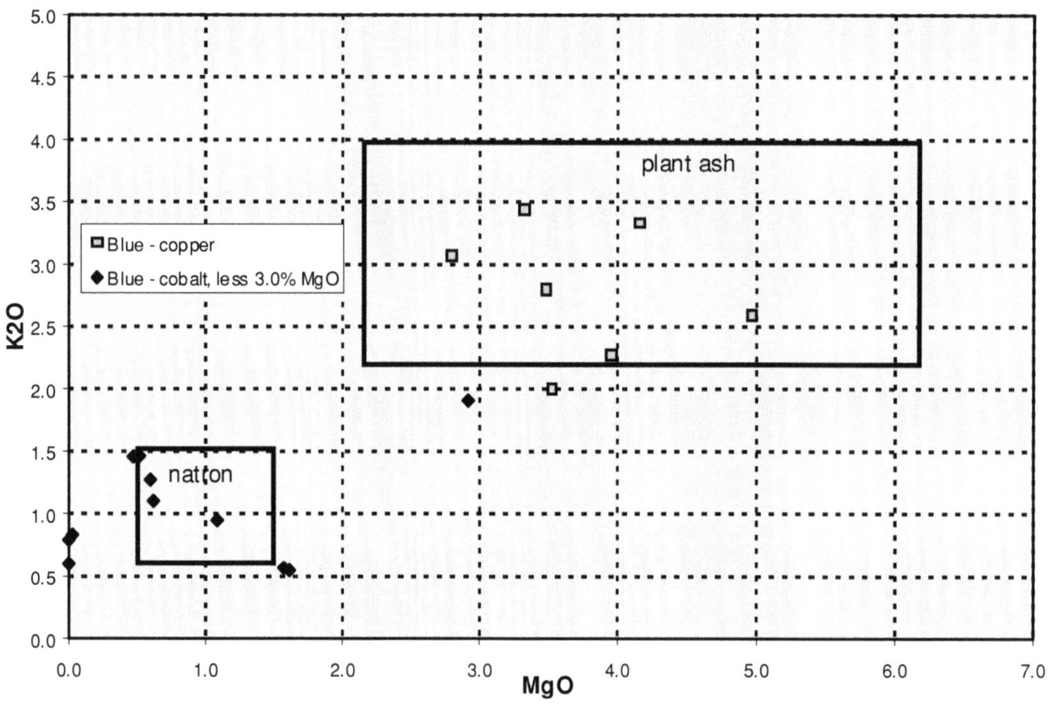

Figure 5-4 : Graph of MgO minus 3.0% against K2O for the glasses. *Data from SEM WDS. In this graph the MgO has been reduced by 3.0% in the cobalt glasses, the amount interpreted as being added by the addition of the alum. Also plotted are the fields in which glass made with natron alkali and plant ash alkali should plot according to Lilyquist and Brill (1995).*

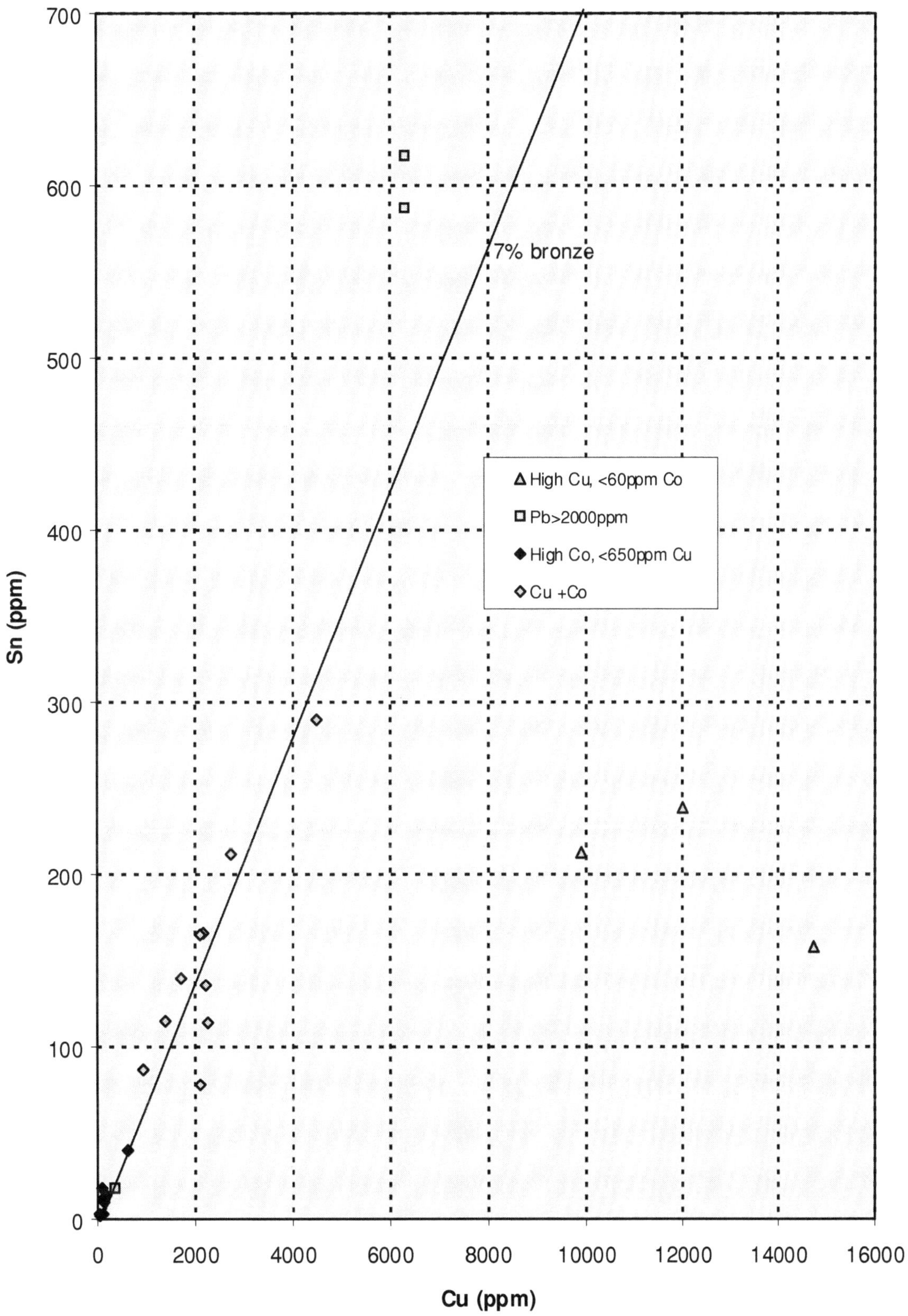

Figure 5-5 : Graph of Cu ppm against Sn ppm *for the glasses in a special study of the colorants, all data from ICP-MS. The line upon which bronzes that are 7.0% Sn would plot is marked.*

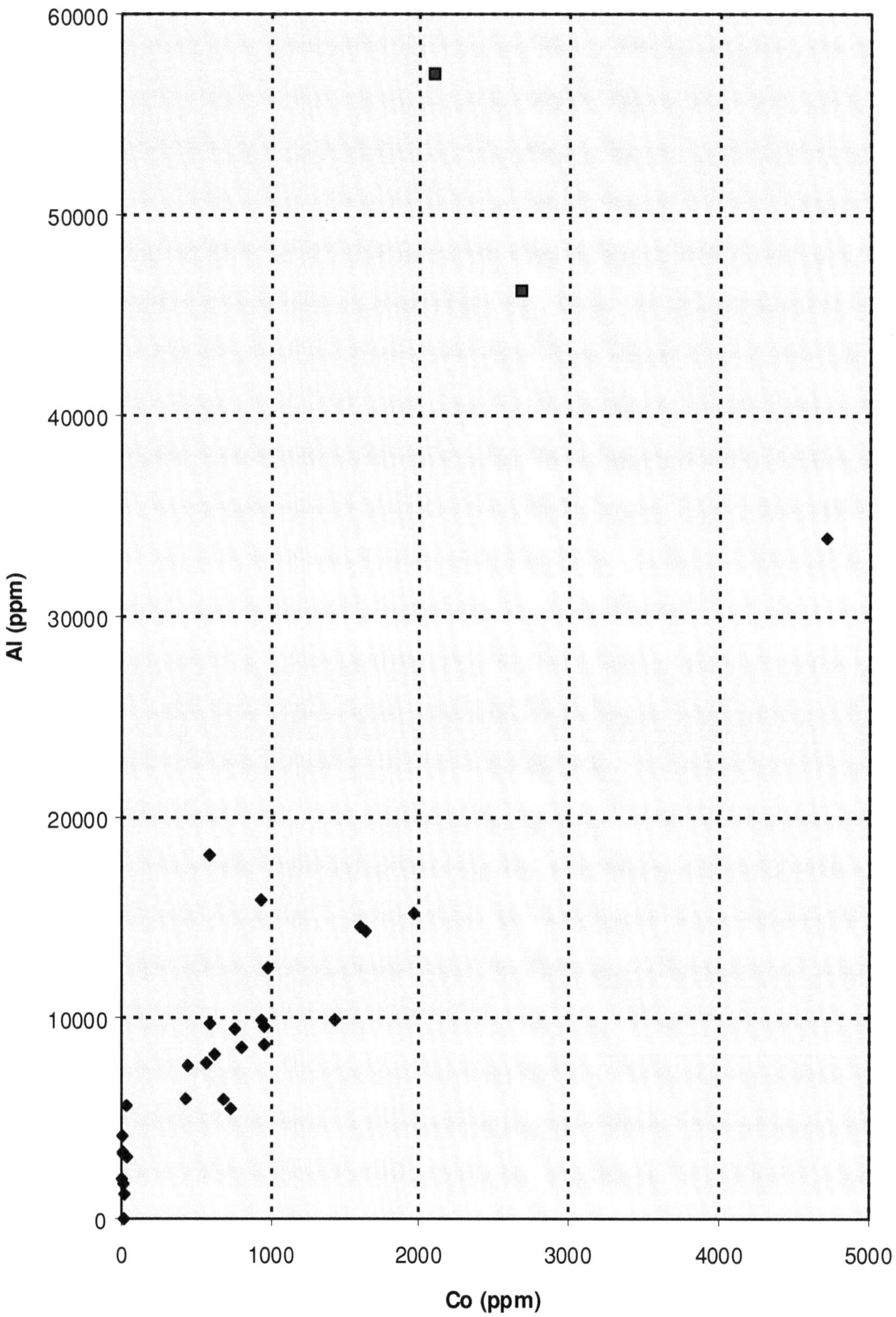

Figure 5-6 : Graph of Co ppm against Al ppm *for the glasses and frits(blue diamonds) and the alums (pink squares) in a special study of the colorants, all data from ICP-MS.*

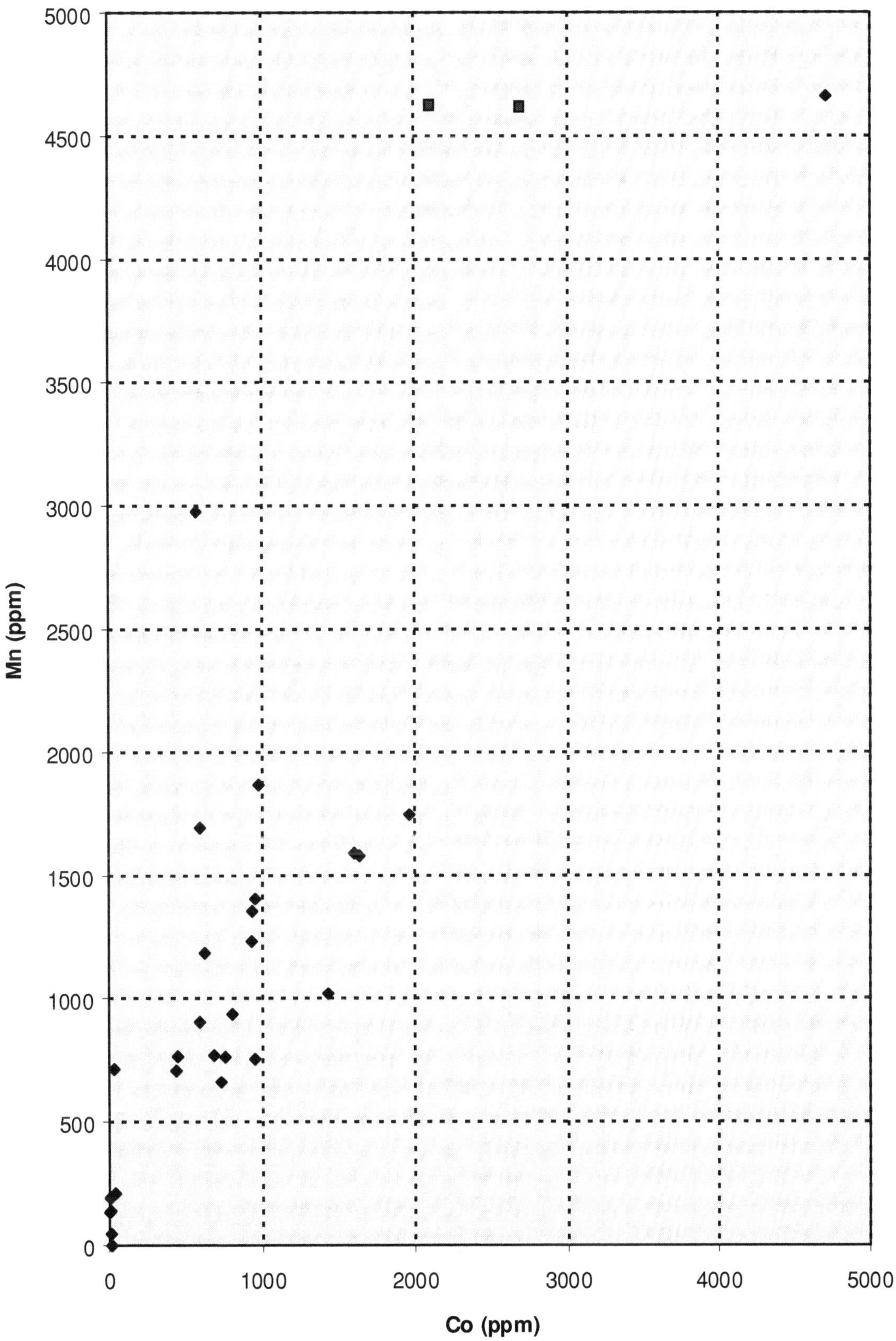

Figure 5-7 : Graph of Co ppm against Mn ppm *for the glasses and frits(blue diamonds) and the alums (pink squares) in a special study of the colorants, all data from ICP-MS.*

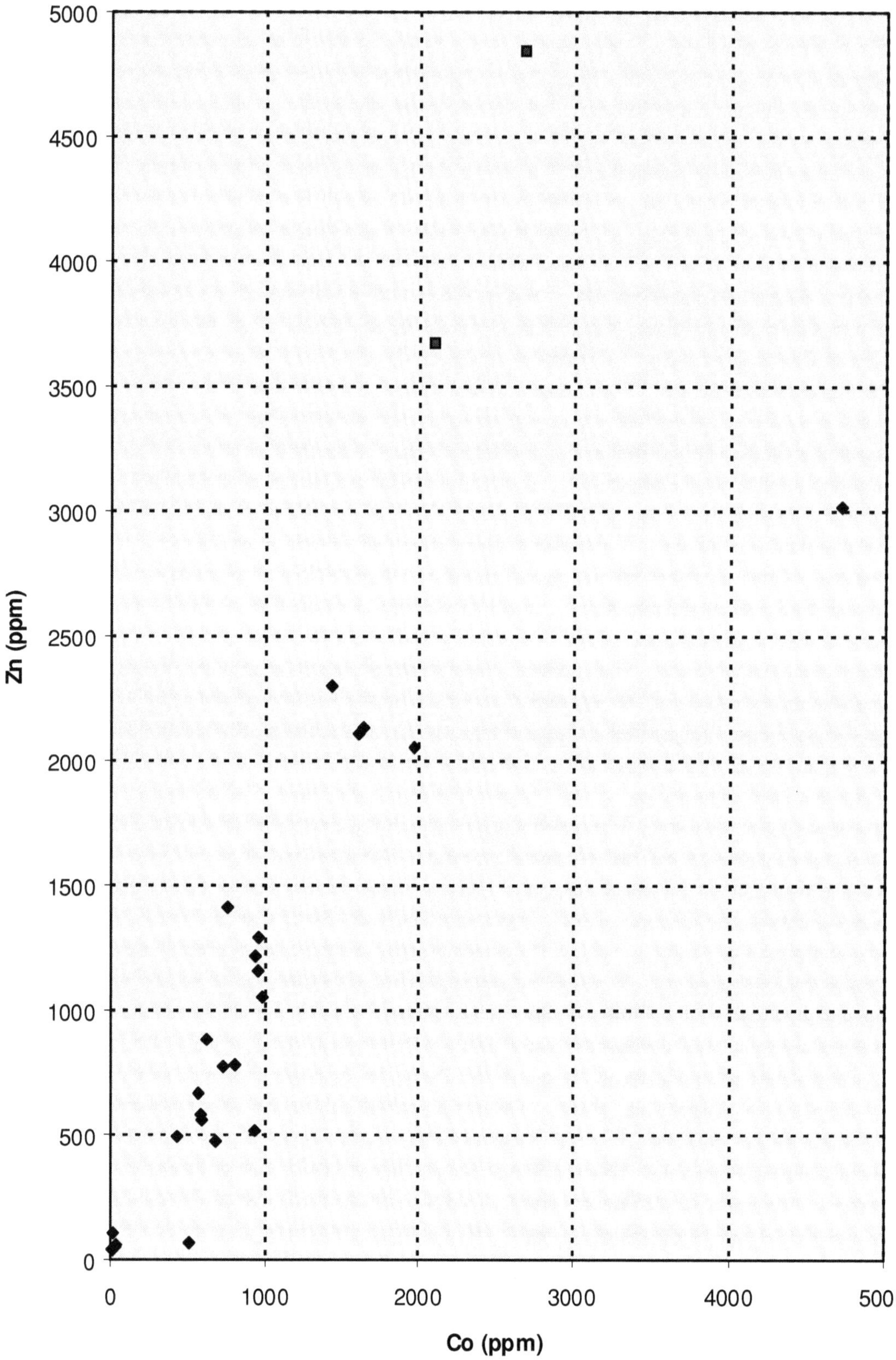

Figure 5-8 : Graph of Co ppm against Zn ppm *for the glasses and frits(blue diamonds, yellow squares with no cobalt are lead coloured) and the alums (pink squares) in a special study of the colorants, all data from ICP-MS..*

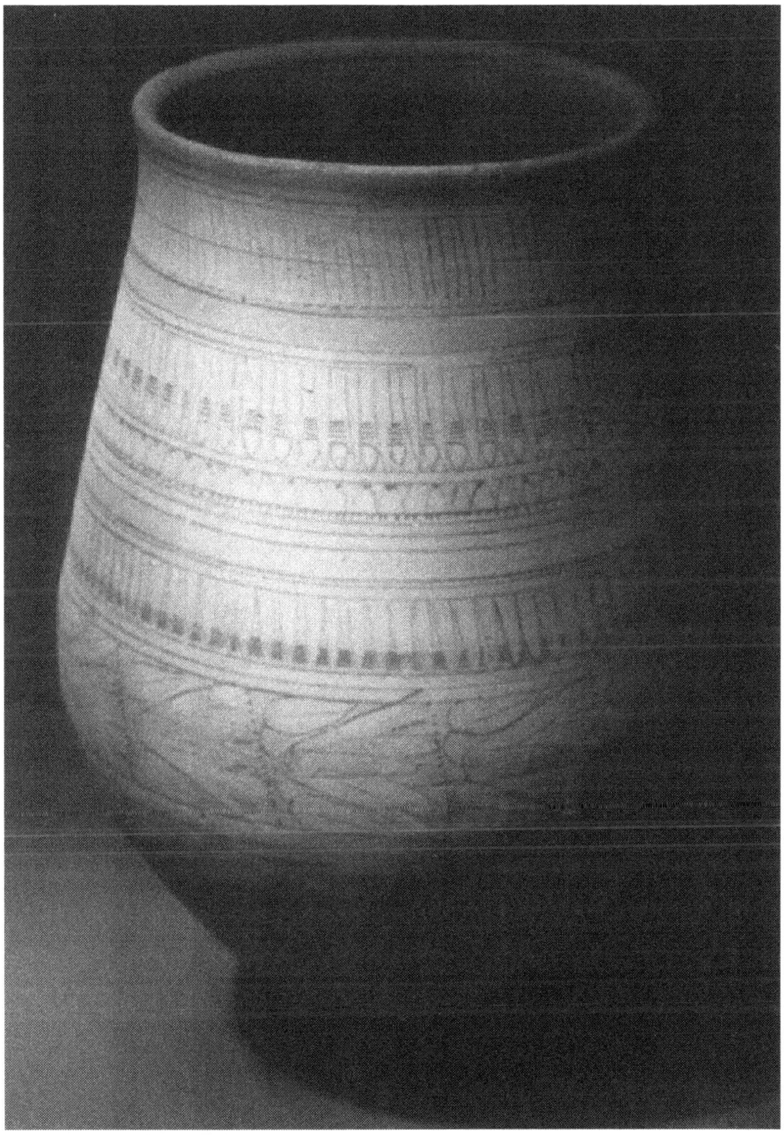

Figure 5-9 : Blue painted ware. *A large example of the blue painted or palace ware from Amarna. Decorated with the typical linear and floral patterns in blue lined with black, on a lime slipped Nile silt ceramic. The example is about 50cm high. Berlin Museum no.22597a, from Amarna, house O49.8.*

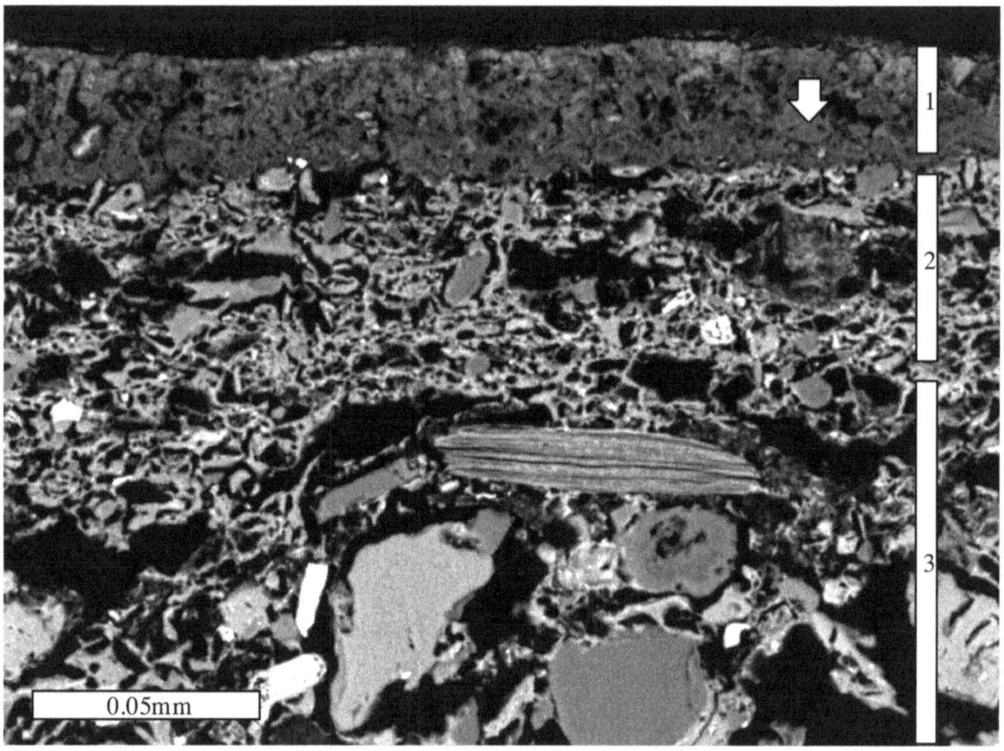

Figure 5-10 : AM21 under SEM. *A cross section of the ceramic showing the various layers. At the bottom of the image is the standard Nile silt ceramic (3), with quite a sharp transition into the lime rich slip (2) above it. Lying on top of that is the blue painted layer (1). Occasional grains can be seen in this layer and one of these is marked. This grain, and others like it, were analysed.*

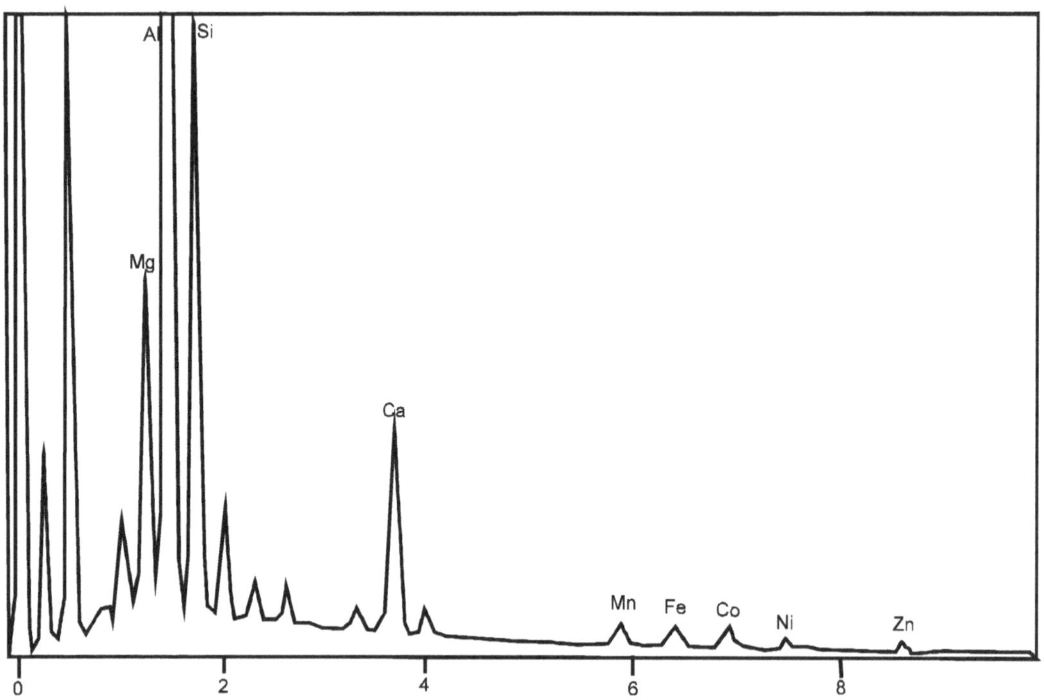

Figure 5-11 : Smoothed EDS spectra of grain from the blue painted layer of AM21.

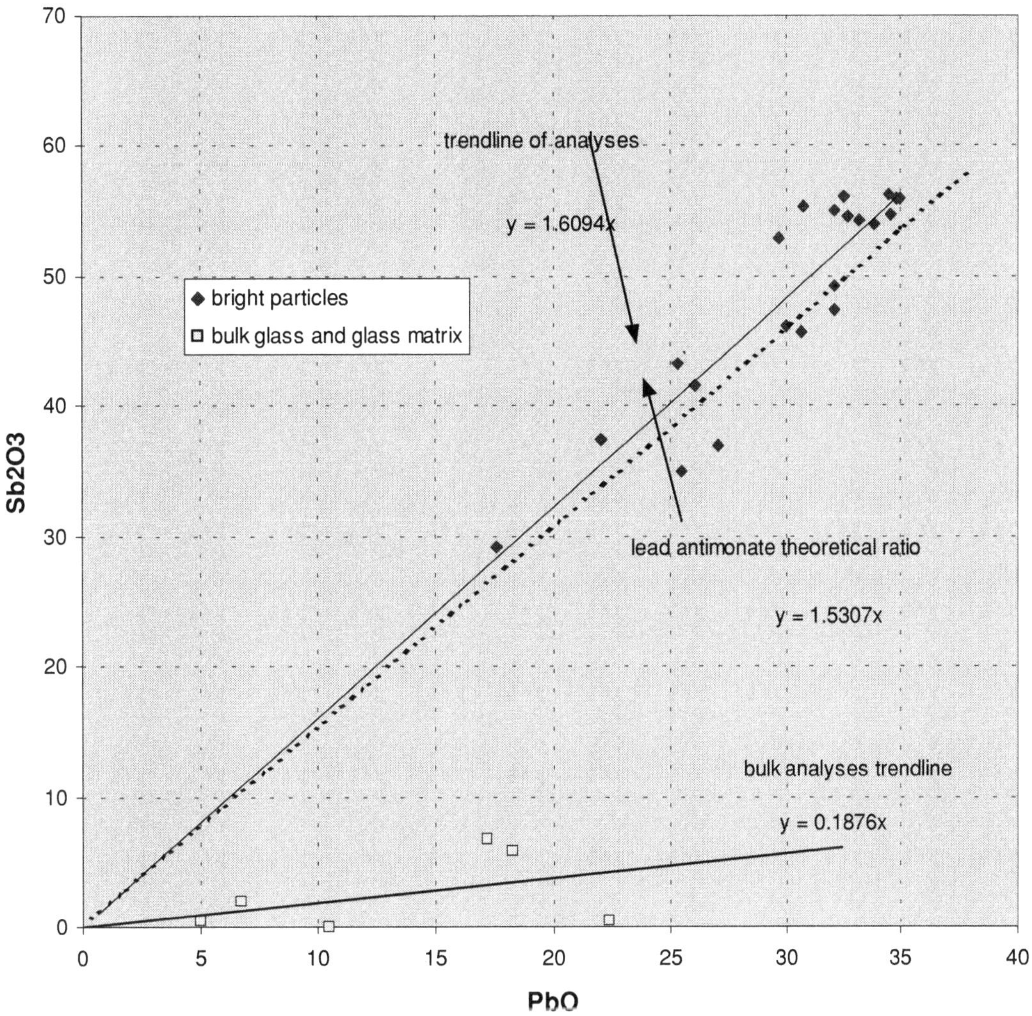

Figure 5-12 : Lead against antimony. *Plot of 21 point analyses of the bright particles from three different samples, two faience (E5032/4 and E5054/1) and one glass (E5613/1). The lead and antimony weight percents are plotted, and the trend line given. The theoretical trend line for ideal lead antimonate is also illustrate as a comparison. Towards the bottom of the graph lie the points for the bulk analyses of yellow glass, faience glaze and glass matrix.*

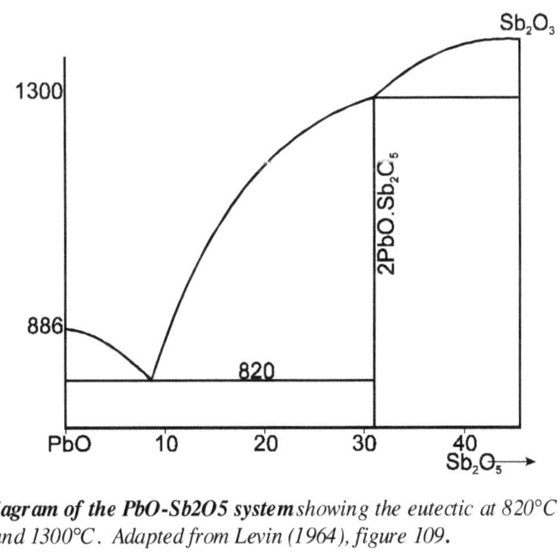

Figure 5-13 : Phase diagram of the PbO-Sb2O5 system *showing the eutectic at 820°C and the melting point of 2PbO.Sb2O5 at around 1300°C. Adapted from Levin (1964), figure 109.*

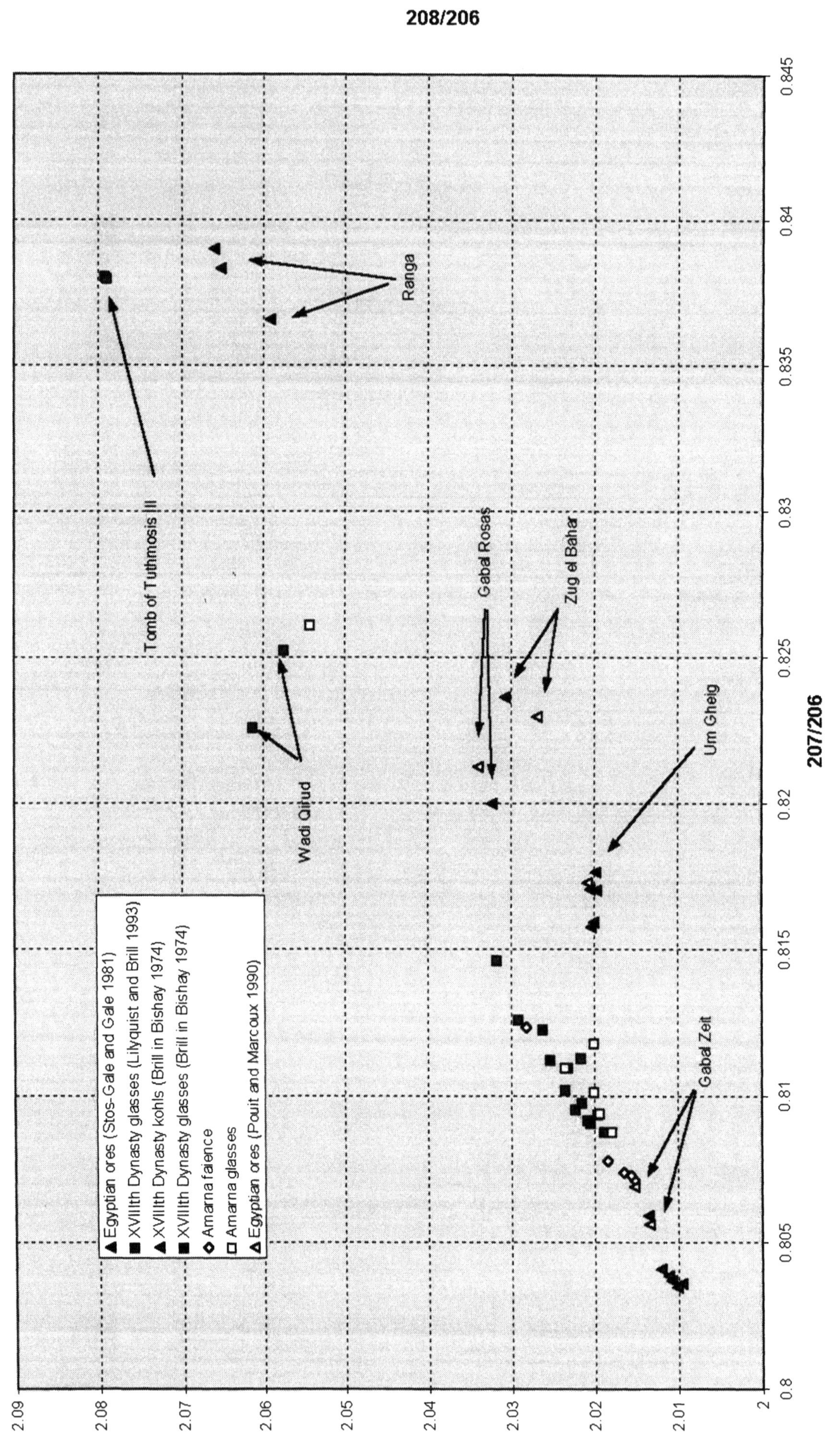

Figure 5-14 : Pb Isotope graph of 208/206 against 207/206. *Data from the current project is compared to that obtained by Brill et al 1974 (glass and kohls), Stos Gale and Gale 1981 (ores) and Lilyquist and Brill 1995 (glass). The sources of the ores are noted next to the fields in which they plot.*

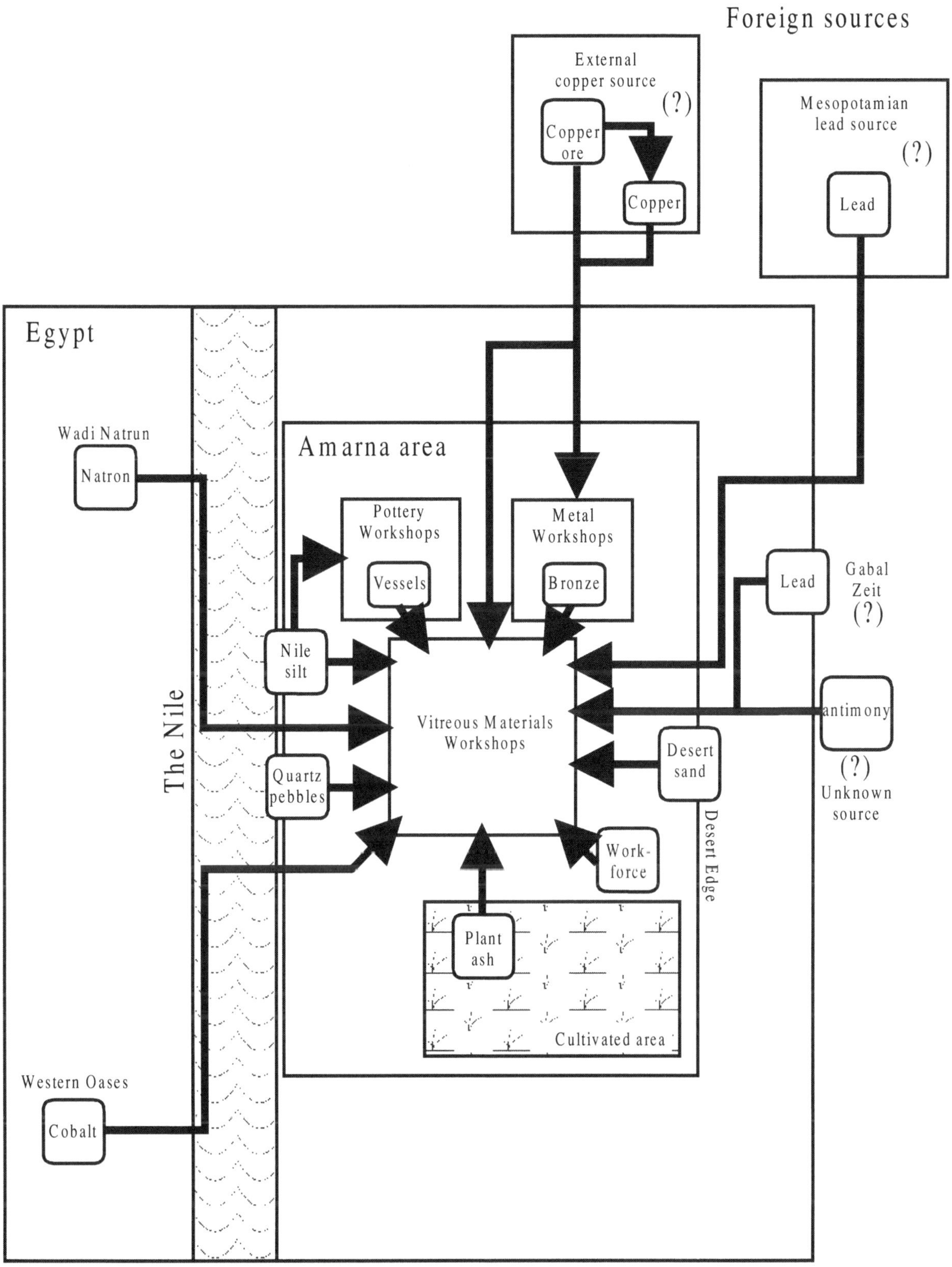

Figure 5-15 : Summary of the sources of the raw materials *for the Amarna vitreous materials workshops. All sources within the square marked "Amarna" are those thought to be within the precincts of Akhetaten, those in the square marked "Egypt" under the control (to a greater or lesser extent) of Egypt, those outside this are foreign imports.*

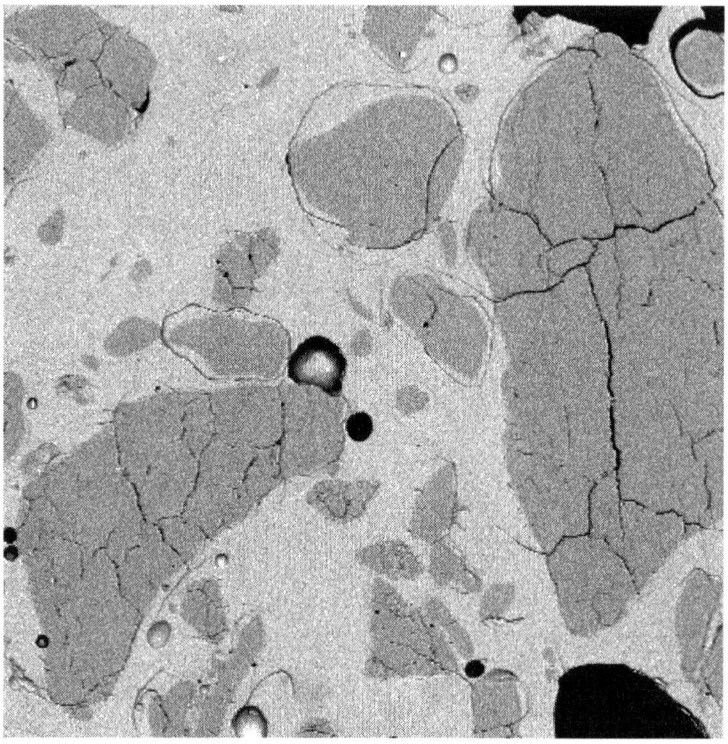

Figure 6-1 : Ancient frit AM34: *SEM image of frit AM34 to compare with the experimental frit below.*

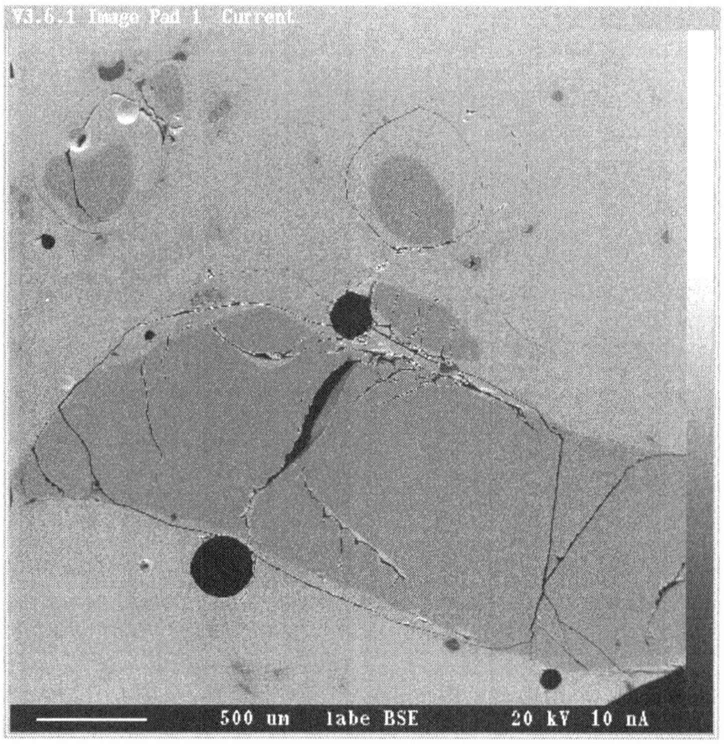

Figure 6-2 : Experimental frit F6. *Frit F6 shows partial melting and rounding of the quartz grains along with beta- to alpha-quartz cracking around the periphery of the grains and internally.*

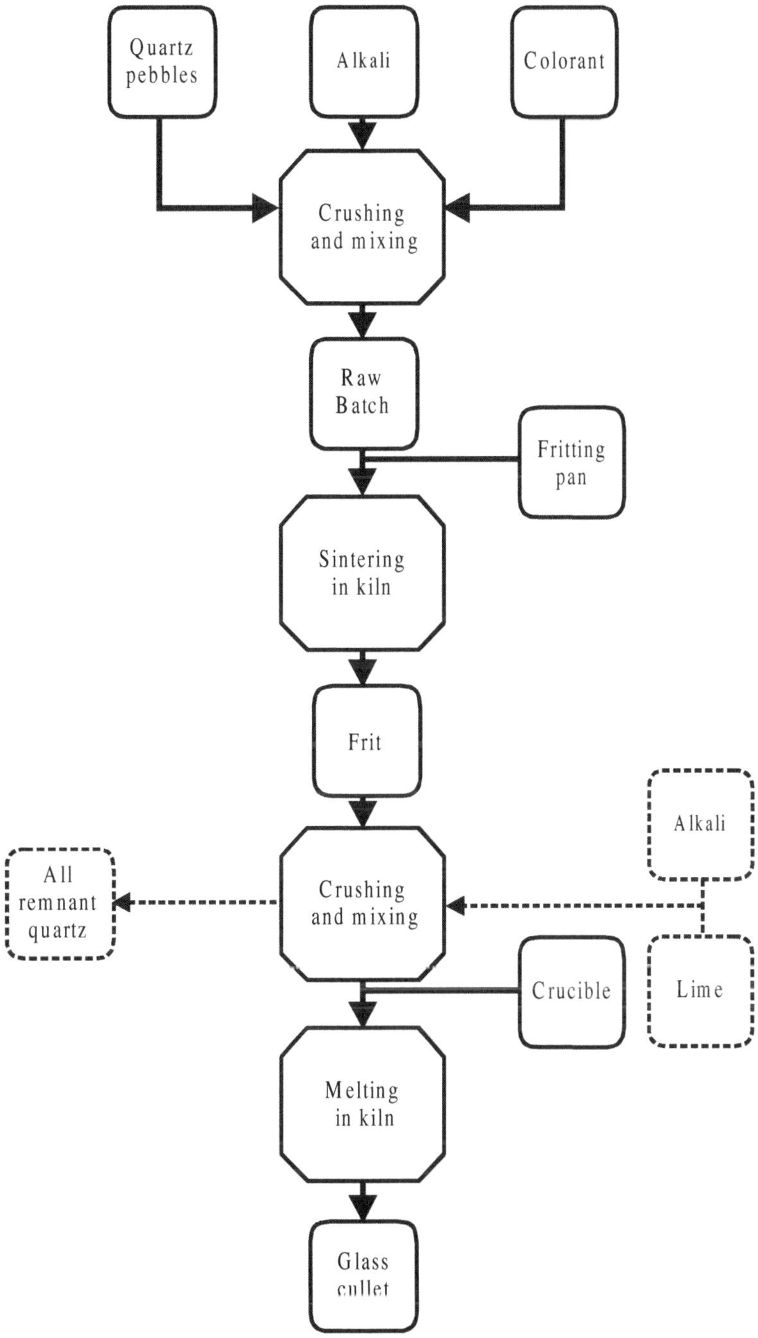

Figure 6-3 : Original explanation of glass production. *Flowchart illustrating the hypothesis of Petrie (1925) and Turner (1954, 1956) explaining the place of the frits in the production of glass. The black outline is the original hypothesis, the dashed addition is an attempt to make the chemistry of the frits closer to that of the glasses.*

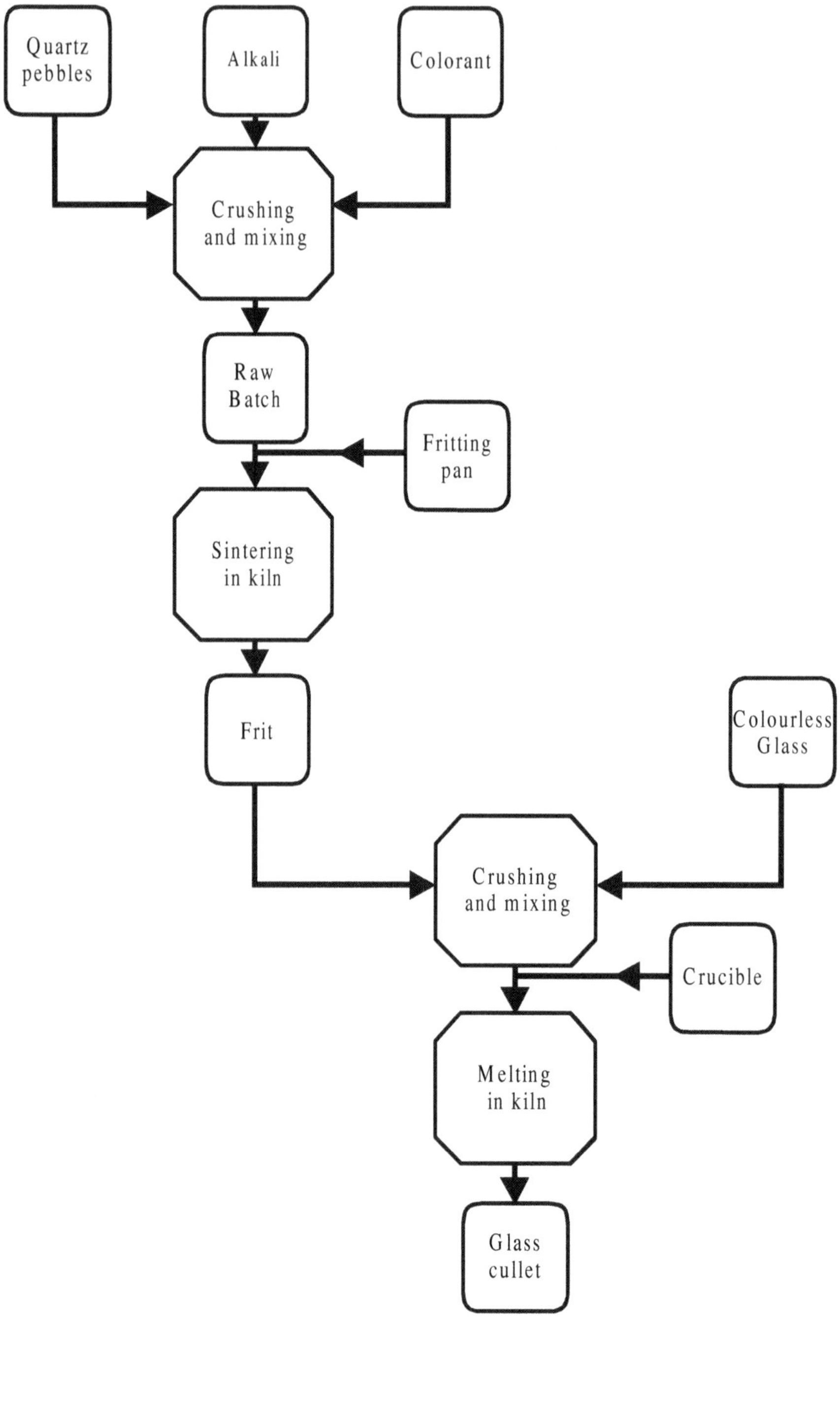

Figure 6-4 : Frits as a colorant, an explanation of glass production. *Flow chart illustrating the hypothesis that the frits are used as a colorant for colourless glass and hence are an intermediate stage in the production of the glasses.*

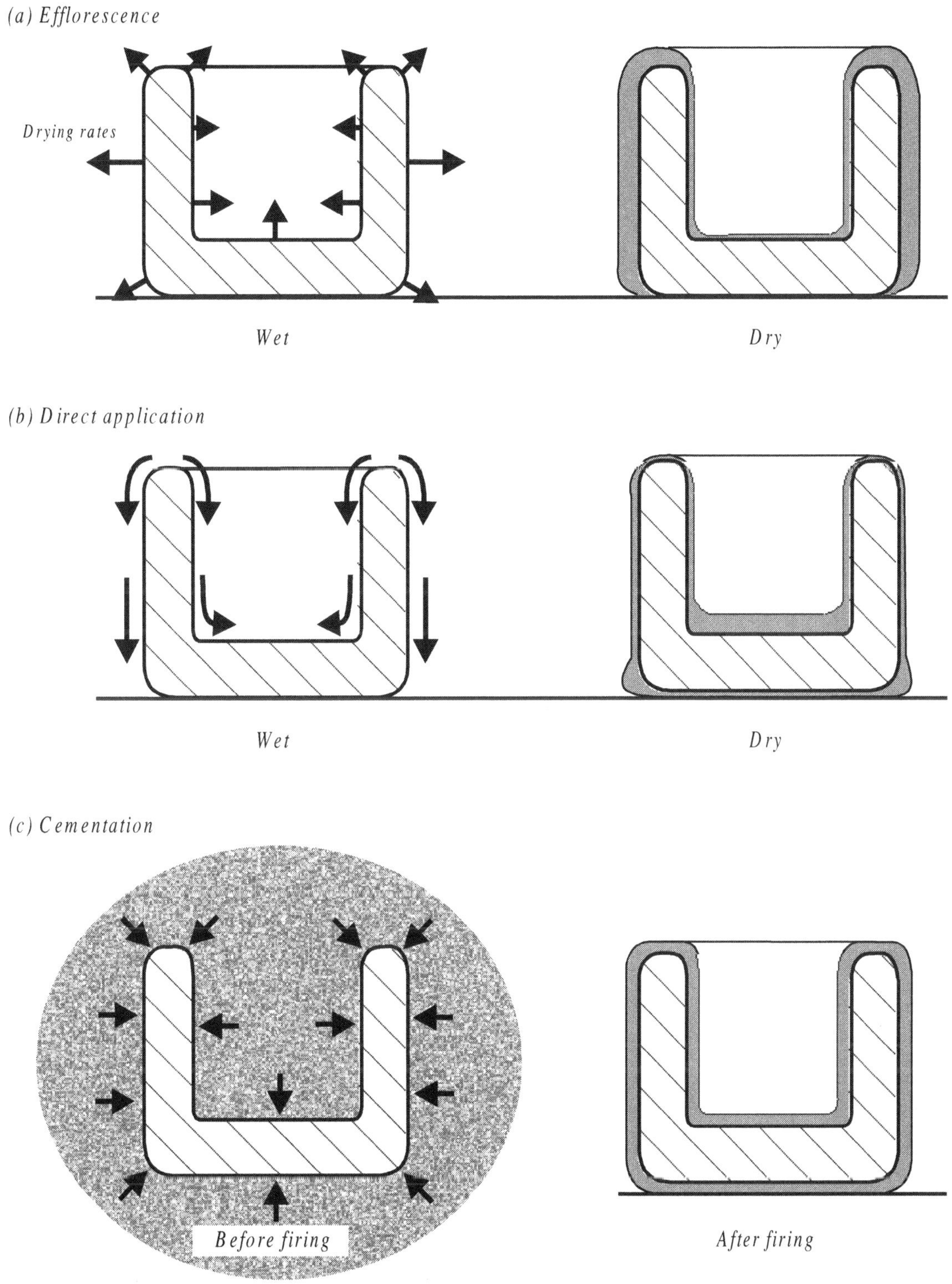

Figure 6-5 : Glazing technology : *The three techniques for the glazing of faience - efflorescence, direct application and cementation. Each method is shown by two schematic faience vessels. The left shows the movement of the unfired glazing material and the right either the dry vessel before firing (a and b) or the completed vessel after firing (c).*

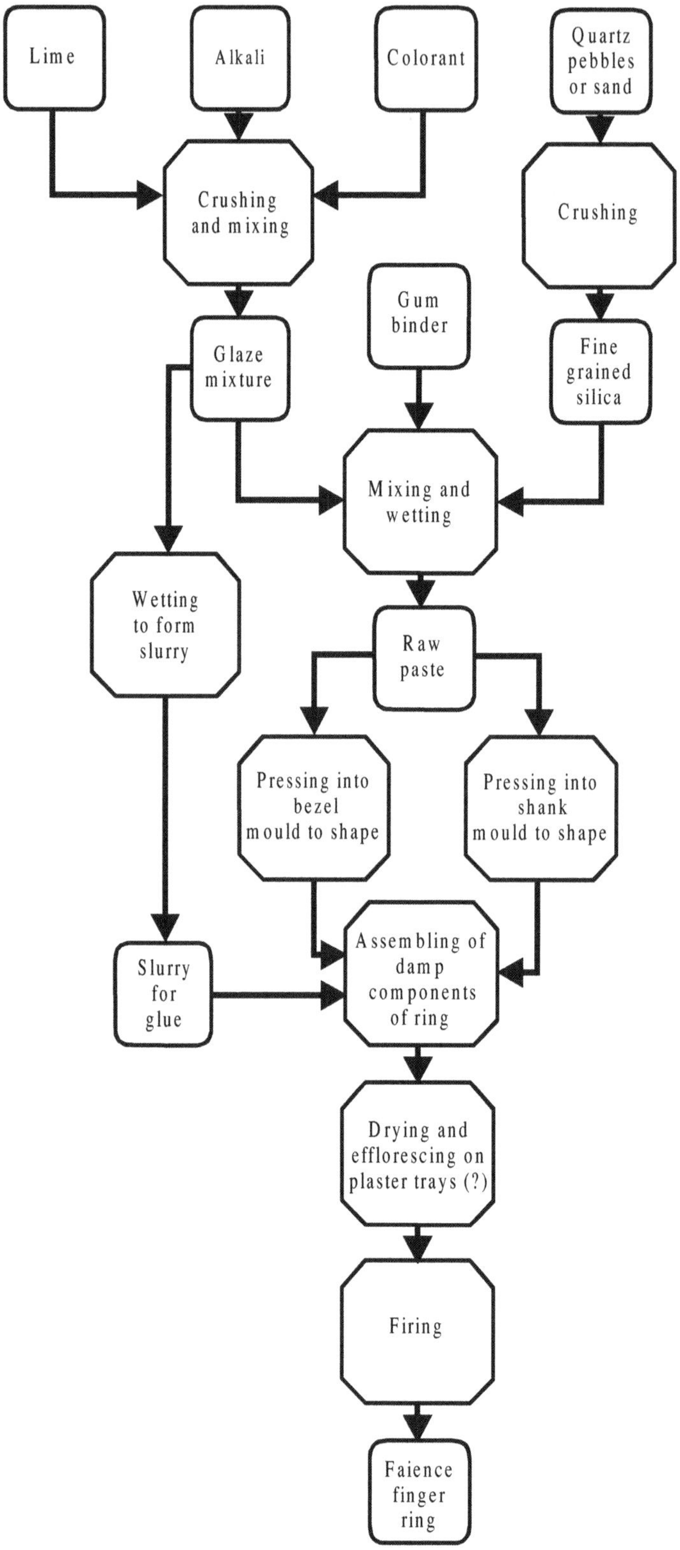

Figure 6-6 : Production of an ordinary faience finger ring. *Flowchart illustrating the theory for the production of small items of ordinary faience (such as rings and amulets) as derived from the observations of Chapters 3 and 5).*

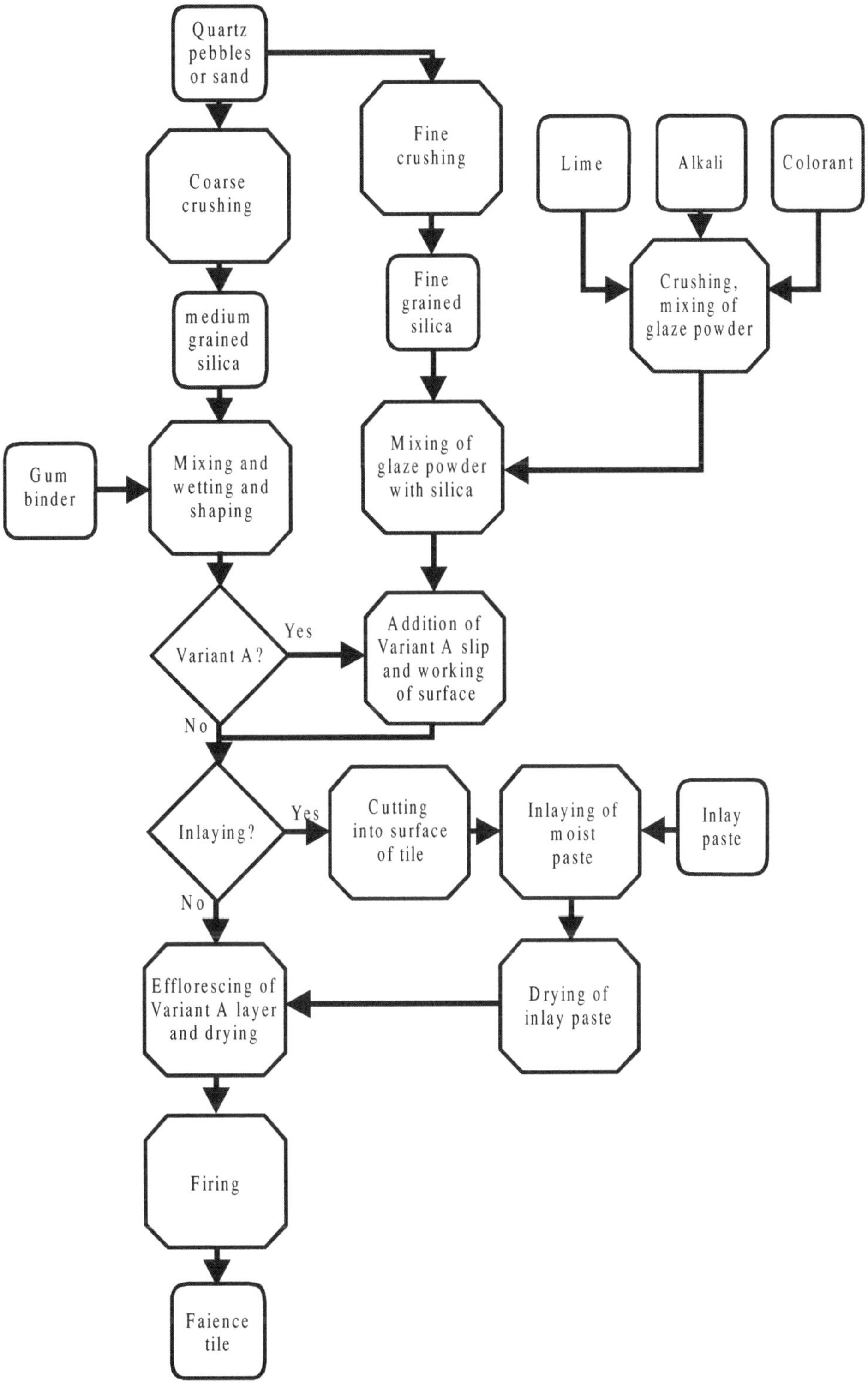

Figure 6-7 : Production of a Variant A faience tile. *Flowchart illustrating production of a Variant A faience tile derived from the information in Chapters 3 and 5 and experimental reproduction.*

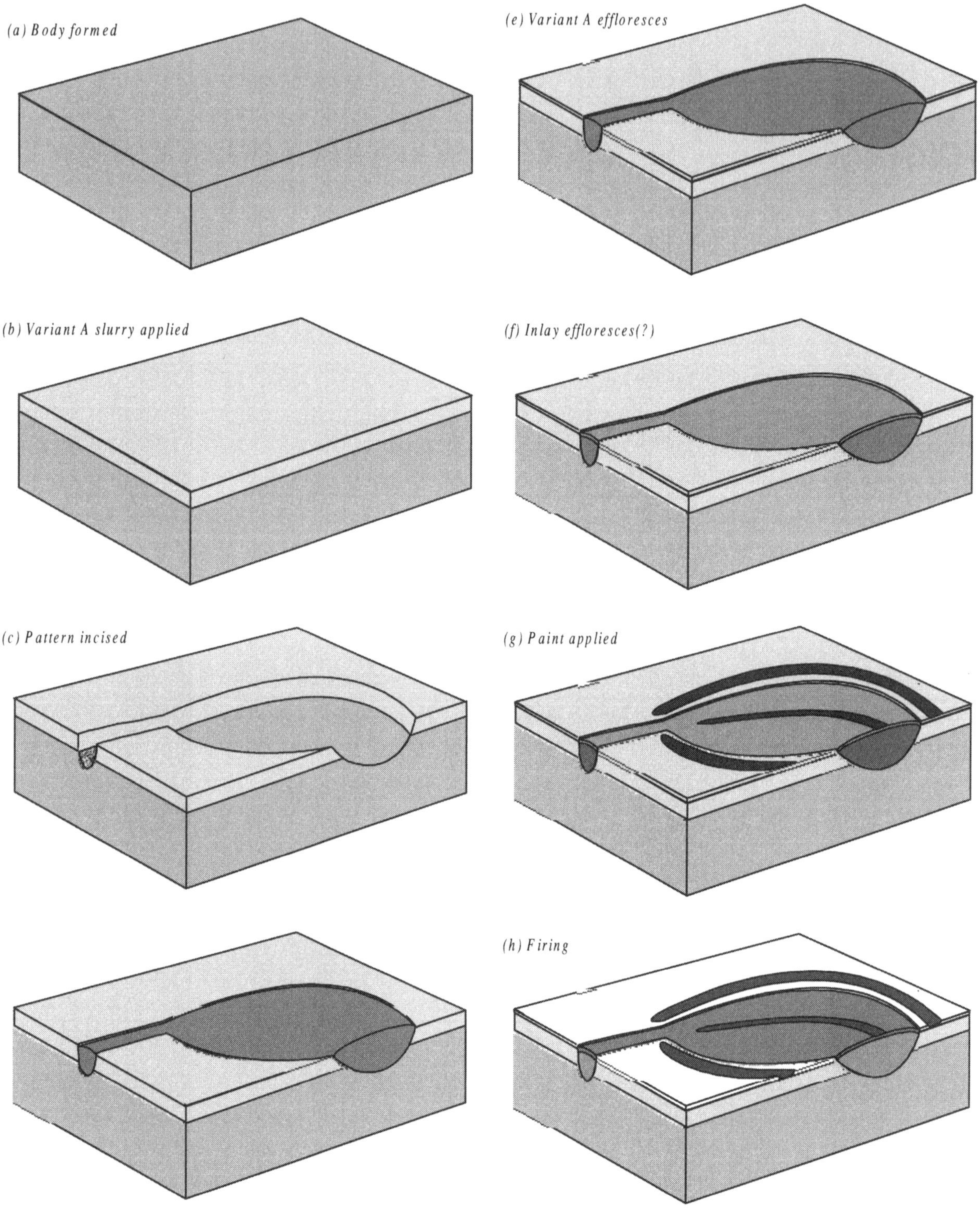

Figure 6-8 : Production of a tile. *Schematic illustration of the stages involved in the production of a tile decorated by both inlaying and painting.*

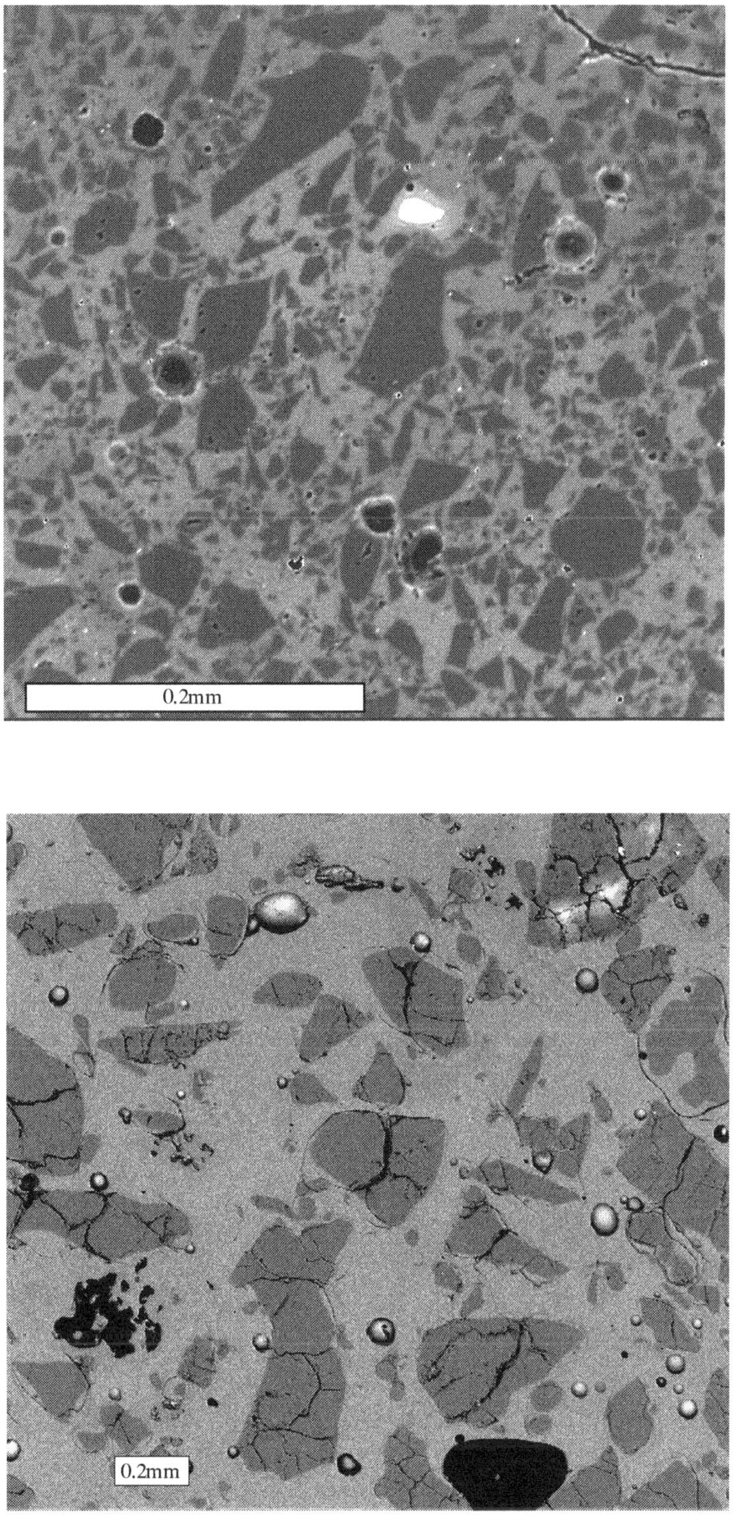

Figure 6-9 : Comparison of frit and Variant D core microstructures. *Top shows the microstructure of the core of Variant D faience sample 3556, while the bottom sample shows frit AM32. Note that the scales are different.*

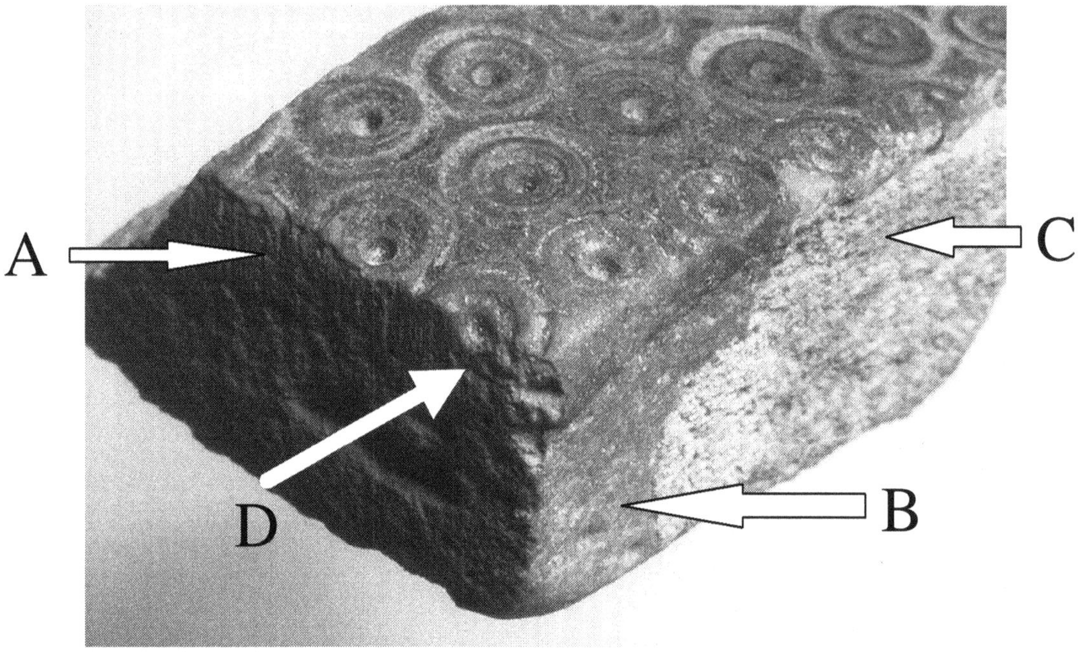

Figure 6-10 : Variant D showing applied glaze structure : *A Variant D faience Khepresh crown fragment (UC23228) showing a Variant A layer with grinding marks ('A') on a coarser base. The green colour of the glaze (also seen and analysed in sample 3556) suggests that it contains copper, while the deep blue of the Variant A layer (and to a lesser extent the lighter blue of the base) indicates the presence of cobalt. The glaze has run off of the top surface and down the side of the piece ('B'), which is strongly indicative of an applied glaze. Other areas, for example 'C', show no glaze development as would be expected if this piece was glazed by efflorescence. 'D' shows glaze preserved at the bottom of the central depression of this set of rings, arguing against post firing drilling.*

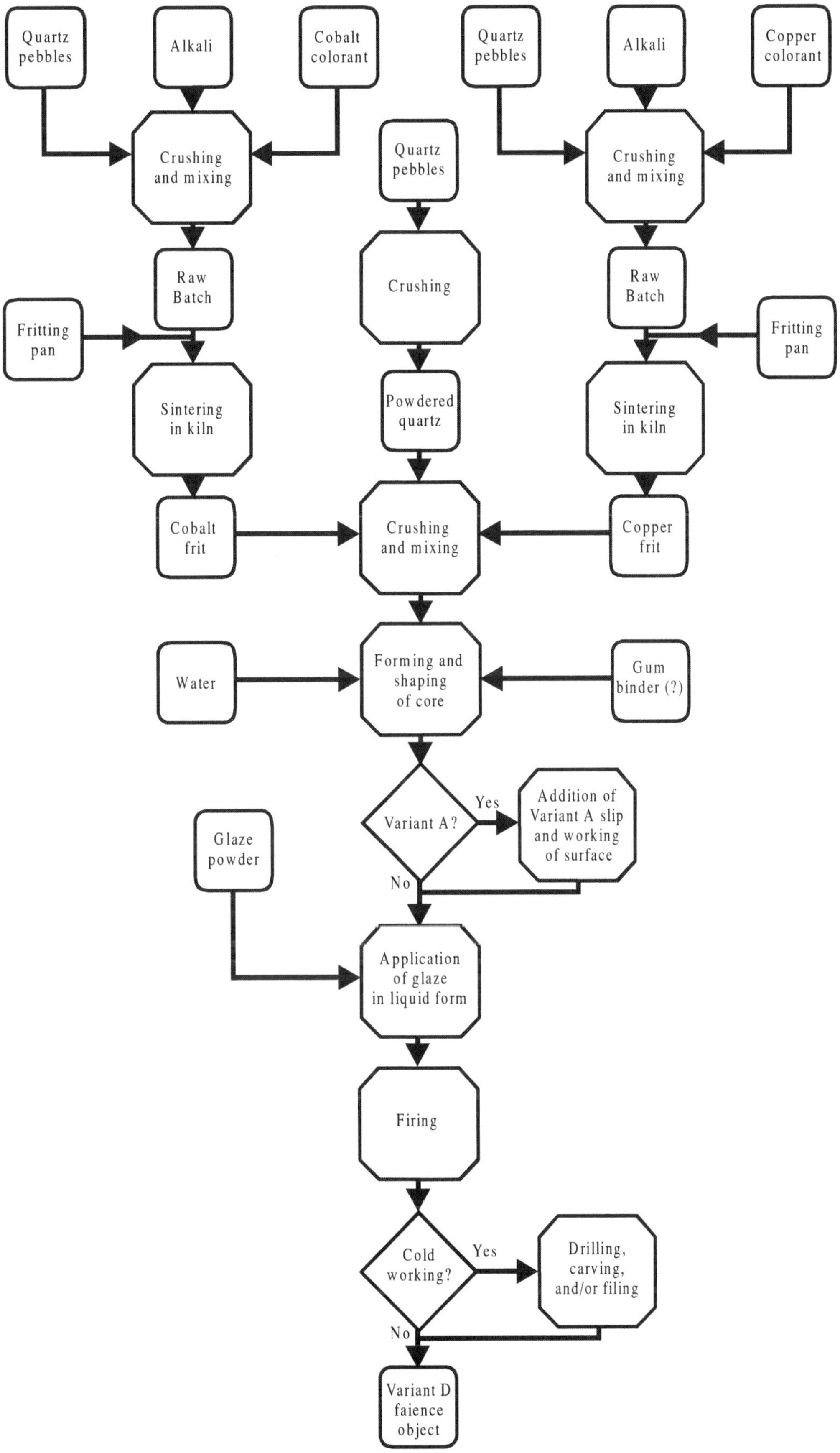

Figure 6-11 : Production of Variant D faience. *Flow chart illustrating the theory for the production of Variant D faience as derived from the observations in Chapters 3 and 5 and confirmed by experimental work.*

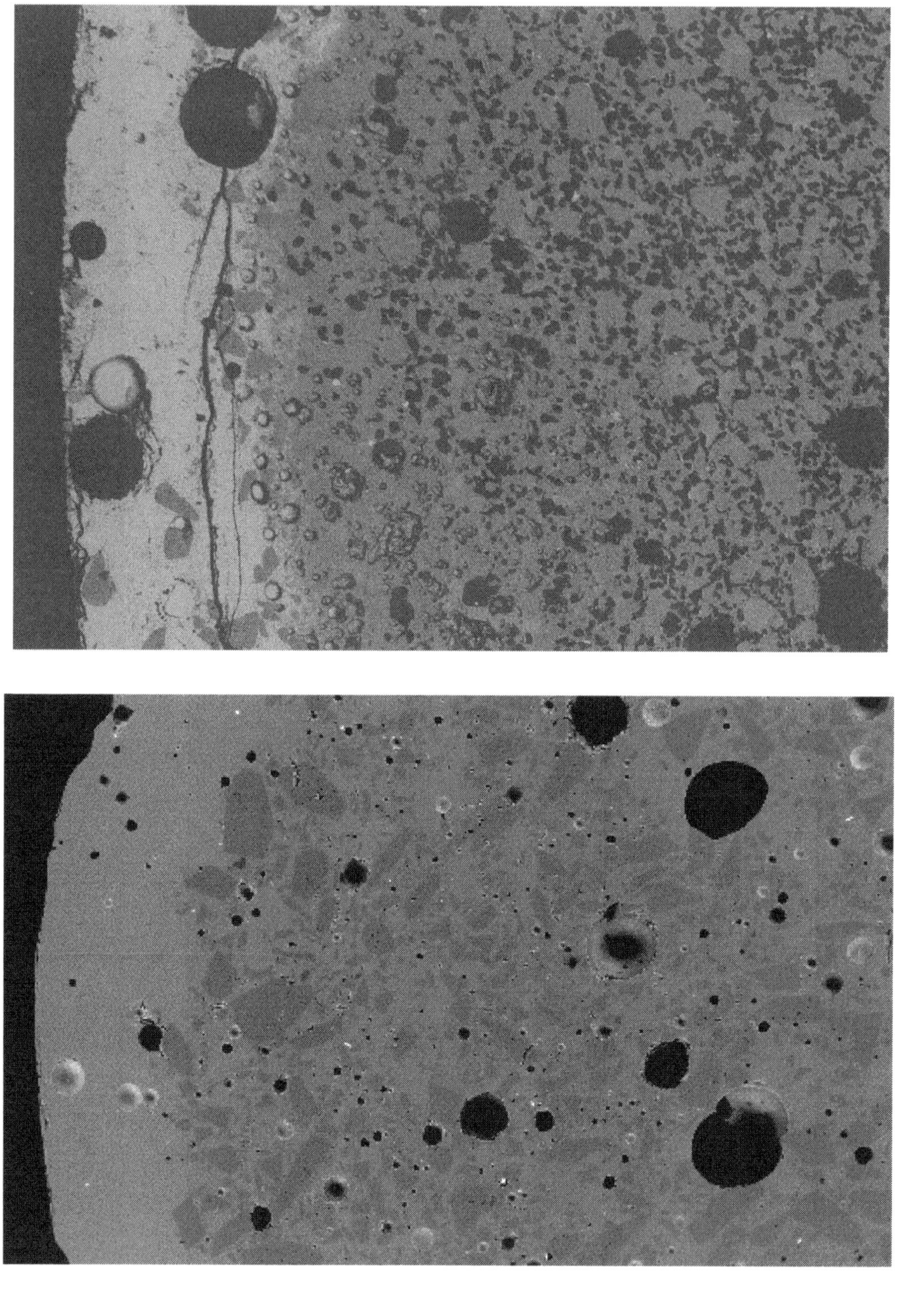

Figure 6-12 : Experimental Variant D core microstructures. *Top sample has a core that was made up from 30% frit F4 and 70% additional quartz, while the bottom sample has a core of 60% frit F6 and 40% additional quartz.*

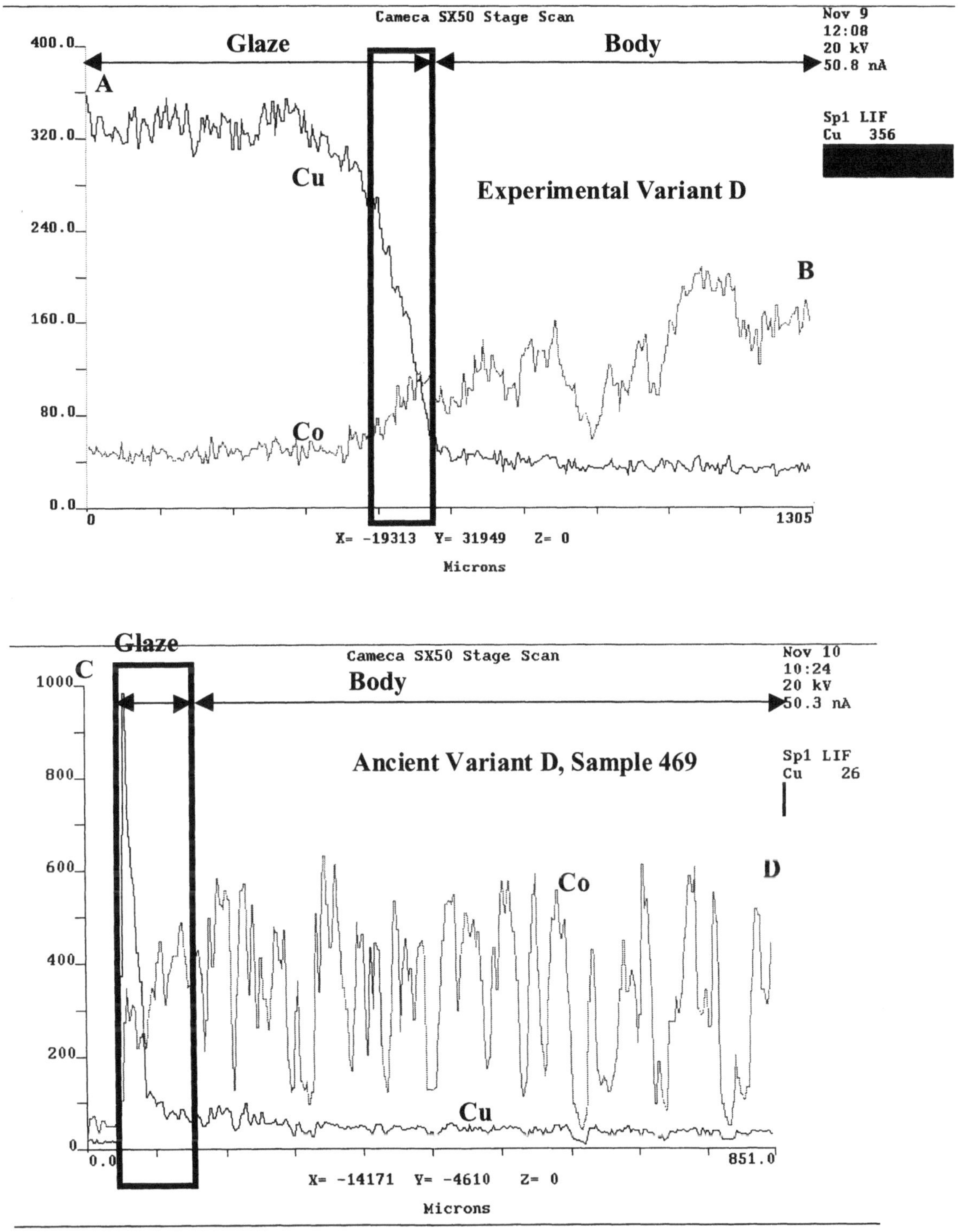

Figure 6-13 : Step scan for Co and Cu. *Top image shows the experimental Variant D step scan (1.305mm long from 'A' to 'B' in Figure 6-14) showing number of counts recorded for Co and Cu given a background level of 40-60 counts. The bottom image is the same scan for the ancient Variant D 469 (0.805mm long from 'C' to 'D' in Figure 6-14). The two zones marked by the squares are equivalent thicknesses at the bottom of the glaze on the glaze/body interface and both show falling Co and rising Cu as the scan goes from the body into the glaze.*

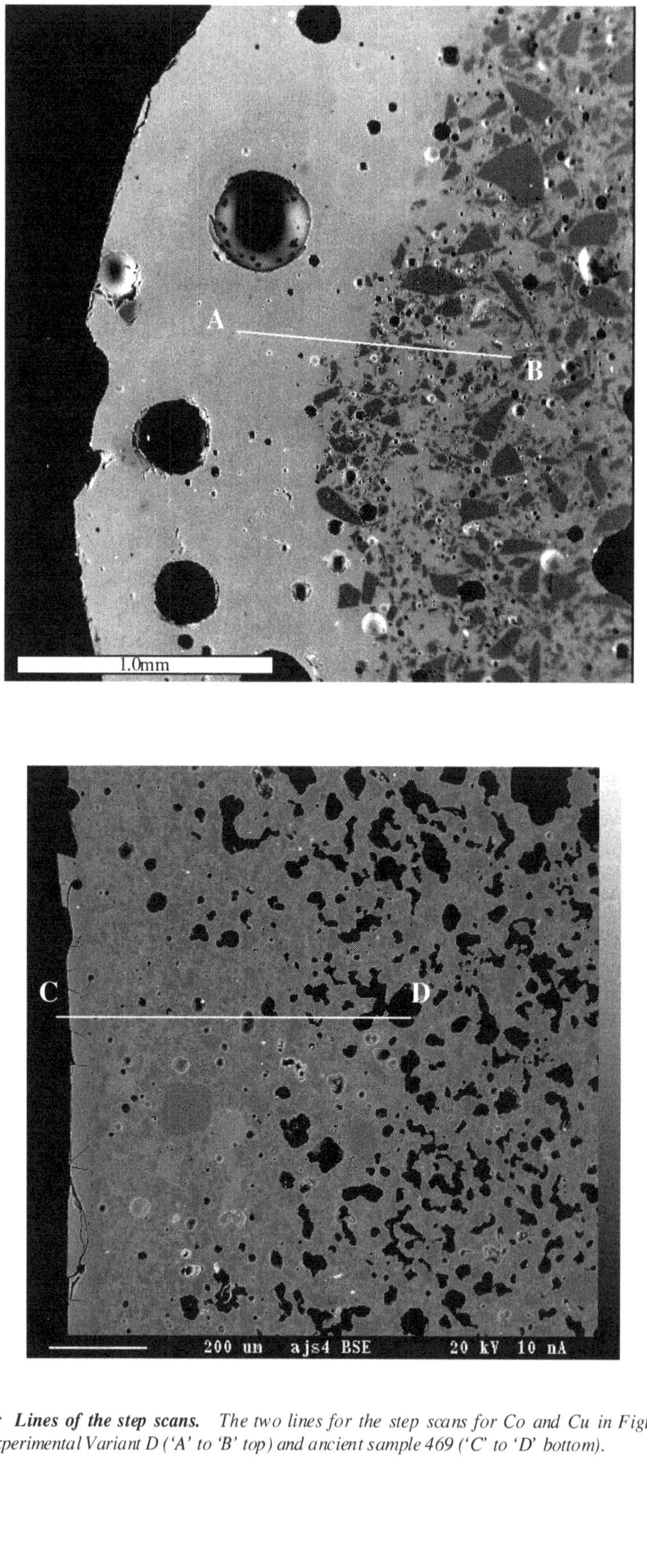

Figure 6-14 : Lines of the step scans. *The two lines for the step scans for Co and Cu in Figure 6-13 through the experimental Variant D ('A' to 'B' top) and ancient sample 469 ('C' to 'D' bottom).*

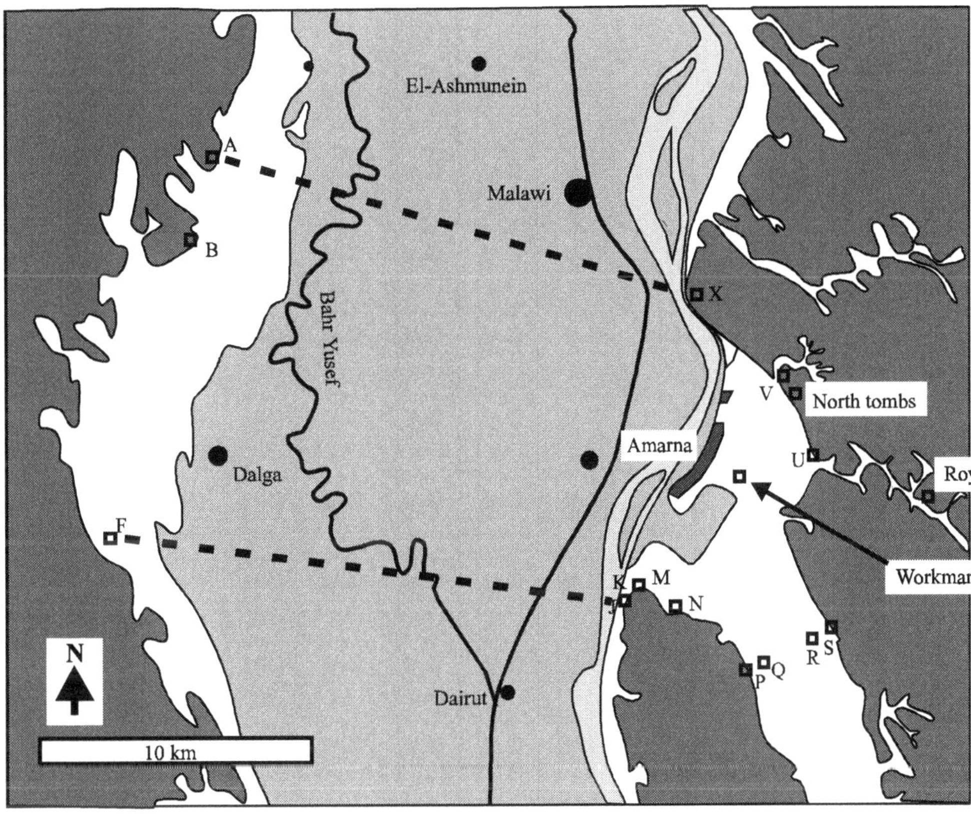

Figure 7-1 : Map of the surroundings of Akhetaten, *showing the positions of the boundary stelae (letters), delineating the extent of the ancient city (A-X and F-J) and tombs. The high desert (dark grey) is also marked along with the area of cultivation (mid grey), the modern courses of the Nile (light grey) and the Bahr Yusef and the various modern villages in the area. Redrawn from Kemp (1989), figure 88.*

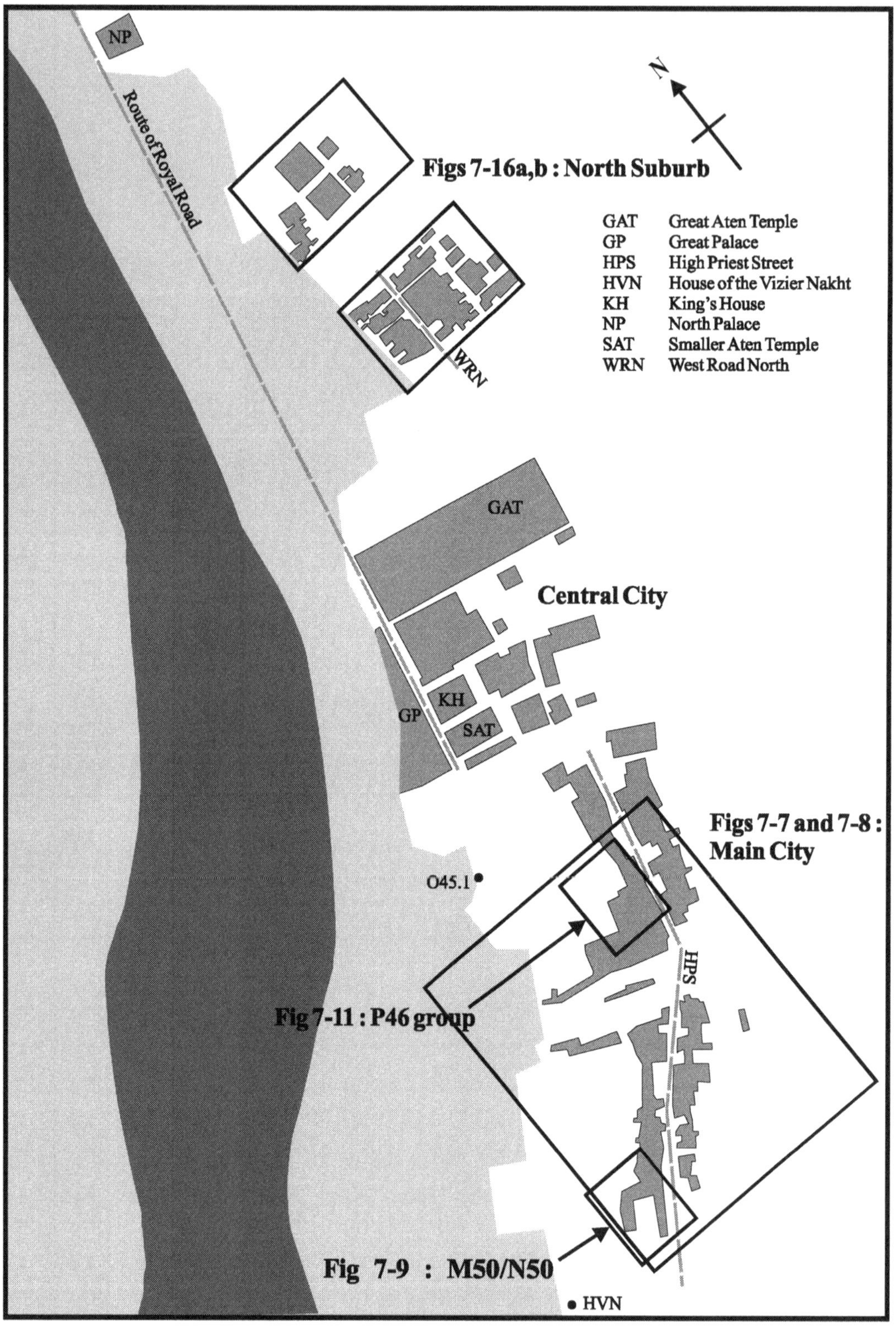

Figure 7-2 : Remains of Amarna - Showing the excavated remains of the various central areas and buildings along with the course of the Nile (dark grey) encroachment of the cultivated area (light grey) onto the site, from Kemp (1989, figure 89). Overlaying this are the areas studied in the mould distribution with the figure numbers that refer to the more detailed maps. Total area shown is 4km across.

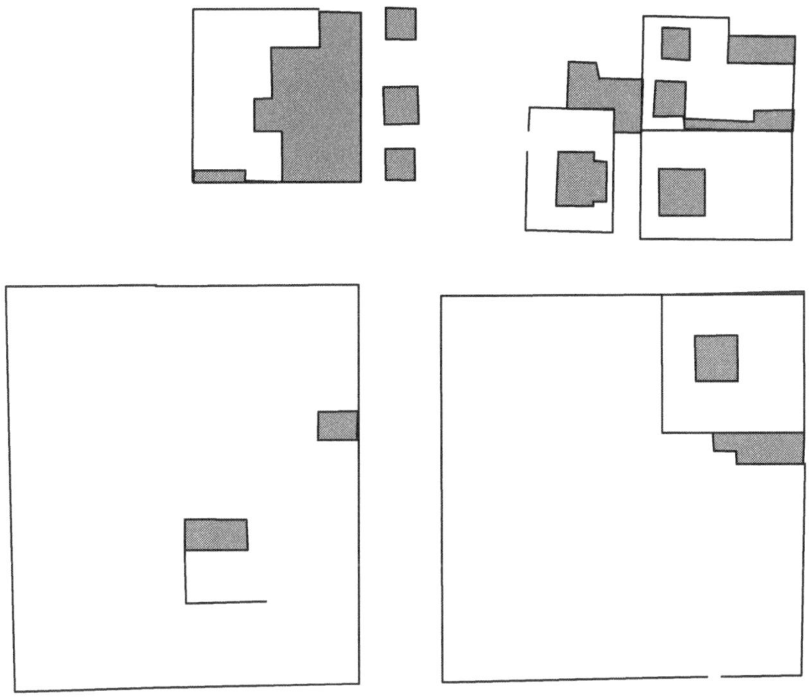

Figure 7-3 : Well spread buildings. *Grid square U33 in the North Suburb at Amarna showing well spaced large properties many set in their own grounds with surrounding walls (adapted from Kemp and Garfi 1993). 200m across, north to the left.*

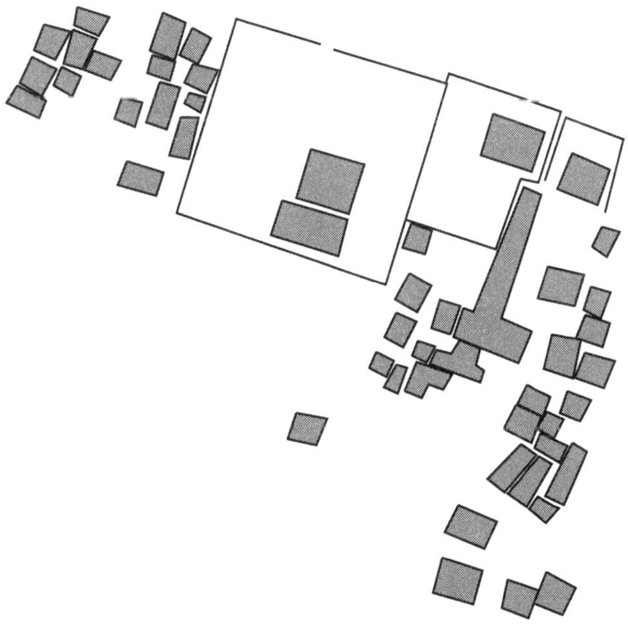

Figure 7-4 : Slums clustering around larger properties. *Grid square Q46 and P46 in the Main City North Suburb at Amarna showing large properties with many small poorly built "slums" clustering around their walls(adapted from Kemp and Garfi 1993). 200m across, north to the left.*

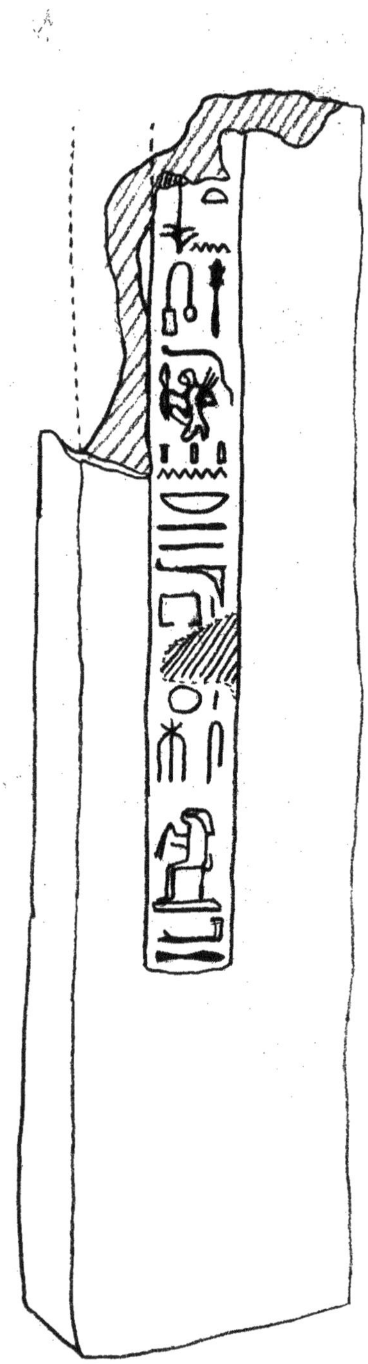

Figure 7-5 : Door frame from the house of Ramose. *This decorated door frame bears the titles and the name of the individual who owned the house, in this case "the King's Scribe, General to the Lord of the Two Lands, the Steward Ramose". (adapted from Borchardt and Ricke 1980)*

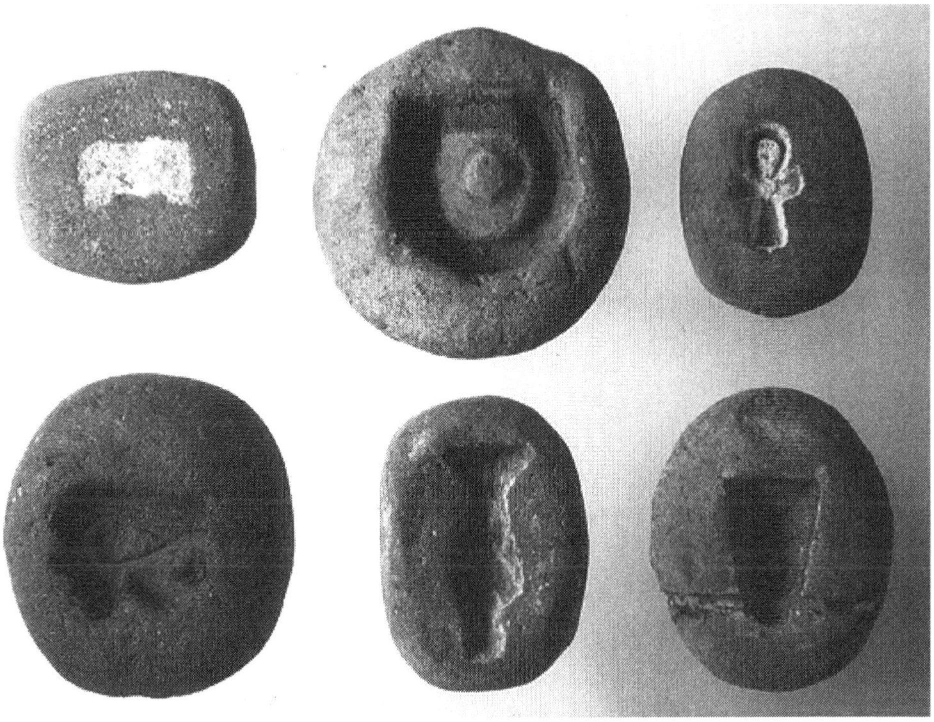

Figure 7-6 : Fired clay moulds for the making of small faience objects. *Moulds such as these were found widely over Amarna by all excavators, although the current study suggests they tend to cluster around faience making areas. Clockwise from top left: a mould still filled with faience paste, mould for a ring shank, for an ankh amulet, a lotus bud, a thistle head and a wadjet eye. Copyright of the Petrie Museum of Egyptian Archaeology, clockwise from top left, UC 23959, UC 25137, UC 24042, UC 24221, UC 23911 and UC 1942.*

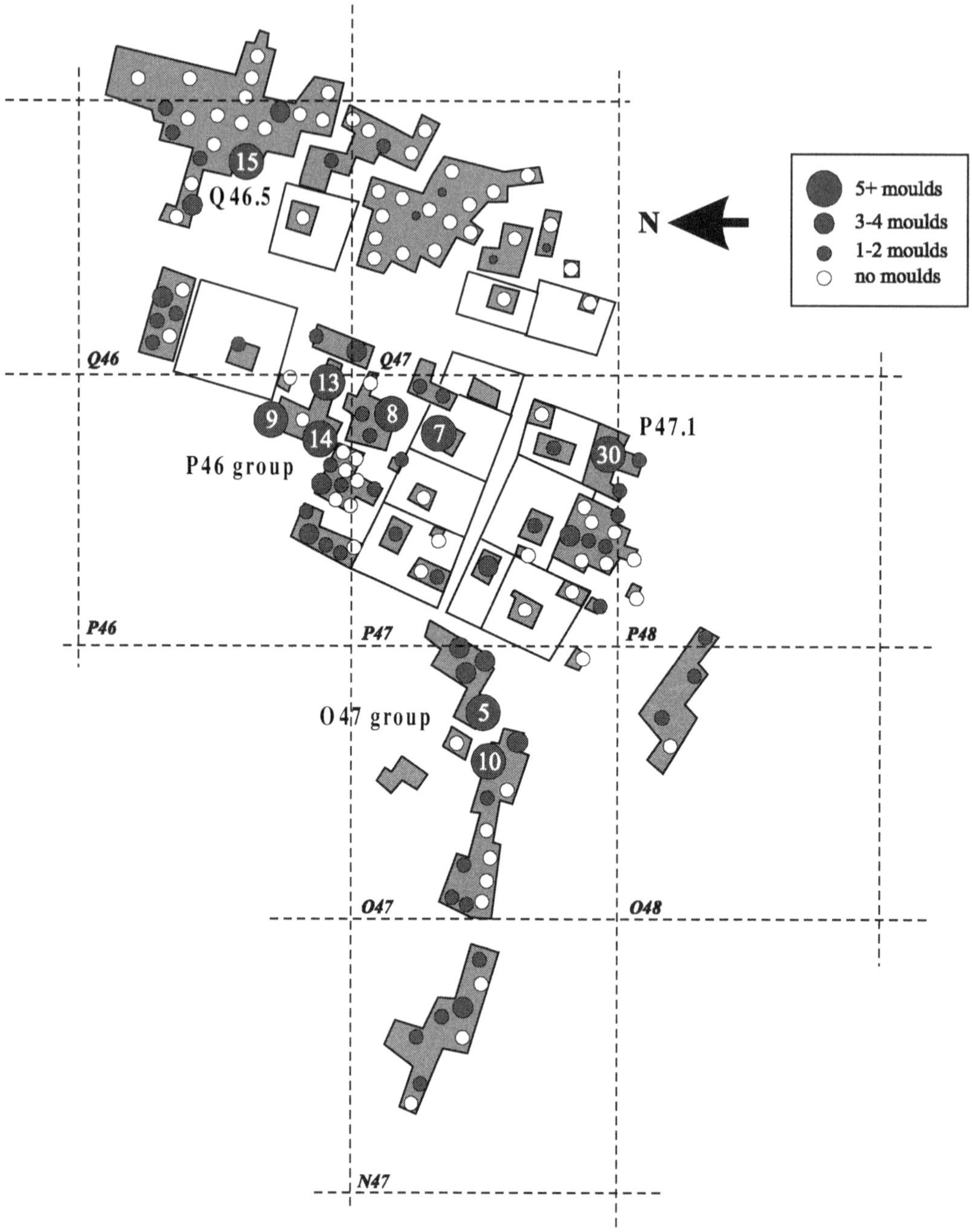

Figure 7-7 : Moulds in Main City North - Results from the search of the excavation reports for the moulds associated with each house. Number of moulds as key.

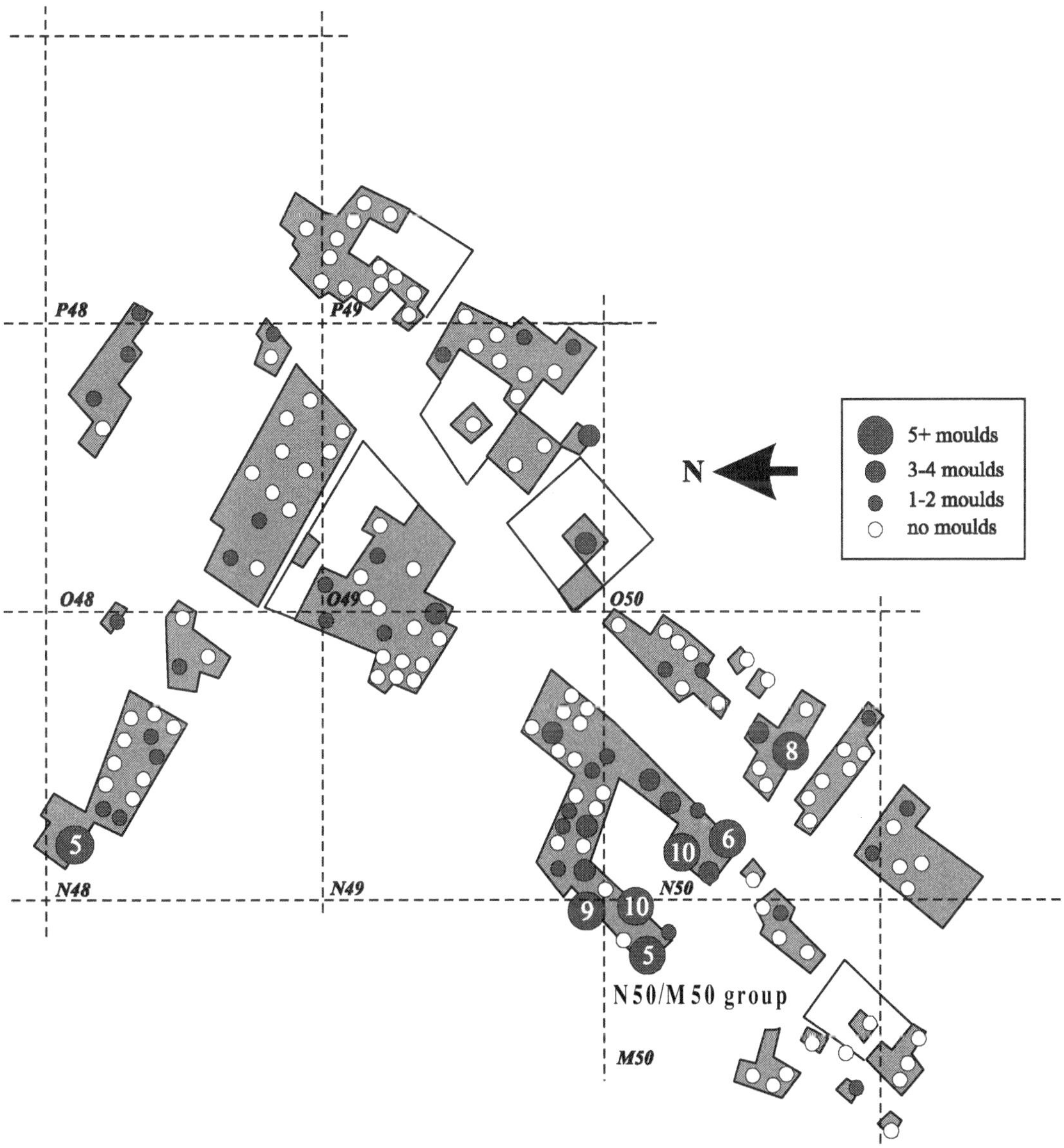

***Figure 7-8 : Moulds in Main City South** - Results from the search of the excavation reports for the moulds associated with each house. Number of moulds as key.*

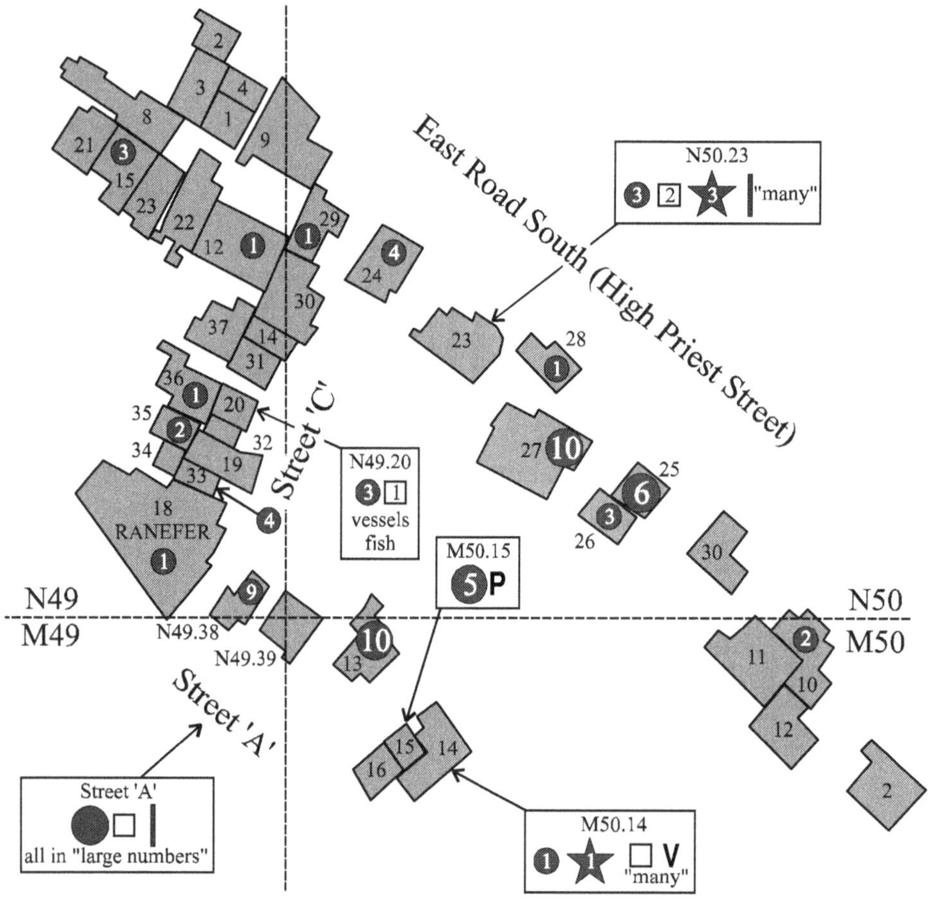

Figure 7-9 : M50/N50 Group- *Results from the search of the excavation reports for the moulds associated with each house. circle = moulds, star = kilns, square = "Werkstuck", rectangle = glass rod, 'V' = cylindrical vessels/crucibles, 'P' = pigment. A number in the symbol, or multiple symbols, indicate the number of that particular item found .*

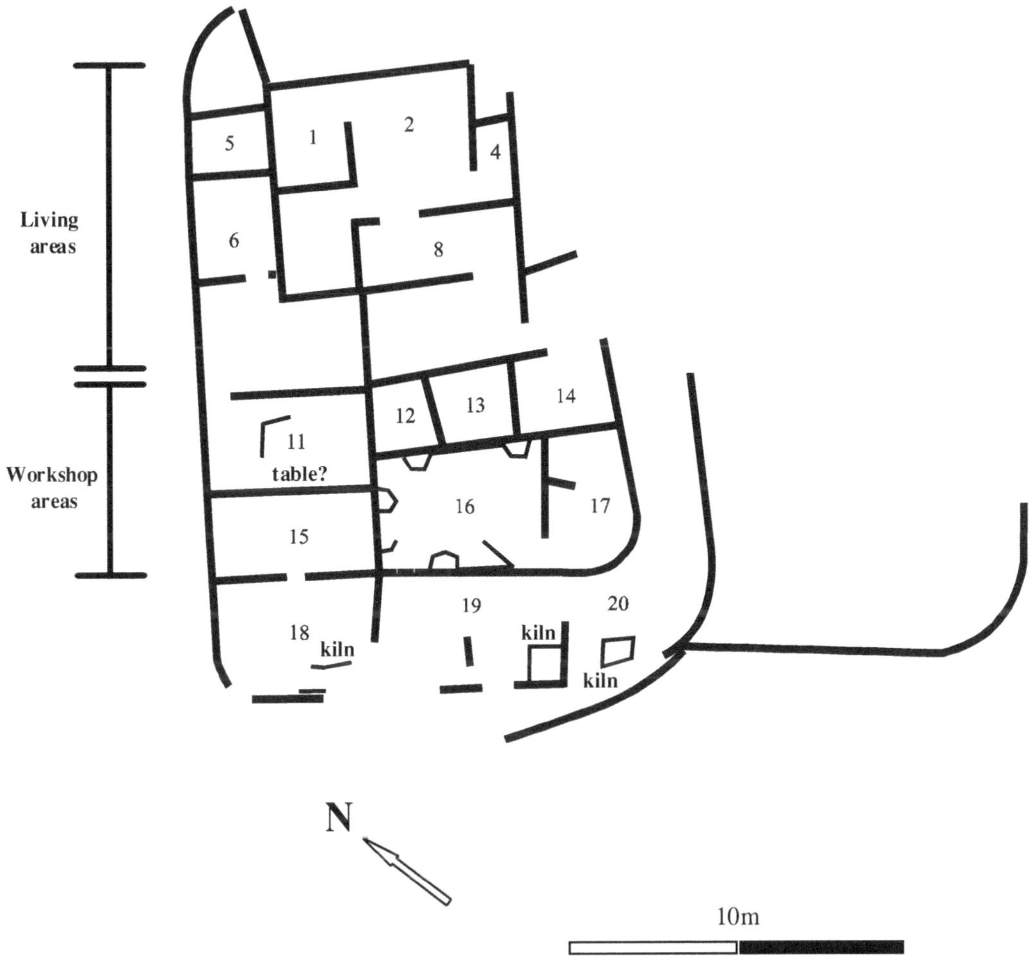

Figure 7-10 : N50.23, a house in the Main City South *with domestic rooms (1-10), workshop areas (11-17) and three kilns, one in each of three small yards near the workshop areas (18-20). Adapted from Borchardt and Ricke (1980)*

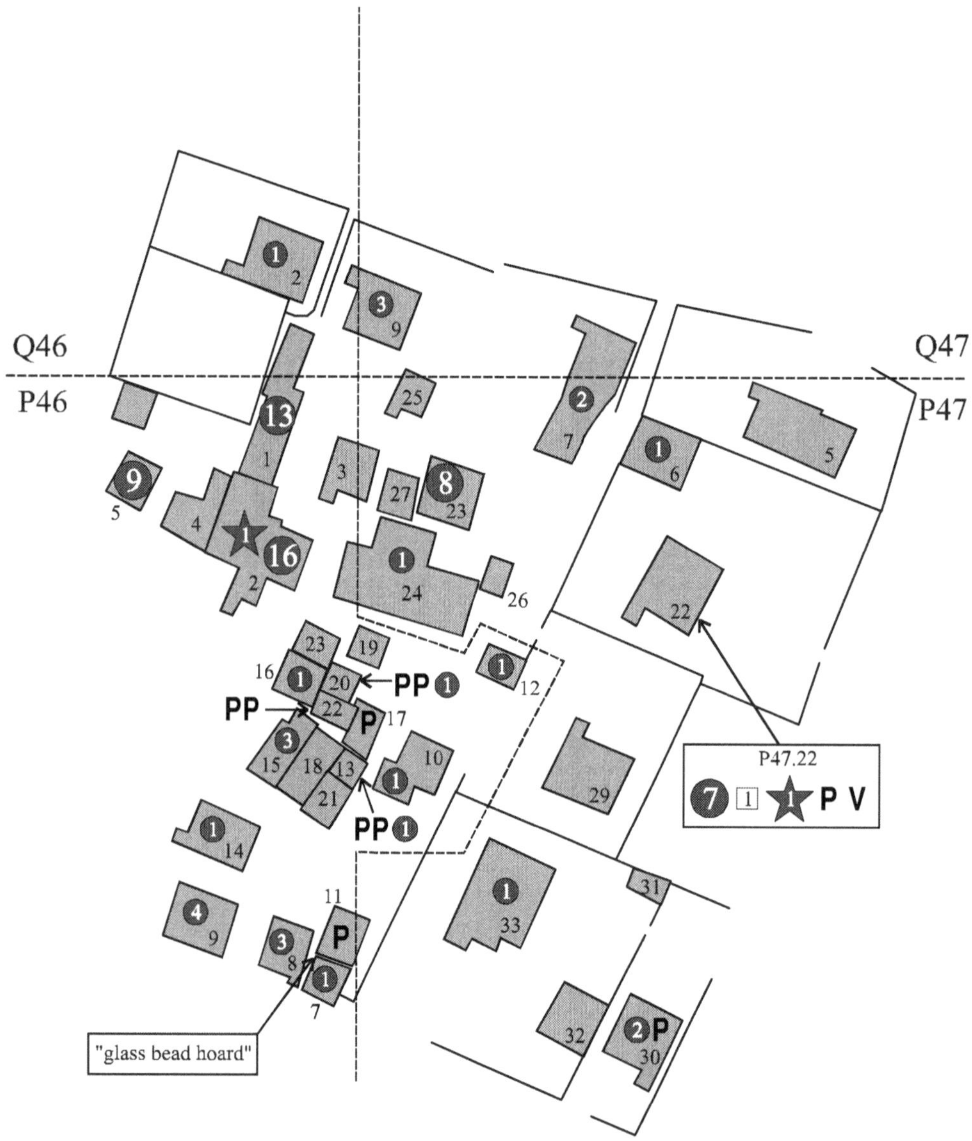

Figure 7-11 : P46 Group- *Results from the search of the excavation reports for the moulds associated with each house. circle = moulds, star = kilns, square = "Werkstuck", rectangle = glass rod, 'V' = cylindrical vessels/crucibles, 'P' = pigment. A number in the symbol, or multiple symbols, indicate the number of that particular item found . Gridlines are redrawn to allow for old numbering system.*

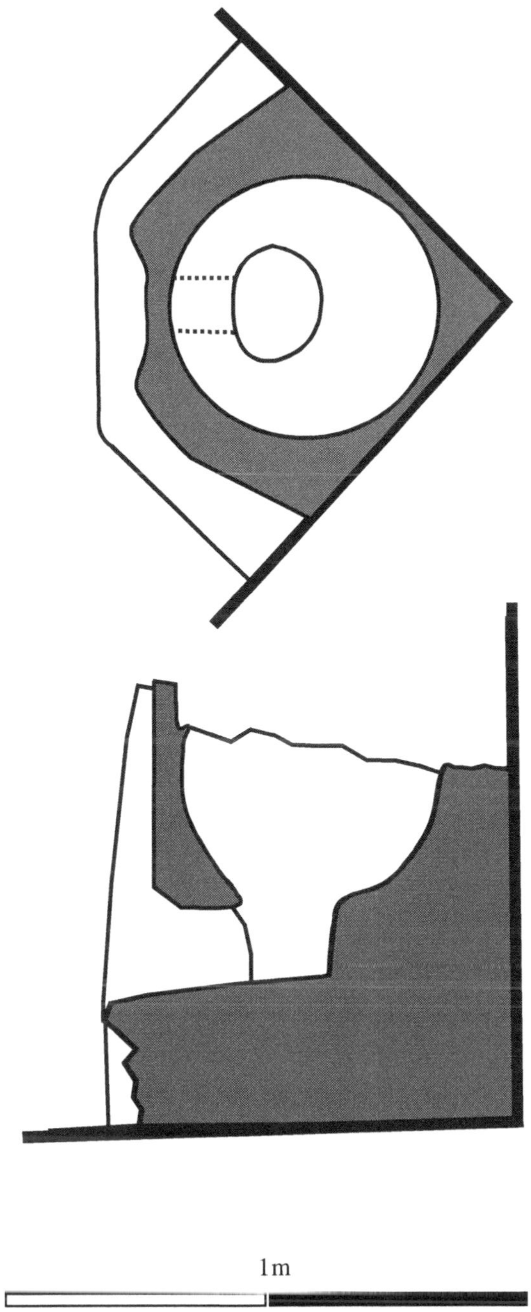

Figure 7-12 : Kiln in Q46.5, *a possible jewellery workshop kiln for the manufacture of small items inlaid with glass? Adapted from Borchardt and Ricke (1980), figure on page 33)*

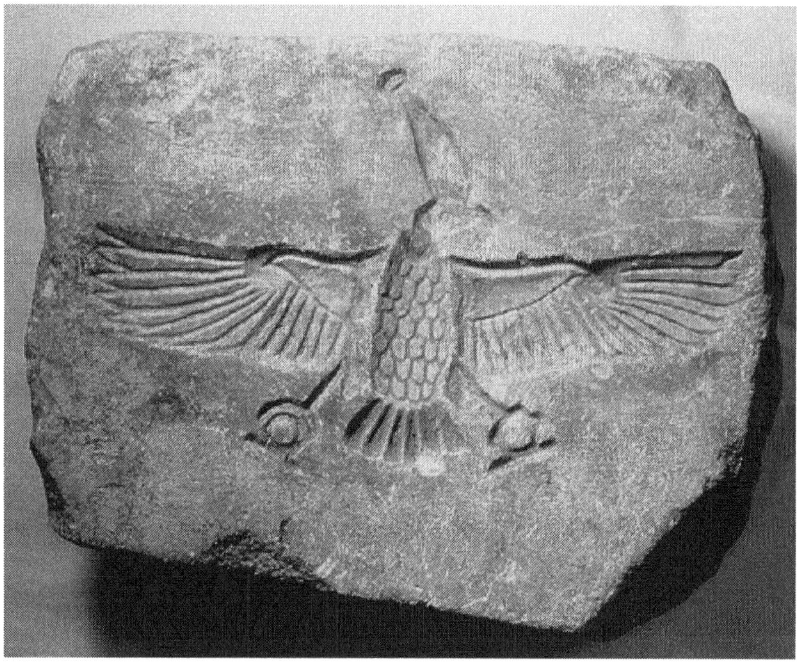

Figure 7-13 : Limestone mould *of the vulture goddess Nekhbet for the shaping of thin gold sheet and found in house Q46.5 associated with clay moulds for faience making, a kiln and some glass making debris, perhaps the remnants of a jewellery workshop (12/13/1234, Borchardt and Ricke 1980). Mould 130mm x 95mm.*

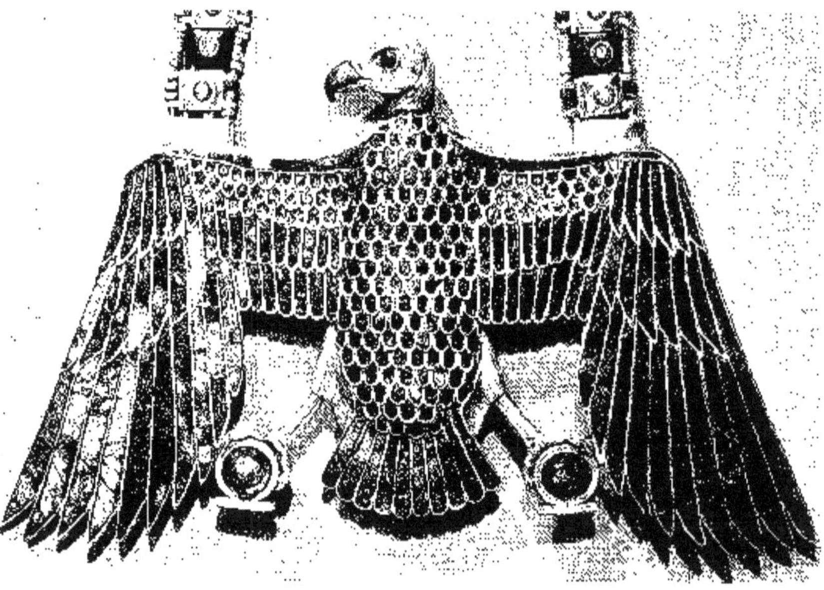

Figure 7-14 : Pendant representing the vulture goddess Nekhbet *of sheet gold inlaid with glass, lapis lazuli and carnelian, from the Tomb of Tutankhamen, Carter and Mace (1923-33), vol II, 124 and pl.LXXXIVA, and British Museum (1972) plate 43. Pendant is 110mm across..*

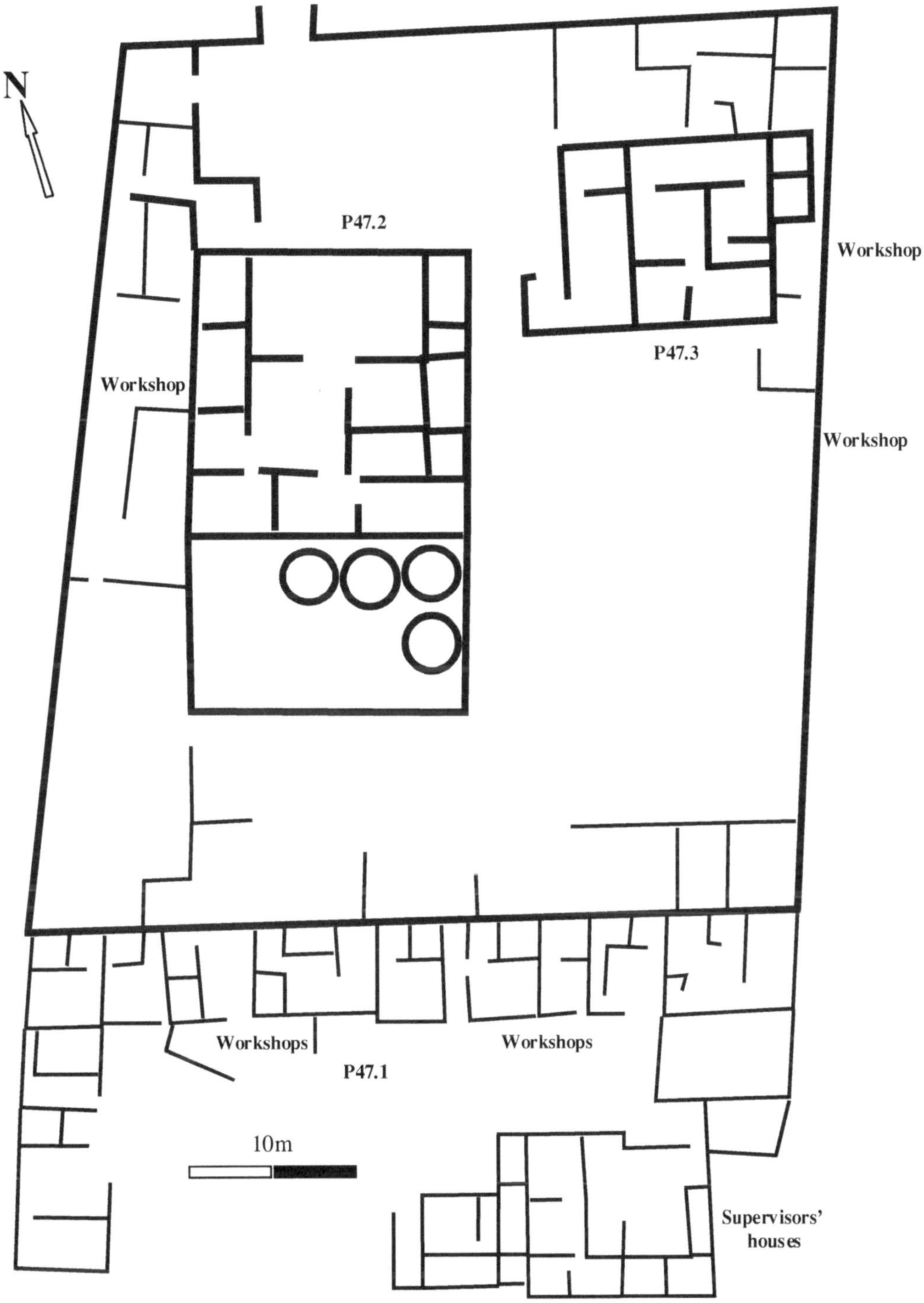

Figure 7-15 : P47.1-3, the house of the sculptor Tuthmose, *a house in the Main City South, showing the house of Tuthmose itself (P47.2), the houses of the younger sculptors (P47.3) and the small combined living and working accommodation of the lower ranking workers (P47.1), where all the faience moulds were found. Also shown is what is thought by Arnold (1997) to be the supervisors' houses. Adapted from Borachardt and Ricke (1980).*

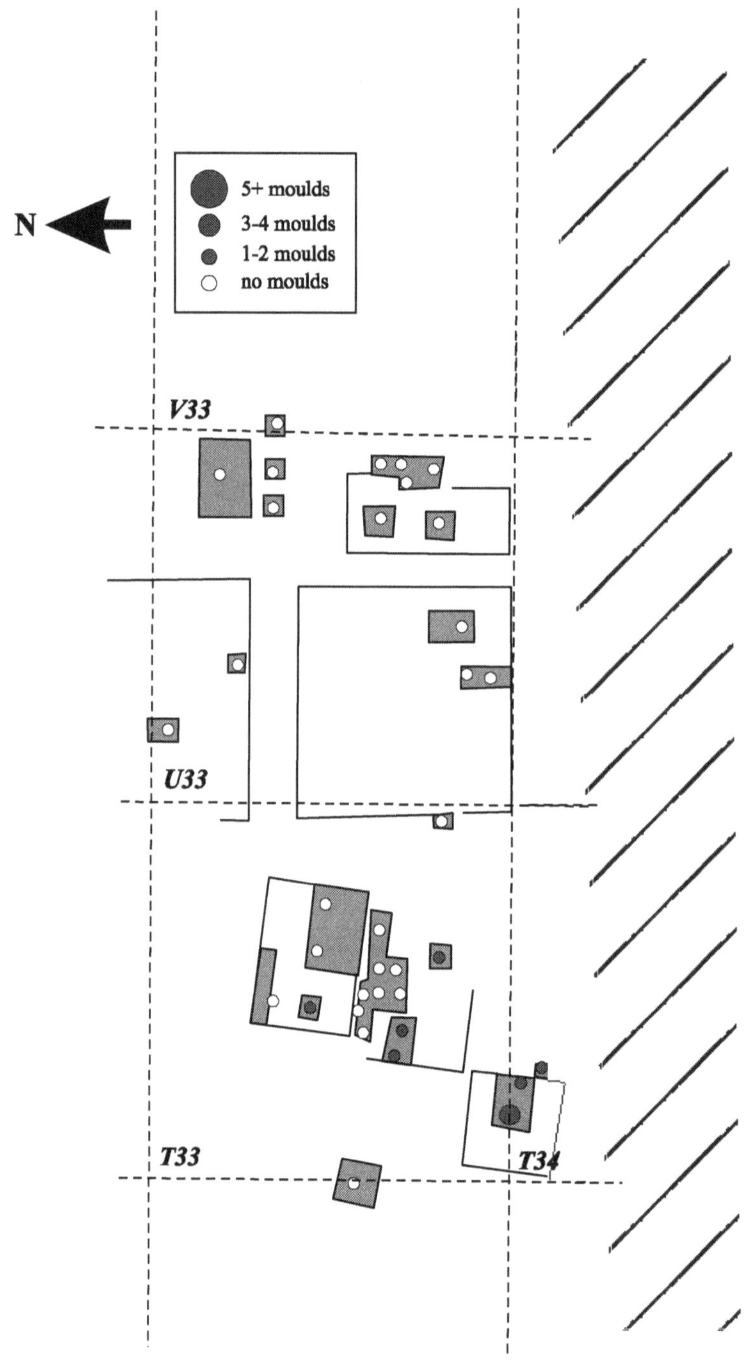

Figure 7-16a : Moulds in North Suburb North *-Results from the search of the excavation reports for the moulds associated with each house. Number of moulds as key.*

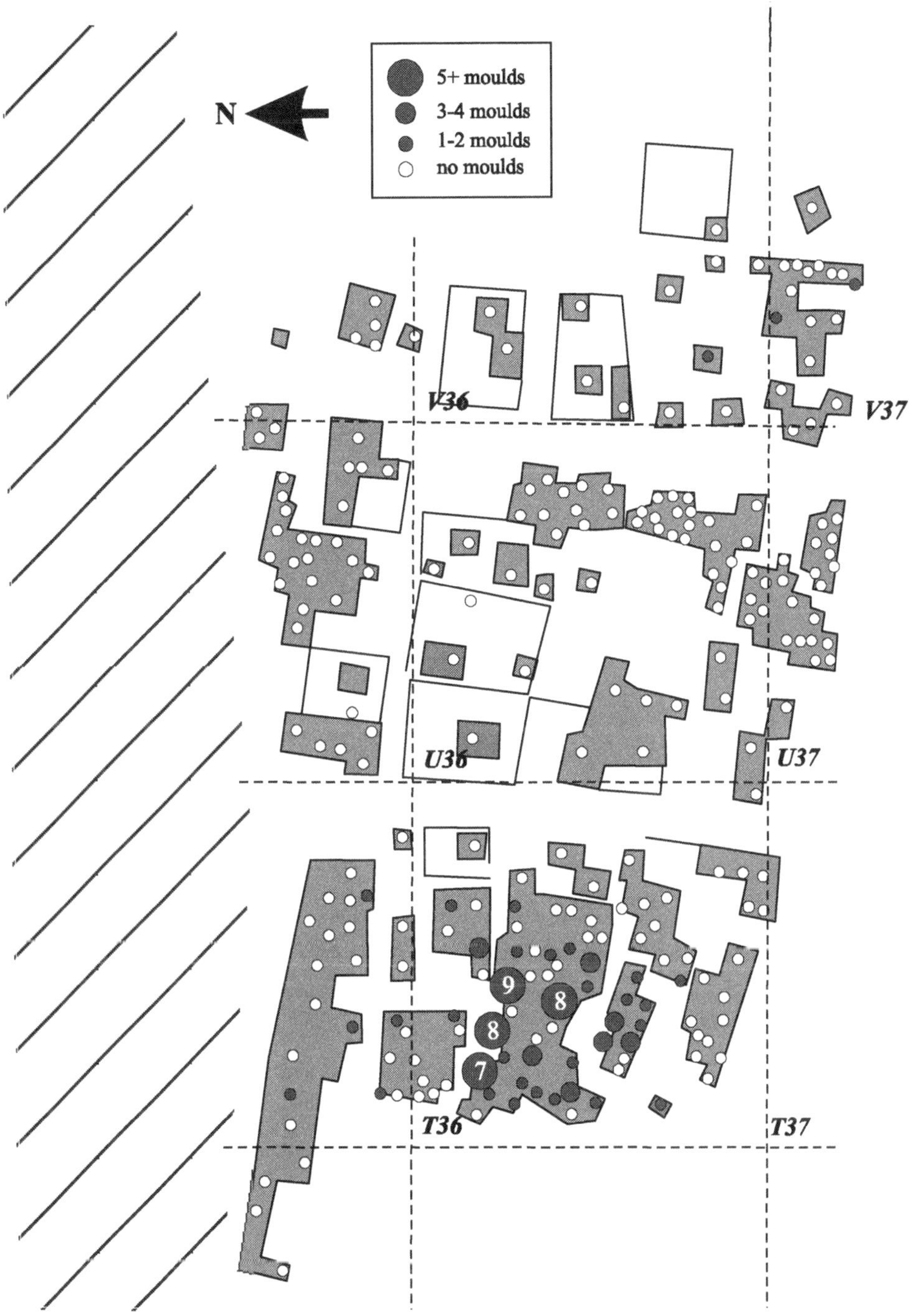

Figure 7-16b : Moulds in North Suburb South *- Results from the search of the excavation reports for the moulds associated with each house. Number of moulds as key.*

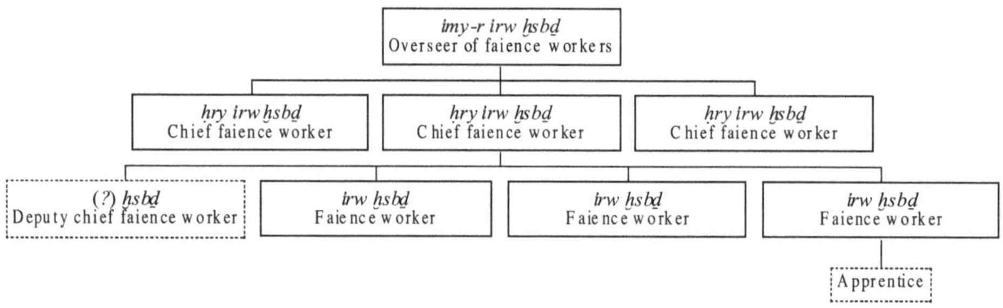

Figure 7-17 : Organisation of workers in faience. *Schematic showing the organisation of faience workers drawing on evidence from the draughtsmen at Deir el Medina (Lesko 1994 and others) and titles of faience workers from outside Amarna, but all drawn from the New Kingdom (see Table 7-4). Deputy chiefs and apprentices are not attested in the record for faience workers, but are present at Deir el Medina, so are shown in dotted outline. Each chief would have a similar collection of workers under him, but for the sake of clarity only one "crew" is shown in detail.*

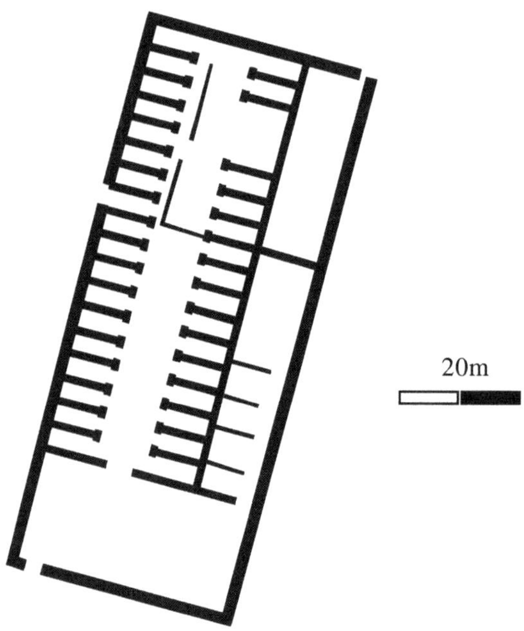

Figure 7-18 : The Bakeries. *The Amarna approach to the problem of mass production was to repeat the normal basic domestic kitchen over and over again, until the required number of loaves were produced. The plan to the left shows one of the bakeries (P43.1) clearly illustrating the many small kitchen units within the larger structure (adapted from Kemp and Garfi, !993).*

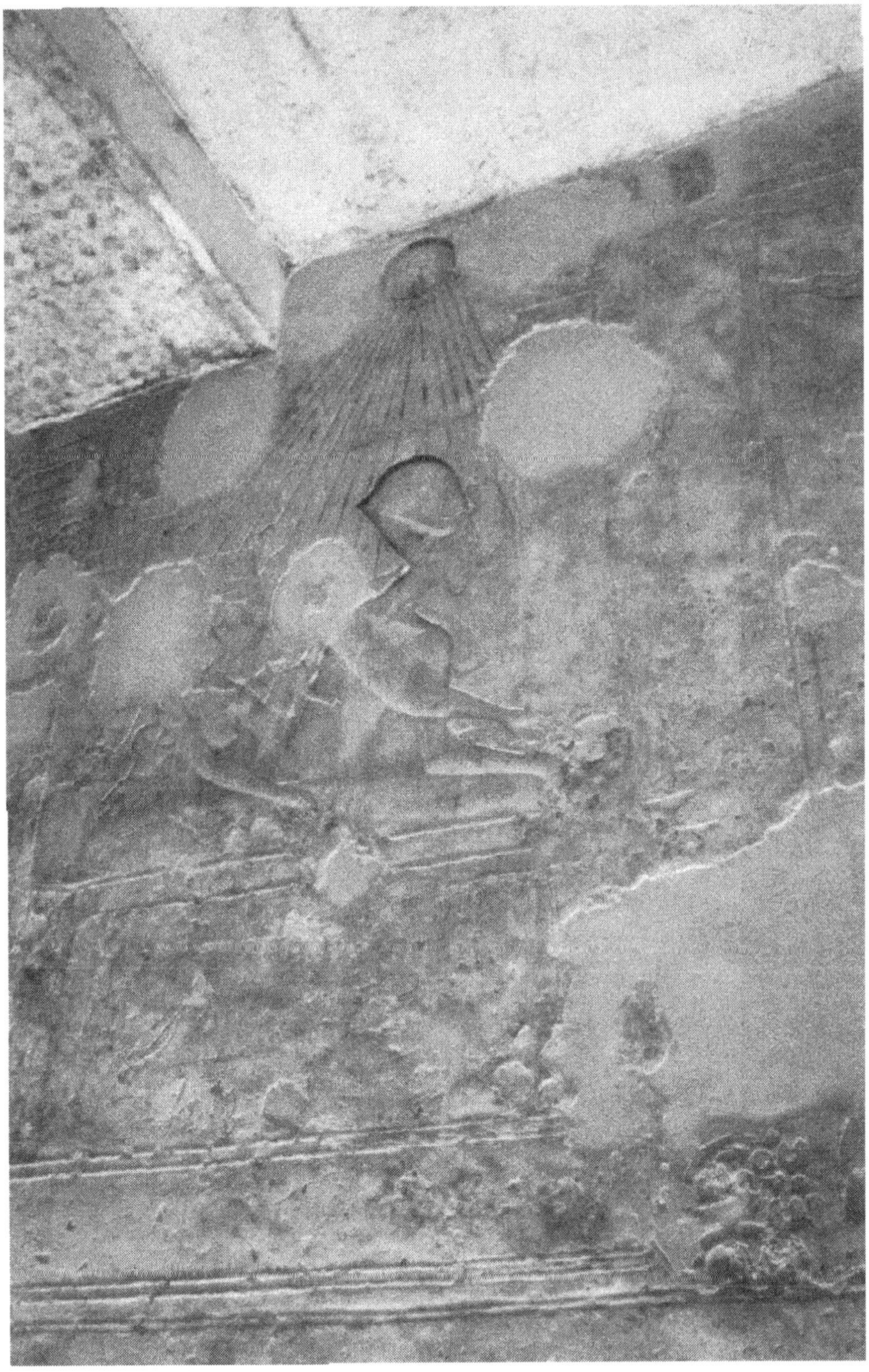

Figure 7-19 : Window of Appearance. *Relief from the tomb of Ramose, in the Tombs of the Nobles at Thebes (TT55), dating from the Amarna period. The King and Queen are shown in the Window of Appearances, the reliefs below which have been identified with reliefs from the King's House. Various items are shown as being given out, including, a range of jewellery in the King's hand and bottom right.*

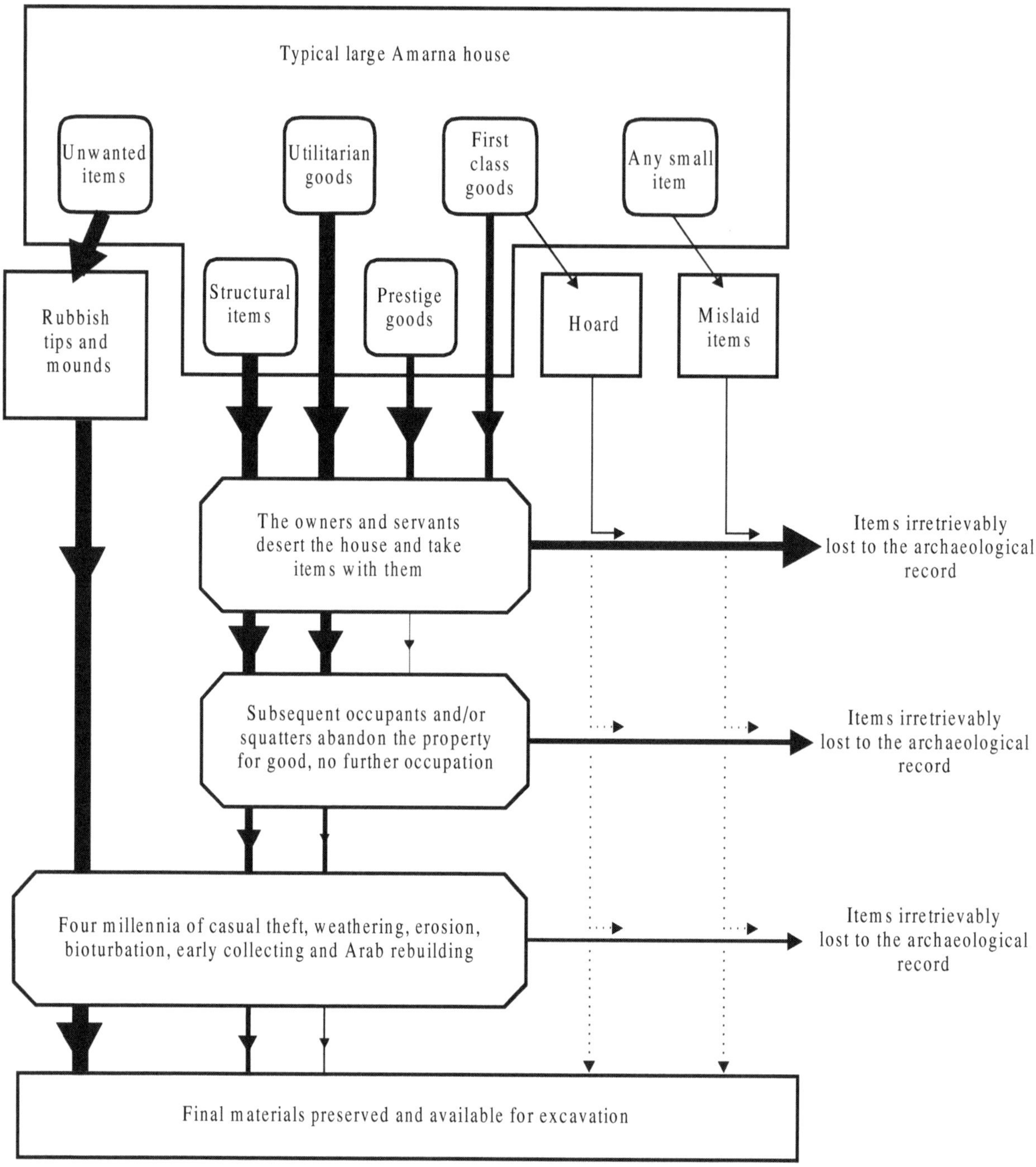

Figure 7-20 : Loss of Material Flowchart : *Flowchart illustrating the loss of material from a typical large house at Amarna through time until final excavation and recovery.*

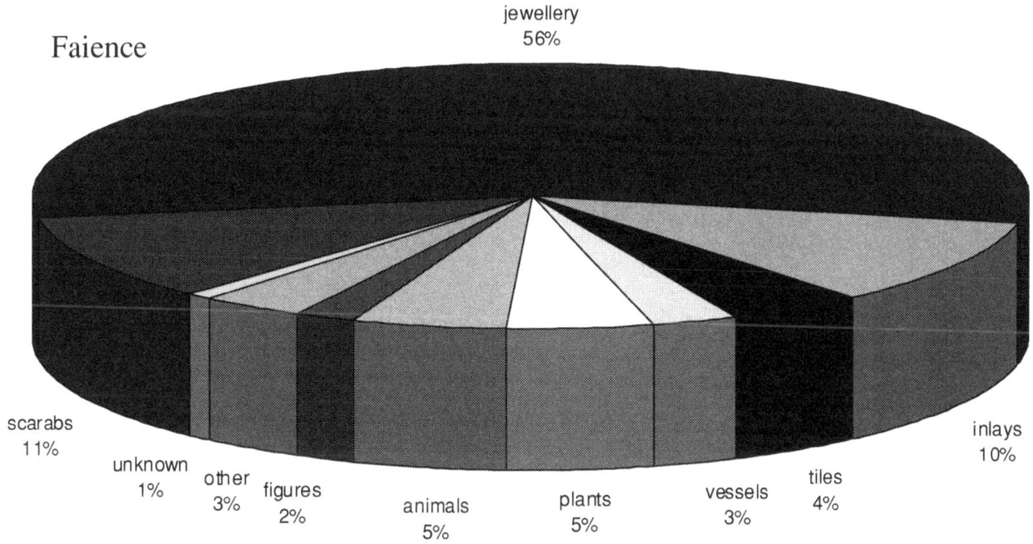

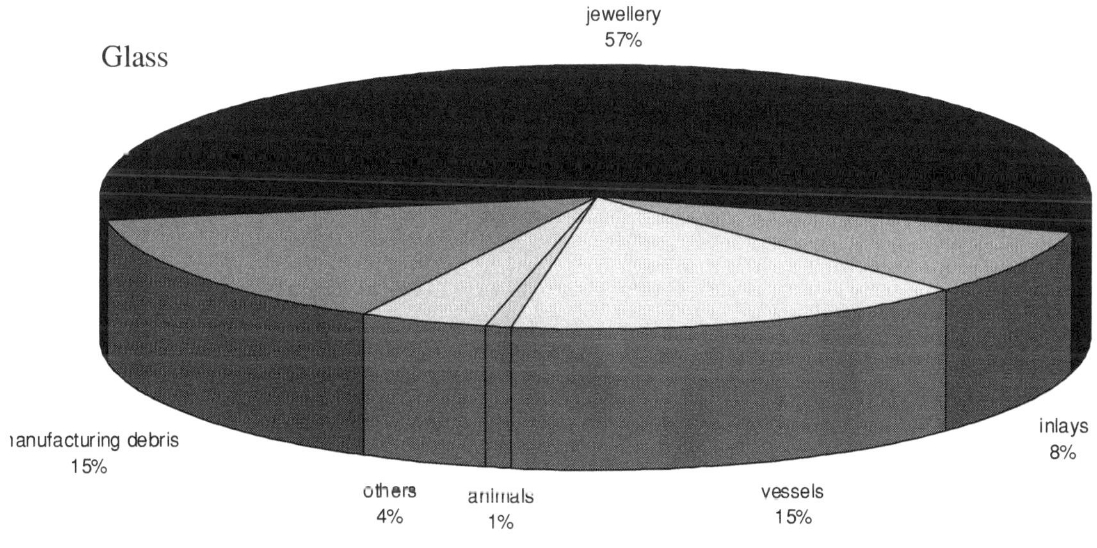

Figure 7-21 : Faience and glass by type. *1610 items of faience (top) and 131 items of glass (bottom) listed in Borchardt and Ricke (1980) classified by type.*

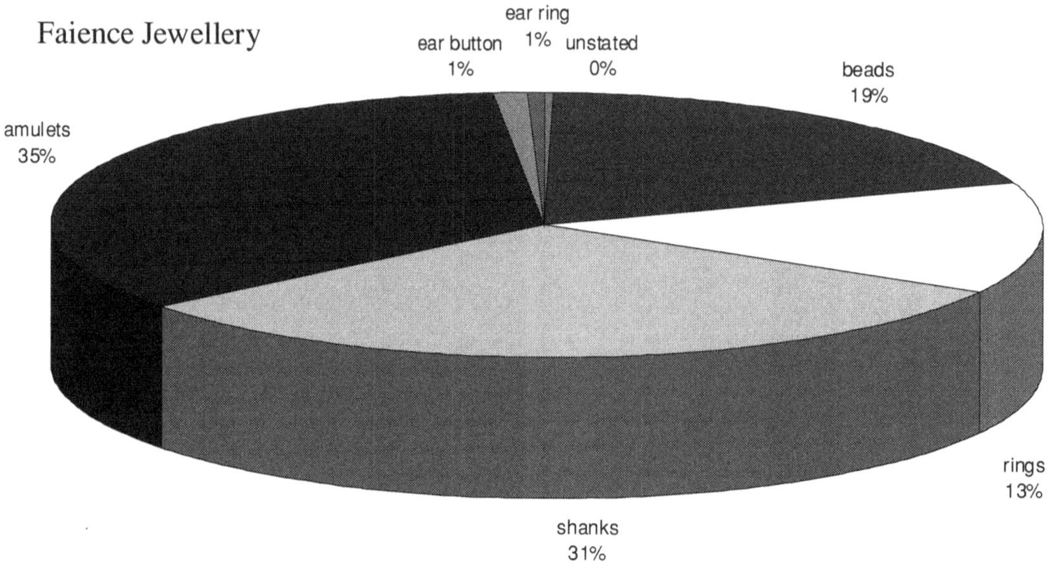

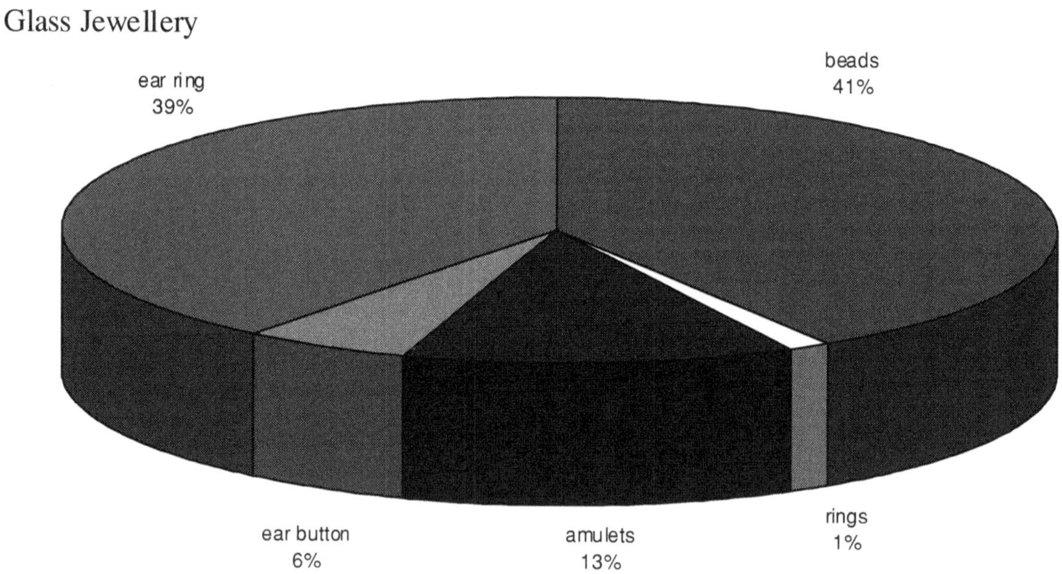

Figure 7-22 : Faience and glass jewellery by type. *874 items of faience jewellery (top) and 71 items of glass jewellery (bottom) listed in Borchardt and Ricke (1980) classified by type.*

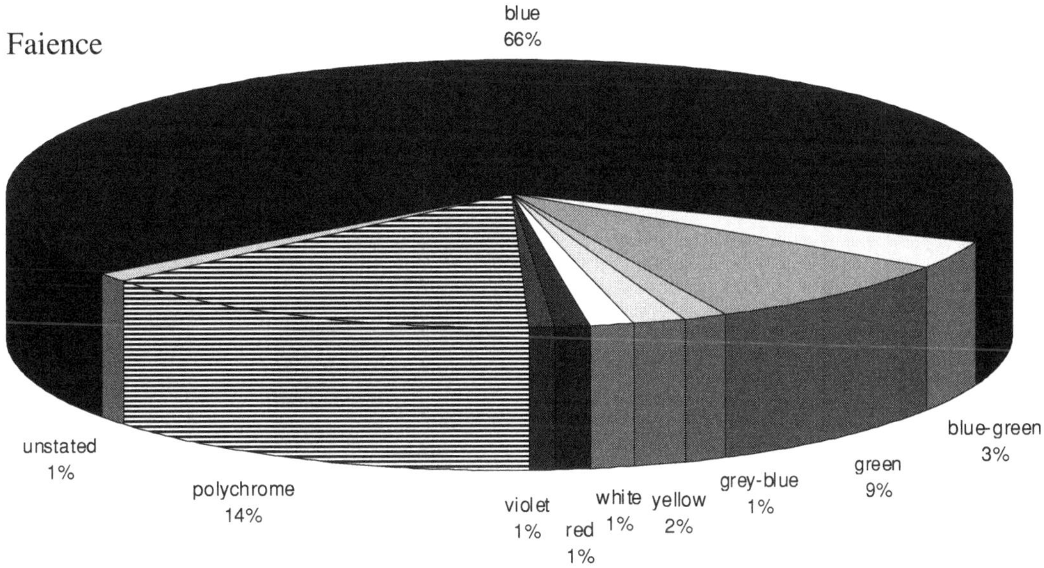

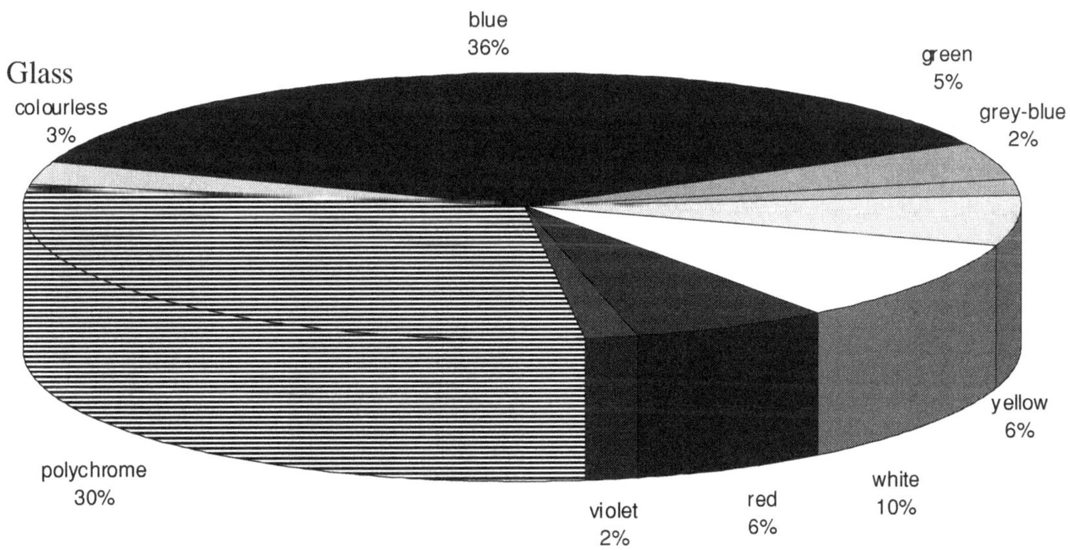

Figure 7-23 : Faience and glass by colour. *1610 items of faience (top) and 131 items of glass (bottom) listed in Borchardt and Ricke (1980) classified by colour.*

Appendix C : Colour plates

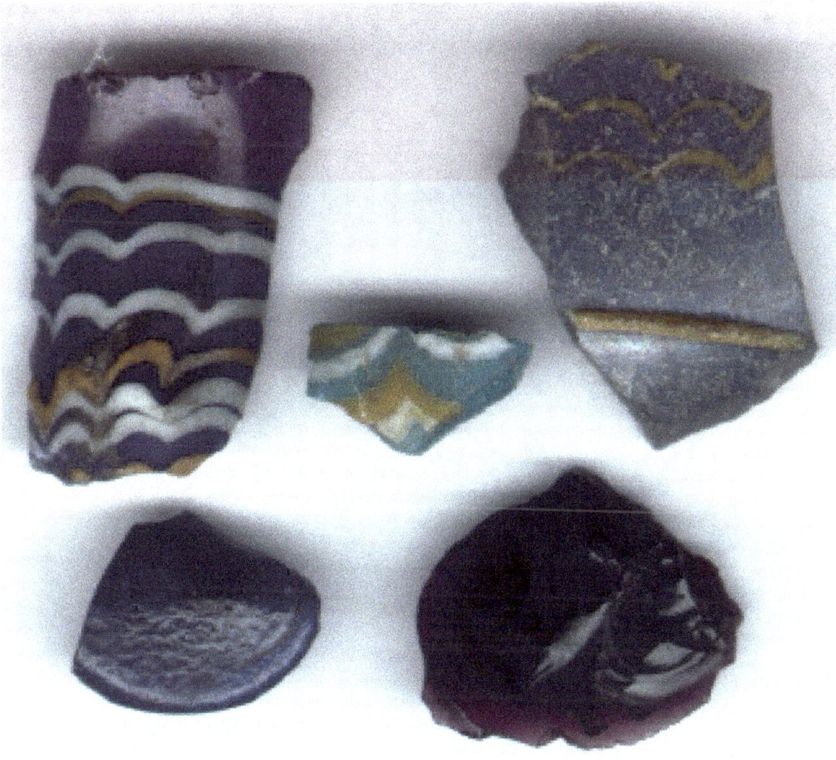

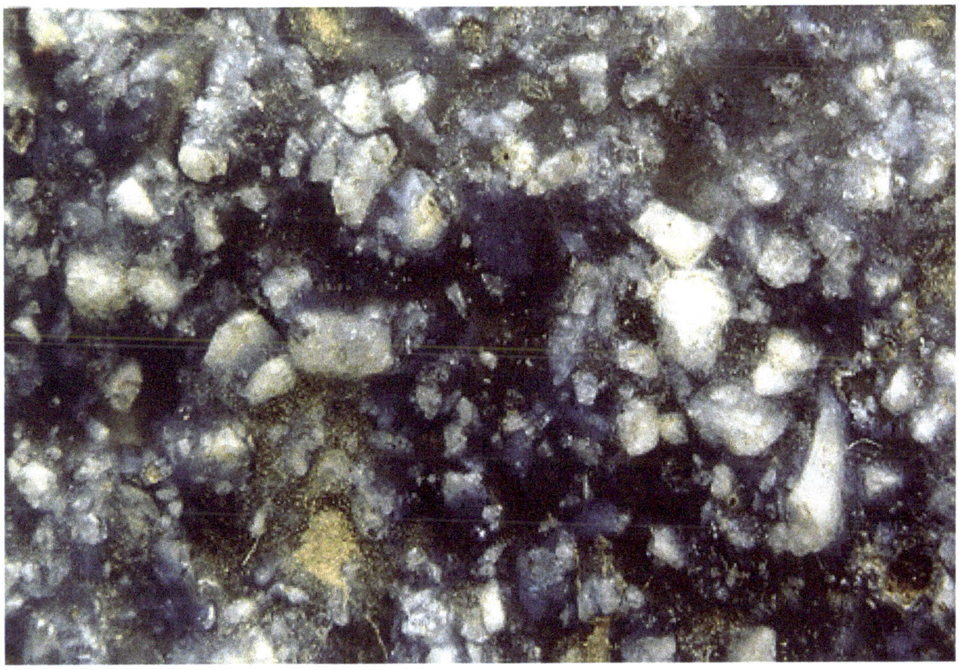

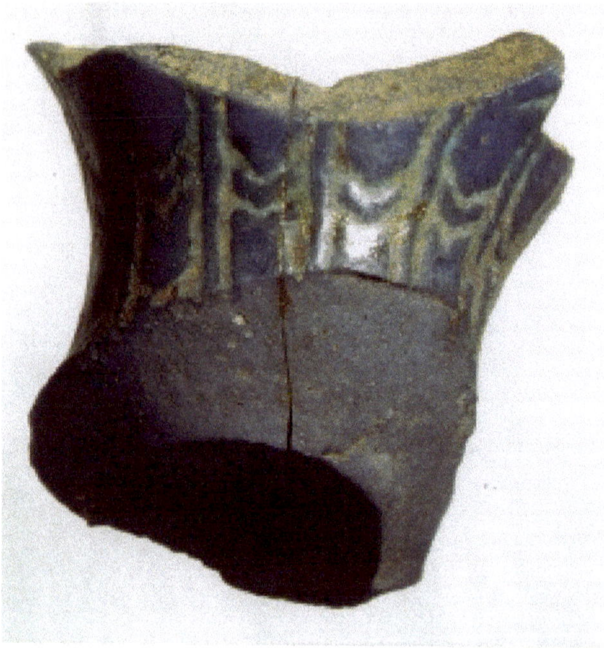
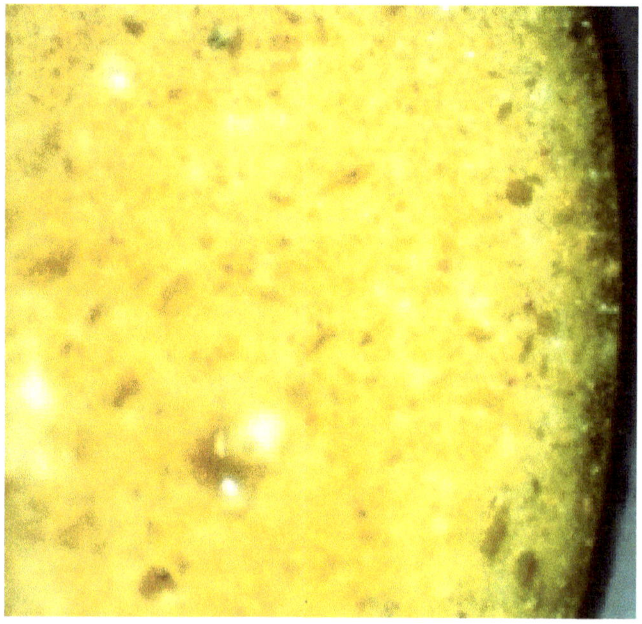
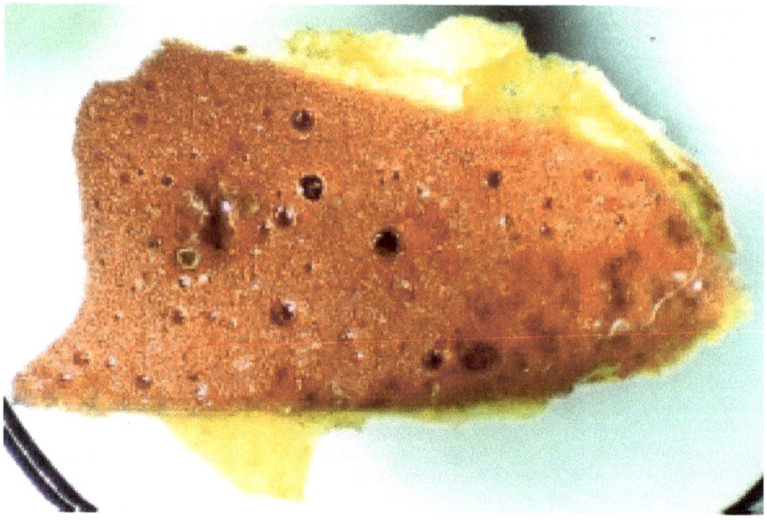
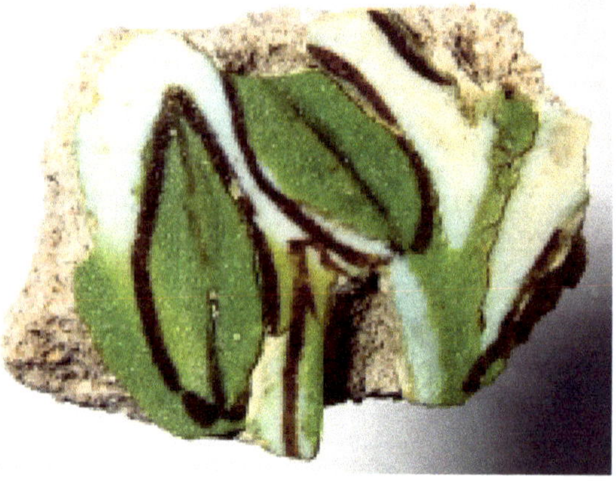
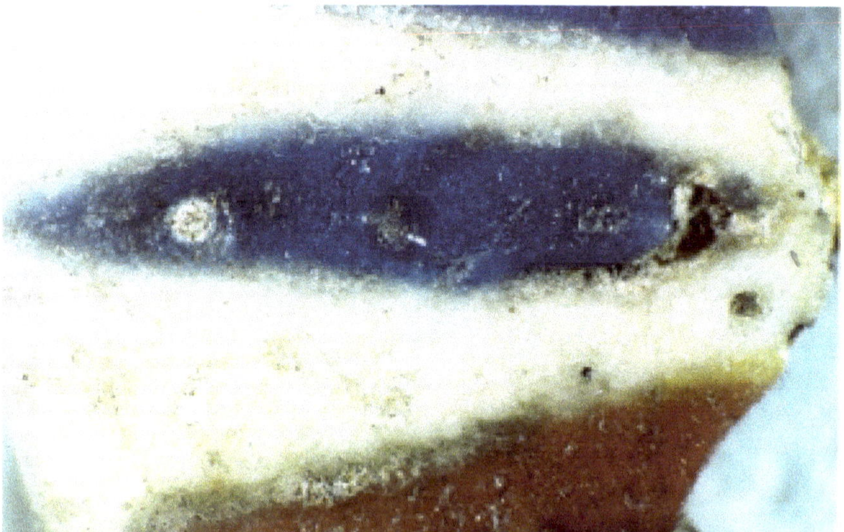